Signal Crosstalk in Plant Stress Responses

Signal Crosstalk in Plant Stress Responses

Edited by

KEIKO YOSHIOKA
Department of Cell and Systems Biology
University of Toronto
Canada

KAZUO SHINOZAKI
Director
RIKEN Plant Science Center
Japan

A John Wiley & Sons, Ltd., Publication

Edition first published 2009
© 2009 Wiley-Blackwell

Blackwell Publishing was acquired by John Wiley & Sons in February 2007.
Blackwell's publishing program has been merged with Wiley's global Scientific, Technical, and Medical business to form Wiley-Blackwell.

Editorial Office
2121 State Avenue, Ames, Iowa 50014-8300, USA

For details of our global editorial offices, for customer services, and for information about how to apply for permission to reuse the copyright material in this book, please see our website at www.wiley.com/wiley-blackwell.

Authorization to photocopy items for internal or personal use, or the internal or personal use of specific clients, is granted by Blackwell Publishing, provided that the base fee is paid directly to the Copyright Clearance Center, 222 Rosewood Drive, Danvers, MA 01923. For those organizations that have been granted a photocopy license by CCC, a separate system of payments has been arranged. The fee codes for users of the Transactional Reporting Service are ISBN-13: 978-0-8138-1963-1/2009.

Designations used by companies to distinguish their products are often claimed as trademarks. All brand names and product names used in this book are trade names, service marks, trademarks or registered trademarks of their respective owners. The publisher is not associated with any product or vendor mentioned in this book. This publication is designed to provide accurate and authoritative information in regard to the subject matter covered. It is sold on the understanding that the publisher is not engaged in rendering professional services. If professional advice or other expert assistance is required, the services of a competent professional should be sought.

Library of Congress Cataloging-in-Publication Data

Signal crosstalk in plant stress responses / edited by Keiko Yoshioka, Kazuo Shinozaki. – 1st ed.
    p. cm.
  Includes bibliographical references and index.
  ISBN 978-0-8138-1963-1 (hardback : alk. paper)
  1. Plants–Effect of stress on. 2. Plant physiology. 3. Cellular signal transduction.
I. Yoshioka, Keiko. II. Shinozaki, Kazuo.
  QK754.S563 2009
  632–dc22

                                                                                     2009009720

A catalog record for this book is available from the U.S. Library of Congress.

Set in 10.5/12 pt Times by Aptara® Inc., New Delhi, India
Printed in Singapore

1   2009

Contents

Contributors		ix
Preface		xiii

1 Role of Abscisic Acid in Disease Resistance 1
VICTOR FLORS, JURRIAAN TON, AND BRIGITTE MAUCH-MANI

Introduction	1
ABA Decreases Disease Resistance	1
ABA Increases Disease Resistance	5
Crosstalk with Other Phytohormones	10
Interactions between Biotic and Abiotic Stress Regulation	15
Prospects and Conclusions	17
Acknowledgments	17

2 Plant Mitogen-Activated Protein Kinase Cascades in Signaling Crosstalk 23
FUMINORI TAKAHASHI, KAZUYA ICHIMURA, KAZUO SHINOZAKI, AND KEN SHIRASU

Introduction	23
MAPK Cascade Components	24
MPK3 and MPK6: Promiscuous MAPKs or Crosstalk Junction?	25
ANP1 as Upstream MAPKKK of MPK3 and MPK6	30
MPK3 and MPK6 in Ethylene Biosynthesis and Signal Transduction	30
MEKK1-MKK1/MKK2-MPK4 Pathway in Biotic and Abiotic Stress	31
Arabidopsis MKK3-MPK6 Pathway in JA Signaling	34
Arabidopsis MKK3 and the Group C MPKs (MPK1, MPK2, MPK7, and MPK14) in Wounding and Pathogen Infection	34
Tobacco NPK1-NtMEK1-NTF6 Cascade in Cell Plate Formation and Defense Responses	35
Protein Phosphatases Negatively Regulate MAPK Cascades	36
Conclusions	37
Acknowledgment	38

3 Transcription Factors Involved in the Crosstalk between Abiotic and Biotic Stress-Signaling Networks 43
YASUNARI FUJITA, MIKI FUJITA, KAZUKO YAMAGUCHI-SHINOZAKI, AND KAZUO SHINOZAKI

Introduction	43
Zinc Finger Transcription Factors in ROS Signaling	43

	MYC and MYB Transcription Factors in JA and ABA Signaling Pathways	45
	NAC Transcription Factors in ABA, JA, and Senescence Signaling Pathways	47
	ERF/AP2 Family Transcription Factors	48
	Transcription Factors Involved in the Biosynthesis and Regulation of Cuticular Components	51
	Conclusions	53
4	**Crosstalk in Ca^{2+} Signaling Pathways** MARTIN R. McAINSH AND JULIAN I. SCHROEDER	**59**
	Introduction	59
	Ca^{2+} Signals in Abiotic and Biotic Stress Responses	60
	Generation of Ca^{2+} Signals	66
	Specificity in Ca^{2+} Signaling	74
	Decoding Ca^{2+} Signals	79
	Conclusions	83
5	**Crosstalk in Pathogen and Hormonal Regulation of Guard Cell Signaling** ELENA BRAY SPETH, MAELI MELOTTO, WEI ZHANG, SARAH M. ASSMANN, AND SHENG YANG HE	**96**
	Introduction	96
	Hormonal Regulation of Stomatal Movements	97
	Are ROS and NO Common Integrators of Hormonal Regulation of Stomatal Aperture?	101
	Pathogen Regulation of Stomatal Movements	102
	Concluding Remarks	107
	Acknowledgments	107
6	**Environmental Sensitivity in Pathogen Resistant *Arabidopsis* Mutants** WOLFGANG MOEDER AND KEIKO YOSHIOKA	**113**
	Introduction	113
	Environmental Effects on *R* Gene-Mediated Pathogen Resistance	114
	Lesion Mimic Mutants and Their Environmental Sensitivity	116
	What Are the Common Aspects in the Described Examples?	121
	Possible Mechanisms for Environmental Sensitivity	122
	Concluding Remarks	128
	Acknowledgments	130
7	**Reactive Oxygen Species, Nitric Oxide, and Signal Crosstalk** STEVEN J. NEILL, JOHN T. HANCOCK, AND IAN D. WILSON	**136**
	Introduction	136
	Synthesis, Turnover, and Interactions of ROS and NO	141
	RS and Signal Crosstalk	151
	Conclusions	156

8	**TORing with Cell Cycle, Nutrients, Stress, and Growth**	**161**
	DESH PAL S. VERMA AND JAYANTA CHATTERJEE	

Introduction 161
Conservation of TOR Pathway in Eukaryotes 161
Presence of TOR Pathway in Plants and Caveat of Plant Resistance
   against Rapamycin 162
Major Players of TOR Pathway in Plants 164
TOR Connections with Plant Cell Cycle and Growth 176
Plant Viral Infections and Possible TOR Connections 179
Nutrient Sensing via TOR 180
Osmotic Stress Signaling and TOR 186
Conclusion 188
Acknowledgment 190

Index   201

Contributors

Sarah M. Assmann
Biology Department
Pennsylvania State University
University Park, PA 16802

Jayanta Chatterjee
Department of Molecular Genetics and Plant Biotechnology Center
The Ohio State University
Columbus, OH 43210

Victor Flors
Biochemistry and Plant Biotechnology Laboratory
Dpt. CAMN
Universitat Jaume I
Castellón 12071
Spain

Miki Fujita
Gene Discovery Research Group
RIKEN Plant Science Center
3-1-1 Koyadai
Tsukuba
Ibaraki 305-0074
Japan

Yasunari Fujita
Biological Resources Division
Japan International Research Center for Agricultural Sciences (JIRCAS)
1-1 Ohwashi, Tsukuba
Ibaraki 305-8686
Japan

John T. Hancock
Centre for Research in Plant Science
Faculty of Health and Life Sciences
University of the West of England, Bristol
Coldharbour Lane
Bristol BS16 1QY
UK

Sheng Yang He Department of Energy Plant Research Laboratory
222 Plant Biology Labs.
Michigan State University
East Lansing, MI 48824

Kazuya Ichimura Plant Immunity Research Group
RIKEN Plant Science Center
1-7-22 Suehirocho
Tsurumi
Yokohama
Kanagawa 230-0045
Japan

Brigitte Mauch-Mani Université de Neuchâtel
Laboratoire de Biologie Moléculaire et Cellulaire
Rue Emile-Argand 11
2009 Neuchâtel
Switzerland

Martin R. McAinsh Lancaster Environment Centre
Department of Biological Sciences
Lancaster University
Lancaster
LA14YQ
UK

Maeli Melotto Department of Biology
University of Texas at Arlington
Arlington, TX 76019

Wolfgang Moeder Department of Cell and Systems Biology
University of Toronto
25 Willcocks Street
Toronto, ON
M5S 3B2
Canada
and
Center for the Analysis of Genome Evolution and
    Function (CAGEF)
University of Toronto
25 Willcocks Street
Toronto, ON
M5S 3B2
Canada

Steven J. Neill	Centre for Research in Plant Science Faculty of Health and Life Sciences University of the West of England, Bristol Coldharbour Lane Bristol BS16 1QY UK
Julian I. Schroeder	Cell and Developmental Biology Section Division of Biological Sciences University of California San Diego La Jolla, CA 92093-0116
Kazuo Shinozaki	Gene Discovery Research Group RIKEN Plant Science Center 3-1-1 Koyadai Tsukuba Ibaraki 305-0074 Japan *and* Plant Functional Genomics Research Group RIKEN Plant Science Center 1-7-22 Suehiro-cho Tsurumi Yokohama Kanagawa 230-0045 Japan
Ken Shirasu	Plant Immunity Research Group RIKEN Plant Science Center 1-7-22 Suehirocho Tsurumi Yokohama Kanagawa 230-0045 Japan
Elena Bray Speth	Department of Energy Plant Research Laboratory Michigan State University East Lansing, MI 48824
Jurriaan Ton	Rothamsted Research West Common, Harpenden Herts, AL5 2JQ UK

Fuminori Takahashi

Gene Discovery Research Team
RIKEN Plant Science Center
3-1-1 Koyadai
Tsukuba
Ibaraki 305-0074
Japan

Desh Pal S. Verma

Department of Molecular Genetics and Plant
    Biotechnology Center
The Ohio State University
Columbus, OH 43210

Ian D. Wilson

Centre for Research in Plant Science
Faculty of Health and Life Sciences
University of the West of England, Bristol
Coldharbour Lane
Bristol BS16 1QY
UK

Kazuko Yamaguchi-Shinozaki

Biological Resources Division
Japan International Research Center for
    Agricultural Sciences (JIRCAS)
1-1 Ohwashi, Tsukuba
Ibaraki 305-8686
Japan

Keiko Yoshioka

Department of Cell and Systems Biology
University of Toronto
25 Willcocks Street
Toronto, ON
M5S 3B2
Canada
and
Center for the Analysis of Genome Evolution
    and Function (CAGEF)
University of Toronto
25 Willcocks Street
Toronto, ON
M5S 3B2
Canada

Wei Zhang

Biology Department
Pennsylvania State University
University Park, PA 16802

Preface

Plants have evolved sophisticated mechanisms to respond to a variety of environmental (abiotic and biotic) stimuli to survive. One can imagine that this ability is more prominent in plants than in animals because plants, as sessile organisms, must continue to adjust their physiology to an ever-changing environment throughout their life cycle after they are rooted in place. It could be several months for some plants, or several hundreds of years for some tree species. In any case, this is an astonishing ability of all plant species.

The question of how plants can defend themselves against various environmental stresses has been one of the prominent topics for plant researchers, including agronomists, because this factor is particularly important for agricultural production. Both abiotic stresses, such as drought, salt, cold, and heat, and biotic stresses from pathogen or insect attack cause a significant negative impact on agricultural production. Therefore, understanding the molecular mechanisms of stress tolerance not only provides scientific knowledge but also contributes various future applications to agriculture.

Over the past two decades, significant effort has been made to understand signal transduction after stress perception, and our knowledge in this area has remarkably increased. For example, the role of the phytohormone abscisic acid (ABA) in drought, temperature, and salt stresses and its signal transduction to induce tolerance to these stresses are increasingly clear. A variety of *Arabidopsis* mutants that show alteration in stress responses was isolated, and some components of the signal transduction cascades, such as transcription factors and kinases, were identified. Furthermore, the phytohormones salicylic acid (SA) and jasmonic acid (JA) are well-known factors that coordinate proper pathogen resistance responses. SA- and JA-dependent signaling pathways are also being studied intensively, and some important factors in these pathways were revealed.

To date, however, these signal transduction pathways have been investigated relatively independently, mainly because it was difficult to consider various stimuli simultaneously. Consequently, we tend to create a model of a linear pathway or, at most, one with several branches from the main pathway. However, considering that plants undergo continuous exposure to multiple stresses under natural conditions, these signal transduction pathways must interconnect to allow plants to coordinate and prioritize their reactions for survival using limited resources. Indeed, recent studies about the crosstalk between these signal cascades are becoming a central theme in various research fields, and we believe that it is time to summarize our current knowledge to understand the complexity of the signaling crosstalk.

Recent groundbreaking findings about the manipulation of phytohormonal crosstalk by pathogenic microorganisms to suppress pathogen resistance responses have been reported (see Chapter 5). Furthermore, a variety of environmentally sensitive *Arabidopsis* mutants that shows alterations in pathogen resistance responses has been reported, providing solid evidence for the crosstalk between abiotic and biotic stress responses (see Chapter 6). These recent discoveries led to the promotion of signal crosstalk research, and we now expect a new era in the field of signal transduction in plants.

In this book, we focus on the forefront of signal crosstalk research in plants. The book comprises eight chapters to cover current important topics, such as phytohormones (Chapter 1), mitogen-activated protein kinase cascades (Chapter 2), transcription factors (Chapter 3), Ca^{2+} cascades (Chapter 4), and reactive oxygen species (Chapter 7). Each chapter was written by a leader in the field and summarizes significant discoveries in recent reports. Because of space limitations, we could not cover other important topics, such as involvement of cytochrome p450 or phosphatases. These factors are currently receiving tremendous attention as players in signal crosstalk. We hope to have another opportunity to cover these topics in the near future. Additionally, we regret being unable to have comprehensive coverage of signal crosstalk between environmental stresses and developmental regulation. However, to cover one of the recent interesting topics in this area, we included a chapter on the TOR (target of rapamycin) signaling pathway, which is well conserved among many living organisms, including plants, as a key element in cell growth control and biotic/abiotic stress responses (Chapter 8).

The complex integration of multiple environmental stimuli is one of the most important abilities for plant survival, and it is now clear that the study of signal crosstalk in environmental stress responses will be a key topic for the next decade. A future challenge is to uncover the key elements that act as nodes of crosstalk and how these work on the molecular level. We are convinced that this book will provide an advanced summary of the key elements involved in signal crosstalk and a stepping stone for further understanding of signal crosstalk in plant stress responses.

Keiko Yoshioka & Kazuo Shinozaki

1 Role of Abscisic Acid in Disease Resistance

Victor Flors, Jurriaan Ton, and Brigitte Mauch-Mani

Introduction

Plants are sessile organisms, and their only means to respond when exposed to biotic stress such as pathogen attack is to activate rapidly a panoply of reactions, including both physical and chemical defenses. These defensive reactions, which result in the death of the pathogen, are the result of highly coordinated sequential changes at the cellular level comprising, among other changes, the synthesis of salicylic acid (SA), jasmonic acid (JA), ethylene (ET), reactive oxygen species (ROS), and nitric oxide (NO), as well as the hypersensitive reaction (HR; Agrios 2005).

Historically, two signaling pathways have been implicated in the amplification and transmission of the signal perceived at the moment a plant recognizes a pathogen until the final defense reactions have been engaged (Kunkel and Brooks 2002). SA has long been recognized as crucial to signal transduction (Ryals et al. 1996), and the importance of a functional JA/ET signaling pathway also rapidly became apparent (Thomma et al. 2001). Although SA signaling in *Arabidopsis* has been shown to be implicated predominantly in defense against biotrophic or hemibiotrophic organisms, JA/ET-dependent signaling seems to be more important when plants are confronted with necrotrophic organisms (Rojo et al. 1999; Thomma et al. 2001).

Because of their involvement and crucial roles in the signal transduction pathways implicated in plant defense against pathogens and insects, SA, JA, and ET have been the most extensively studied hormones. In contrast, the role of other hormones, such as auxins, gibberellins (GAs), abscisic acid (ABA), and brassinosteroids (BR), has been neglected in this context. An increasing body of evidence has recently pointed to interesting and important roles for these hormones in plant defense against biotic stresses (for review, see Robert-Seilaniantz et al. 2007). In this chapter, emphasis is placed on the role of ABA in basal and induced resistance and in priming phenomena against biotic stresses.

ABA Decreases Disease Resistance

The development of recent techniques for simultaneous analysis of phytohormones has highlighted the relevance of classical hormones such as indole acetic acid, cytokinins, GAs, and ABA in plant-pathogen and plant-insect interactions (Durgbanshi et al. 2005; Jakab et al. 2005; Robert-Seilaniantz et al. 2007; Schmelz et al. 2003; Flors et al. 2008). Interestingly, certain saprophytic and parasitic fungi, such as *Botrytis, Ceratocystis, Fusarium*, and *Rhizoctonia*, are also able to produce ABA (Robert-Seilaniantz et al. 2007). Plant hormones do not induce significant developmental alterations in pathogens, and thus it is probable that pathogens produce plant hormones to modulate the hormonal balance in their host, leading to interference with plant defenses (Robert-Seilaniantz et al. 2007).

ABA has also been shown to play a role during plant-insect interactions. Investigations into the relationship between plant stress and resistance to herbivory led to the hypothesis that herbivores should have increased performance on mildly stressed plants (White et al. 1974). ABA may be involved in defense against insects after aphid infestation of celery plants (aphid susceptible) or sorghum and wheat (aphid resistant). During these interactions, several genes that share sequences with those involved in the biosynthesis of or that are activated by ABA are upregulated (Divol et al. 2005; Park et al. 2006; Boyko et al. 2006). However, as with plant reactions toward pathogens, ABA also has various regulatory functions against insects. Voelckel et al. (2004) found that a sequence encoding 3-hydroxy-3-methyl glutaryl coenzyme A reductase, which is involved in ABA and GA biosynthesis, is downregulated in *Nicotiana attenuata* infested by *Myzus nicotianae*. In *Schizaphis graminum*–resistant sorghum plants, Park et al. (2005) identified several highly upregulated genes under ABA control that are involved in cell-wall strengthening (Smith and Boyco 2007). ABA levels are not altered by corn infestation by the ear worm caterpillar; however, in tomatoes, ABA increases significantly after wounding (Schmeltz et al. 2003).

Leaf humidity surrounding stomata is reduced under water stress conditions. This situation can alter the leaf environment for insects. Low humidity reduces hatching of lepidopteran eggs (Godfrey and Holtzer 1991), spider mite population growth (Perring et al. 1984), and growth of many species of insect larvae (Mattson and Scriber 1987). Consequently, ABA-deficient tomato plants (*sitiens* and *flacca*) have a reduced resistance to *Spodoptera exigua* (Thaler and Bostock 2004). These results show that ABA differentially regulates pathogenic or insect responses and is highly dependent on the challenge and the plant species.

Many reports support the role of ABA in disease promotion during pathogenic infection of plants (Audenaert et al. 2002; Mohr and Cahill 2003; Thaler and Bostock 2004; Koga et al. 2004; Robert-Seilaniantz et al. 2007). Treatment of potato plants with ABA before infection with *Phytophthora infestans* or *Cladosporium cucumerinum* suppresses the accumulation of phytoalexins and significantly decreases plants resistance (Henfling et al. 1980). It is generally accepted that ABA often displays negative crosstalk with other plant defense hormones (see the section "Crosstalk with Other Phytohormones" later in the chapter), but it is difficult to attribute a direct role to ABA in disease promotion because the observed phenotypes may be side effects of the downregulation of other plant defenses. This could be the case for ABA-treated soybean plants that display susceptibility to nonvirulent strains of *Phytophthora sojae* (McDonald and Cahill 1999). During incompatible interactions between nonpathogenic strains of *P. sojae* and soybean, ABA levels decrease (Cahill and Ward 1989). ABA appears to induce a strong repression of phenylalanine ammonia lyase (PAL) at the transcriptional level, which directly interferes with the plant's ability to stop the pathogen.

Rezzonico et al. (1998) showed that ABA also downregulates an antifungal β-1,3-glucanase (also known as PR2) in tobacco cell suspensions. Such a downregulation might lead to a reduction in disease resistance in tobacco after ABA treatment; however, this has never been confirmed. *Magnaporthe grisea* infection is also favored by ABA treatments followed by cold stress (Koga et al. 2004). Accordingly, inhibition of ABA synthesis prevents cold-induced susceptibility. In *Arabidopsis*, ABA increases after exogenous treatment or drought stress induction and leads to a decrease in resistance to avirulent bacteria (Mohr and Cahill 2003). Treatment of *Arabidopsis* with increasing concentrations of ABA before inoculation with *Pseudomonas syringae* pv *tomato* (*Pst*) 1065 led to the development of necrosis in a concentration-dependent manner (Mohr and Cahill 2003). In the same report, the authors showed that the endogenous concentration of ABA and time of pathogen challenge can be important for the plant's susceptibility. Accordingly, the ABA insensitive mutant *abi1-1* has no altered response to *Pst*, and the ABA biosynthetic mutant *aba1-1*, which produces less ABA, is less susceptible to *Hyaloperonospora arabidopsis*. However, it cannot be generalized that ABA has an inhibitory effect on plant defenses in all situations.

Pst DC3000 induces auxin biosynthetic genes and a rise in the ABA content (Robert-Seilaniantz et al. 2007). This increase in ABA levels seems to be related to the ability of the bacteria to secrete effectors through the Type III secretory system (T3SS). Induction of ABA during *Pst* infection is T3SS-dependent, and full virulence is only achieved when the ABA biosynthetic pathway is intact (de Torres-Zabala et al. 2007). There is evidence that coronatine is the effector responsible for the ABA increase upon *Pst* infection (Schmelz et al. 2003). Here, the authors suggested that ABA increases are small and something like a secondary response to biotic stress; however, de Torres-Zabala et al. (2007) demonstrated that they play a direct role in the penetration process of *Pst*. In a compatible interaction leading to disease, the induction of *NCED3* (a key regulator enzyme in the ABA biosynthesis) takes place between 4 and 12 hours post infection (hpi) and follows T3SS effector (T3E) delivery (de Torres-Zabala et al. 2007). At 18 hpi, a dramatic increase in ABA content is observed following inoculation with virulent *Pst* but not with an *hrpA−* strain (avirulent). The increase in ABA is proportional to the increase of the bacterial population. Together with ABA, *Pst* DC3000 also induces JA and SA. The avirulent strain, however, only induces SA. This shows clearly that virulence is linked to ABA increases, but the fact that exogenous application of ABA enhanced the multiplication of *Pst* DC3000 and *hrpA−* is even more convincing. In accordance with these experiments, the ABA-insensitive mutants *abi2-1* and *abi1-1* are more resistant to *Pst* DC3000.

The ABA-deficient mutant *aba3* has been reported to be more susceptible to *Pst* DC3000 when the bacteria are inoculated by spraying (Robert-Seilaniantz et al. 2007). Supporting a positive role for ABA against *Pst* infection, ABA can antagonize *Pst* penetration by inducing a closure of the stomata (Melotto et al. 2006). A possible explanation for these apparently contradictory data is that the timing of the infection determines the role of ABA. ABA controls stomatal closure; therefore, it can control bacterial entrance into the plant. Once inside the plant, the bacteria need to maintain a high water potential to establish disease (Robert-Seilaniantz et al. 2007). In this situation, ABA indeed helps *Pst* progression. A possible reason for the enhanced susceptibility to *Pst* induced by ABA is the repression of several genes that are involved in defense, such as *FRK1* and *NHO1*, and are not downregulated in *abi2-1* and *abi1-1*. In addition to downregulation of these genes, ABA can attenuate early callose deposition after *Pst* inoculation. In sum, ABA is a repressor of basal defenses against *Pst*. The bacterial effector *AvrPtoB* stimulates a rise in ABA levels, which is probably due to a secretion by the bacterium that allows it to take control of plant metabolism.

Although there is an apparent contradiction with other reports that attribute to ABA a positive effect concerning callose deposition against *Blumeria graminis*, *Alternaria brassicicola*, and *Plectosporium cucumerina* (Thaler and Bostock 2004; Ton and Mauch-Mani 2004), this illustrates that, as is known for SA or JA/ET, effects of ABA on defense are dependent on the pathogenicity mode of the respective pathogen and probably on the nature of pathogen-associated molecular patterns (PAMPs) carried by each microbe (de Torres-Zabala et al. 2007).

Involvement of ABA in mediating *Arabidopsis*-necrotrophic fungus interactions is beyond doubt; however, ABA can be a positive or a negative signal depending on the necrotroph that is attacking the plant. Against the soil-borne fungus *Fusarium oxysporum*, ABA acts as a negative regulator of defense through its antagonistic interaction with the JA/ET signaling pathway (Anderson et al. 2004). In the interaction between tomato and the necrotrophic pathogen *Botrytis cinerea*, ABA is also necessary to develop disease (Audenaert et al. 2002). It is important to note that in the tomato–*B. cinerea*, plant defenses are controlled by SA instead of JA, as is the case in *Arabidopsis*. Leaves of the tomato ABA mutant *sitiens* remain more resistant, showing a considerable decrease in spreading lesions produced by *B. cinerea*. The *sitiens* mutation can be complemented by applying increasing concentrations of ABA. At 10 µM of ABA, *sitiens* plants become as susceptible as the wildtype, showing that ABA requires a threshold concentration to influence resistance in tomato (Audenaert et al. 2002). Interestingly, *sitiens* shows increased PAL activity, suggesting that ABA might induce susceptibility in tomato due to negative crosstalk with SA-dependent defenses. More recent findings on the influence of ABA on resistance have shed light on the tomato–*B. cinerea* interaction. It is well known that ABA is linked to ROS production—in particular to H_2O_2 accumulation during abiotic stress. *Arabidopsis* responds to avirulent strains of *B. cinerea* by inducing a minor oxidative burst 18 hours after infection and a second stronger burst at 33 hours postinoculation (Unger et al. 2005). However, when *Arabidopsis* plants are infected with virulent fungal strains, the second burst is suppressed by the pathogen, thus hindering a proper buildup of defense barriers. Recently, a role similar to H_2O_2 has been proposed for ABA in tomato–*B. cinerea* interactions (Asselbergh et al. 2007). The ABA-deficient tomato *sitiens* is more resistant to this necrotroph, and this resistance seems to be dependent on a faster mRNA accumulation of genes involved in cell-wall modifications, together with an enhanced accumulation of H_2O_2 in the epidermal cell walls 1 day after infection. This enhanced production of ROS leads to an increase in phenolic compounds and to cytoplasmic aggregation, both hallmarks for HR (Heath 2000). Accordingly, when this fast oxidative burst is suppressed by treating tomato plants with scavengers such as ascorbic acid or catalases, the fast accumulation of H_2O_2 is abolished, and, accordingly, the enhanced resistance of *sitiens* reverts to wildtype (Asselbergh et al. 2007). ABA application to *sitiens* leads to a loss in resistance, together with a loss of diaminobenzidine (DAB) staining in the anticlinal cell walls, indicating that this mutant is no longer able to accumulate H_2O_2 rapidly. In summary, the lower ABA levels in *sitiens* result in an early-alert state of defense, allowing the mutant to respond earlier and more strongly to pathogen challenge. This strongly suggests that *sitiens* is primed for defense responses against *B. cinerea*.

sitiens also shows a disease phenotype different from wildtype tomatoes when inoculated with the necrotrophic plant pathogenic bacterium *Erwinia chrysanthemi*

(Asselbergh et al. 2007). Although wildtype plants show soft rot symptoms caused by tissue maceration by pectinolytic enzymes produced and secreted by *E. chrysanthemi*, *sitiens* shows only minimal maceration symptoms that never go beyond the infiltration zone of the bacteria. Interestingly, a single application of ABA before infection renders both wildtype and *sitiens* plants very susceptible. Following infection, *sitiens* shows a faster and stronger activation of defense responses, such as H_2O_2 accumulation, peroxidase activation, and cell-wall strengthening. This again points to a constitutively primed state of *sitiens* plants that also operates against bacterial infection.

Despite all this evidence, no clear link between ABA deficiency and enhanced resistance can be established. *Pti5* is a gene that, once overexpressed in tomato, induces resistance specifically against biotic stress and shows a similar gene expression profile to that of *sitiens* (Asselbergh et al. 2007). However, the consideration that *sitiens* is a constitutively stressed plant may be the reason for the observed priming in these mutants because abiotic stress can prime for biotic stress defenses (Conrath et al. 2002).

ABA Increases Disease Resistance

Many reports also show a positive correlation between ABA and disease resistance, but far from being a contradiction, this again shows that ABA can play different roles depending on the plant-pathogen interaction in which it is studied. ABA not only plays a role in modulating plant defenses against pathogens, it can also act directly as an antimicrobial compound. It can inhibit growth and sporidia and teliospore formation of the wheat pathogen *Neovossia indica* (Singh et al. 1997).

Treatment of *Commelina communis* with an inhibitor of plant phospholipase C (U-73122) shows that ABA-induced cytoplasmic calcium, $[Ca^{2+}]cyt$, is suppressed transiently in guard cells that are preincubated with this inhibitor (Staxen et al. 1999). This and later findings by Klüsener et al. (2002) suggest that inositol 1,4,5 triphosphate–triggered Ca^{2+} release mechanisms may contribute to ABA-induced increases in $[Ca^{2+}]cyt$ in *Arabidopsis*. Recently, Ton et al. (2005) characterized a T-DNA insertion mutant called *ibs2* (*insensitive to BABA-induced sterility2*) that is mutated in the 5'-UTR of the *AtSAC1/IBS2* gene. This mutation causes phenotypes disrupted in the *PtdIns*-dependent signaling pathway and correlates with a defect in the priming for ABA-inducible gene expression, indicating that *AtSAC1/IBS2* regulates sensitization for ABA-dependent defenses.

Leaf vitamin C contents may also regulate defense transcripts that affect plant resistance (Pastori et al. 2003; Barth et al. 2004). Interestingly, ascorbic acid (AsA) deficiency results in an increase of transcript levels of 9-cis-epoxycarotenoid dioxygenase (*NCED*), an AsA and Fe^{3+}-dependent dioxygenase that catalyzes the formation of xanthoxin, the precursor of ABA (Figure 1.1; Schwartz et al. 1997). This suggests that low levels of AsA in *vtc1* (vitamin c-1) mutants may decrease the flux through the dioxygenase reaction. The authors hypothesized that elevated *NCED* transcript levels in *vtc1* compensate for the decreased AsA (cofactor) availability, resulting in increased ABA biosynthesis (Pastori et al. 2003). Infection with either virulent *P. syringae* or *H. arabidopsis* results in largely reduced bacterial and hyphal growth in *vtc1* compared with control plants (Barth et al. 2004). *vtc1* mutants have elevated ABA levels, together with elevated PR-5 and PR-1 levels, which could explain the enhanced resistance to

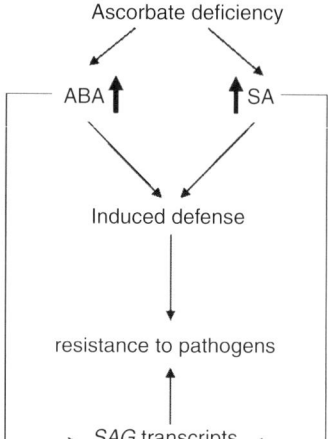

Fig. 1.1 Ascorbate deficiency results in the accumulation of abscisic acid (ABA) and salicylic acid (SA) leading to the induction of defense against pathogens, either directly or through an induction of a subset of senescence associated genes (SAG). (Adapted from Barth et al., 2004).

these pathogens. This is one of the few exceptions in which SA induction and ABA upregulation are present simultaneously, resulting in a resistant phenotype to *Pst*.

Additional partners for ABA different from the classical defense hormones (SA, JA, ET) have been found in pepper plants (Jung et al. 2004). The *CALRR1* gene, a leucine-rich repeat (LRR) motif, is strongly induced by *Xanthomonas campestris*, *Phytophthora capsici*, *Colletotrichum coccodes*, and *Colletotrichum gloesporoides* in pepper leaves. Application of resistance inducers such as BTH, MeJA, BABA, SA, or ET does not induce *CALRR1* transcription. However, it is induced by high salinity, ABA, and wounding. Hybridization studies showed that *CALRR1* mRNA was localized in phloem tissues of leaves and green fruits during pathogen infection and ABA treatment. Therefore, it seems that location and timing pattern of *CALRR1* may play a role in resistance against biotic stress, and ABA could be involved in the signal transduction pathway for its induction (Jung et al. 2004).

ABA also plays an active role in the regulation of papillae formation, consisting mainly of callose, a β-1,3-glucan polymer (Flors et al. 2005). Callose can form a barrier to viral spread through plasmodesmata and hinders cell-wall penetration by *A. brassicicola* and *P. cucumerina* (Ton and Mauch-Mani 2004). Because β-1,3-glucanases are able to degrade callose, their induction can weaken plant defense. A good example for a positive role of ABA in defense is the increased resistance of ABA-pretreated tobacco plants against tobacco mosaic virus that might result from an increased callose deposition in response to infection (Lucas and Wolf 1993) because ABA inhibits the transcription of the basic β-1,3-glucanase (Rezzonico et al. 1998). In such situations, ABA increases resistance to viruses (Mauch-Mani and Mauch 2005) and necrotrophs (Ton and Mauch-Mani 2004; Flors et al. 2008). In *H. arabidopsis-Arabidopsis* interactions, callose also plays a role in defense (Zimmerli et al. 2000); however, in these interactions, ABA does not visibly induce resistance because there are other mechanisms influencing the interaction that contribute more to resistance than callose alone.

Characterization of *leaf wilting* (*lew*) mutants revealed that defects in secondary cell-wall formation have a strong influence on ABA synthesis and proline-synthesis-related genes—and consequently, an influence on drought stress resistance (Chen et al. 2005). A map-based cloning approach revealed that *lew2* mutants are alleles of the *AtCesA8/IRX1* gene. Recently, Hernandez-Blanco et al. (2007) have shown a positive role of ABA in defense based on the *irregular xylem5* (*irx5*) mutant. *IRX5* codes for a subunit of cellulase synthase (*CESA4*) that is required for secondary cell-wall formation (Taylor et al. 2003; Somerville et al. 2004; Brown et al. 2005; Persson et al. 2005). *irx5* and the allelic *irx1* were less susceptible to *P. cucumerina* and strongly resistant to *Ralstonia solanacearum* (Berrocal-Lobo et al. 2002; Hernandez-Blanco et al. 2007). In addition, these mutants showed reduced susceptibility to *B. cinerea* and to the powdery mildew *Erysiphe cichoracearum* USC1 (Hernandez-Blanco et al. 2007). Interestingly, the primary cell-wall mutants *ixr1* and *ixr2* were also more resistant to *B. cinerea* and *E. cichoracearum*. By using the double mutant *irx1-(NahG, ein5, coi1)* it was demonstrated that *irx* resistance is independent of SA and JA/ET signaling (Hernandez-Blanco et al. 2007). A possible explanation for this resistance lies in the constitutive expression of ABA-responsive and defense-related genes in these mutants. This upregulation of ABA-dependent genes also confers drought and osmotic stress resistance to *irx1*. Although the positive role for ABA is consistent in the interaction in *Arabidopsis-R. solanacearum*, it is rather confusing in the case of *P. cucumerina* because *irx1* and *irx5* (with upregulated ABA signaling) are more resistant and *abi1-1*, *abi2-1* and *aba1-6* display reduced susceptibility. This could be explained by the fact that ABA alterations in these mutants result in the activation of separate pathways that seem to regulate distinct sets of defense genes. *IRX* represses ABA-mediated antimicrobial peptides (lipid transfer proteins; Figure 1.2; Molina et al. 1993) or regulators of secondary metabolite biosynthesis (*ATR1*; Grubb and Abel 2006), whereas in the *aba* and *abi* mutants ABA impairment shows that ABA is a

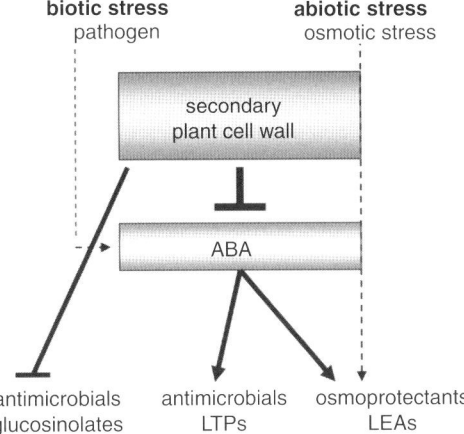

Fig. 1.2 Alterations in the secondary plant cell wall can trigger abscisic acid (ABA) signaling and subsequently lead to the accumulation of antimicrobial substances, such as glucosinolates and lipid transfer proteins (LTPs) and osmoprotectants such as late embryogenesis-associated proteins (LEAs). (Adapted from Hernandez-Blanco et al., 2007).

negative regulator of JA/ET-dependent defense responses but also a positive modulator of JA-dependent defense genes (Anderson et al. 2004; Lorenzo et al. 2004; Adie et al. 2007).

In the interaction between *Arabidopsis* and *Leptosphaeria maculans*, ABA also plays a role in defense (Kaliff et al. 2007). *aba1-3* and *abi1-1* are more susceptible to *L. maculans*; however, *abi2-1*, *abi3-1*, and *abi5-1* (in Ws background) remain resistant to the pathogen. In these resistant mutants, lack of ABA results in high camalexin (a phytoalexin in *Arabidopsis*) levels, creating a resistant phenotype. This was confirmed by double-crossing *aba2-1* and *abi3-1* with *pad3*. The double mutants display enhanced susceptibility toward the pathogen. In this interaction, ABA signaling is dependent on the $RLM1_{Col}$ pathway (Kaliff et al. 2007). The differences in resistance phenotypes among ABA-insensitive mutants seem to be related to the role of *ABI1*, which is involved in a feedback that modulates $RLM1_{Col}$, whereas *ABI2* is not implicated in this pathway. Interestingly, the susceptible phenotype of *abi1-1* and *aba3-1* is associated with a reduction of callose deposition during the infection. Although this reinforces the link between callose and ABA, it is not the only ABA-regulated pathway involved in resistance; the callose deficient mutant *pmr4-1* can also be protected by ABA treatments against *L. maculans* through a different mechanism (Kaliff et al. 2007). Additionally, ABA-mediated callose deposition needs a functional $RLM1_{Col}$ pathway because $rlm1_{col}$-*pad3* is impaired in ABA-induced resistance.

During the interaction between *Arabidopsis* and *Pythium irregulare*, ABA acts as an important regulator of defense gene expression. Analysis of *cis*-regulatory elements revealed an overrepresentation of ABA response elements in promoters of *P. irregulare*-responsive genes. *P. irregulare* is an oomycete that has neither typical necrotrophic nor biotrophic characteristics. During its interaction with *Arabidopsis*, both the JA/ET and SA pathways play a role in pathogen resistance. During infection, a rise in ABA contents is observed 6 and 12 days after inoculation. Using ABA-deficient or -insensitive mutants, it was shown that the ABA pathway needs to be functional for defense against this oomycete, and it acts as a positive element to regulate resistance (Adie et al. 2007). It seems that ABA activation of *Arabidopsis* defenses against *P. irregulare* affects JA biosynthesis. During *P. irregulare* infection, ABA levels increase up to 12hpi. Moreover, the ABA-impaired mutants *aba2-12*, *aba3-2*, and *abi4-1* are more susceptible to *P. irregulare*. In a similar set of experiments with the necrotrophs *B. cinerea* and *A. solani*, a very different pattern of defense is observed. Whereas ABA-defective or ABA-insensitive mutants are more resistant to *B. cinerea*, the same mutants behave more susceptibly to *A. solani*. Therefore, ABA is not restricted to influence against *P. irregulare* but is also a positive signal for defense against *A. solani*, but somehow diminishes resistance against *B. cinerea* (Adie et al. 2007). Moreover, treatment with ABA protects *Arabidopsis* against *A. brassicicola* and *P. cucumerina* (Ton and Mauch-Mani 2004), further indicating that ABA is necessary and sufficient to enhance defense responses against several necrotrophic pathogens. Results regarding ABA-induced resistance against *A. brassicicola* must be considered carefully because once the senescence process has started, ABA is no longer effective in stopping the pathogen. Interestingly, the callose-impaired mutant *pmr4-1* is also susceptible to *P. irregulare*. Similarly, *pmr4-1* is also more susceptible to *A. brassicicola*, although this enhanced susceptibility seems to be controlled by a complicated hormonal interplay (Flors et al. 2008).

To summarize, ABA is a component of the complex defense signaling network that activates resistance against some (but not all) necrotrophic pathogens. ABA seems to enhance resistance through two pathways, priming callose deposition upon attack or regulating defense gene expression and activating JA biosynthesis (Ton and Mauch-Mani 2004; Flors et al. 2008). Although it is widely recognized that ABA negatively affects *Arabidopsis* defenses against *Pst*, as we briefly commented earlier, it can also have a positive role in defense during early penetration stages of the bacterium (Melloto et al. 2006). Far from being passive natural openings for pathogen penetration, stomata play a key role in the innate immune system of plants (Melotto et al. 2006). PAMPs derived from *Pst*, such as *flg22* or the LPS, lead to a stomatal closure within 1 hour of infection. The basis for stomatal participation in innate immune responses is based on a gene-for-gene resistance mediated by *avrRpt2/RPS2* that has a positive effect on promoting stomatal closure. The mutants *ost1 kinase* and ABA-deficient *aba3-1* are unable to perceive the *flg22* or the LPS signal to induce stomatal closure; therefore, it is clear that PAMP-induced stomatal closure is linked to ABA signaling in guard cells (Melotto et al. 2006). However, pathogens have developed a variety of virulence factors to circumvent plant defenses. Virulent races of *Ps* secrete pathogen effectors that inhibit PAMP-triggered immunity (PTI; Jones and Dangl 2006). When virulent strains of *Pst* are inoculated on *Arabidopsis* leaves, a reopening of the stomata can be observed 3 hours after inoculation; accordingly, *Pst cor* mutants or avirulent races are unable to induce such a reopening. Thus, *cor* seems to be the candidate for effector triggered susceptibility (ETS) by inhibiting PAMP-induced ABA signaling in the guard cells. This was further confirmed using *Arabidopsis coi1* mutants that showed a deficient reopening of stomata in the presence of virulent *Pst*. Furthermore, this also establishes a link between JA and ABA signaling in resistance to *Pst*. Interestingly, *COR* does not inhibit NO production in response to PAMPs or ABA. Therefore, *COR* may act downstream of both signals to block PAMP effect on stomatal behavior (Melotto et al. 2006).

A particular case of stomatal control by pathogens occurs during *Plasmopara viticola* grapevine infection that overtakes ABA control for its profit. Allègre et al. (2006) recently reported that stomata of infected grapevine remain open during darkness and water deficit. They presented evidence that this stomatal opening is not due to mechanical forces but seems to be locally confined to those cells with growing hyphae in the substomatal cavity. An increase in the stomatal conductance and water loss is observed between 5 and 8 days after inoculation. The timing is different from the process observed after *Pst* infection in *Arabidopsis*, which leads to stomatal reopening a few hours after infection (Melloto et al. 2006). Stomatal opening induced by *P. viticola* is not different in uninfected leaves during the day. During the process of stomatal control, guard cells remain active; thus, it seems obvious that the pathogen provokes a deregulation of the stomatal closure process. Application of exogenous ABA, darkness or drought stress does not close the stomata in the infected leaf area. Therefore, an unknown, nonsystemic pathogen effector may interfere with ABA signaling to maintain stomata opened (Allègre et al. 2006).

Recently, Ramirez and coworkers (personal communication, January 7, 2007) found that the OCP3 protein, a transcriptional regulator of the homeodomain (HD) family, is regulated by ABA and MeJA and plays a significant role in *Arabidopsis* resistance to *B. cinerea*. *ocp3* shows a highly resistant phenotype to *B. cinerea* and also to drought stress. *ocp3* constitutively expresses *PDF1.2*, which could be the reason for the resistant phenotype to the necrotroph (Coego et al. 2005). Accordingly, when *ocp3* is crossed with *coi1*, the resistant phenotype is lost, but it is still present in crosses with *abi1*. However, *ocp3* is also resistant to drought stress and this phenotype can be reverted by crossing *ocp3* with *abi1* but not with *coi1*. In addition, Ramirez et al. demonstrated that OCP3 interacts with ABI1 by using the yeast two hybrid system (personal communication). Therefore, OCP3, through phosphorylation and desphosphorylation changes, could give flexibility to the fine-tuning of plant responses to various kinds of stress.

Crosstalk with Other Phytohormones

With the use of the model plant *Arabidopsis*, molecular genetic analysis allowed better determination of the signal transduction pathway components involved in hormone signaling (McCourt 1999). A screening procedure designed to pinpoint early signaling components through the identification of specific genes by mutant analysis rapidly showed that theses genes often control the behavior of several plant hormones at the whole plant level. Thus, it became clear that the linear representation of hormone signaling pathways is not accurate and that hormones or hormone signaling pathways interact with each other, as a more or less intensive crosstalk between them that is now a subject of intense research (Rojo et al. 1999; Moller and Chua 1999; Swidzinski et al. 2002; Mahalingam et al. 2003; Gazzarrini and McCourt 2003; De Paepe et al. 2004).

ABA is well known to be a crucial player in the regulation processes governing growth and development and their response to abiotic stress in plants (Srivastava and Lalit 2002; Shinozaki et al. 2003), but until recently, not much was known about the crosstalk of ABA with other plant hormones in which plant pathogen interactions are concerned.

ABA and SA

One of the first reports pointing to possible crosstalk of ABA with the SA-mediated defense pathway was described for the pathosystem soybean-*Phytophthora megasperma f. sp. glycinea* (*Pmg*; Ward et al. 1989). It was shown that when the hypocotyls of etiolated-resistant soybeans are treated with ABA before infection with an avirulent race of *Pmg*, they became susceptible to the oomycete. Concomitantly, the accumulation of transcripts for *PAL*, as well as PAL enzymatic activity, are repressed in ABA pretreated hypocotyls. The levels of the soybean isoflavonoid phytoalexin glyceollin are also lower than in the untreated controls. The authors concluded that the biosynthesis of the phytoalexin may be controlled at the transcriptional level by changes in ABA concentrations caused by infection. From today's point of view, the observed downregulation of PAL activity is likely also to interfere with the SA pathway. The accumulation of SA usually observed after infection with *Pmg* (Shirasu et al. 1997) leads to the accumulation of PR proteins and finally to resistance. Thus, blocking PR protein accumulation, lowering phytoalexin synthesis, and most likely also inhibiting lignification through ABA treatment should lead to the susceptible phenotype observed in such treated soybeans.

The reduction in SA and lignification has been shown to strongly influence the outcome of an infection of *Arabidopsis* by the hemibiotrophic *Pst* (Mohr and Cahill 2007). In this pathosystem as well, lignin and SA increase during the incompatible interaction, but this increase is reduced in case of a preceding treatment with ABA. Interestingly, this was shown to be independent from the development of a HR (Lorenzo et al. 2003). Gene expression in this system was analyzed using the ATH1 genomewide *Arabidopsis* Affymetrix Genechip. ABA treatment turned out to lead to a suppression of many defense-related genes, including members of the phenylpropanoid biosynthesis pathway and genes coding for PR proteins.

Although ABA has been clearly shown to negatively regulate SA signaling, there is no evidence for downregulation of ABA by SA. In a study on the role of PMR4, a pathogen inducible callose synthase of *Arabidopsis*, on the interplay among the SA, JA, and ABA signaling and callose deposition in basal resistance and BABA-IR against hemibiotrophic bacteria (*P. syringae*) and a necrotrophic fungus (*A. brassicicola*), it was shown that the mutant *pmr4-1* overproduces SA; however, there is no reduction in ABA accumulation in this mutant (Flors et al. 2008).

ABA and JA/ET

In a thorough study on the interplay of various defense pathways, Anderson and colleagues (2004) were able to demonstrate a complex crosstalk between the ABA and JA/ET signaling pathways. They showed that treatment of plants with ABA suppresses the JA/ET-activated transcription of defense genes such as *PDF1.2*, *CHI*, *PR4*, and *LEC*. In this interaction, *AtMYC2* (a positive regulator of ABA signaling) overexpression resulted in a repression of *PDF1.2*. However, *AtMYC2* is a dispensable signal for JA repression when ABA is applied exogenously, probably because of multiple control points to define the crosstalk. On the contrary, the ABA biosynthetic mutants, *aba1* and *aba2*, show a constitutive upregulation of JA/ET responsive defense genes. The *jin1/myc2* mutant and the *aba2-1* mutant both display an increase in resistance toward the necrotrophic fungus *Fusarium oxysporum* and an upregulation of JA/ET responsive defense genes, showing that ABA deficiency can positively affect disease resistance to *F. oxysporum*. These results suggest that *AtMYC2* differentially regulates different stress responses, because it can also upregulate the wound-responsive gene *VSP2*. In this context, ABA and its positive regulator (*MYC2*) can interfere with defense signals upstream of *ERF1* that integrates signals of JA and ET pathways (Lorenzo et al. 2003).

Taken together, these results strongly point to a modulation of defense- and stress-responsive gene expression in response to biotic and abiotic stresses through antagonistic interactions between multiple components of ABA and the JA/ET signaling pathways. The more recent finding that AtMYC2 is a common transcription factor of light, ABA, and JA signaling pathways in *Arabidopsis* (Yadav et al. 2005) suggests that light may be involved as an additional component in these already-complex interactions.

An additional, controversial aspect was introduced by Adie et al. (2007). On the basis of a meta-analysis of transcriptomic data, they were able to show that ABA upregulated approximately one-third of the genes induced by *P. irregulare* in *Arabidopsis*. This is in contrast to the results of Anderson et al. (2004), which show that ABA actually downregulates defense genes. The authors speculated that this discrepancy between results is probably explained by the fact that ABA represses only a small, specific group of JA/ET-dependent genes; however, the vast majority of ABA-specific and ABA/JA-related defense genes are clearly upregulated.

ABA and Auxin

Recently, the involvement of auxin and auxin-dependent processes has been repeatedly described in plant-pathogen interaction, especially in the case of bacterial pathogens. Levels of auxin were observed to change rapidly following infection of *Arabidopsis* with virulent strains of either *Pseudomonas or Xanthomonas* (O'Donnell et al. 2003). In addition to the rise in auxin levels after *Pseudomonas* infection, a concomitant rise in ABA was observed (Schmelz et al. 2003). These changes in auxin levels seem to be due to the action of bacterial type III effectors, which have been clearly shown to affect the expression of auxin-responsive genes. An elegant proof of auxin's role in plant response to pathogens was demonstrated by Navarro et al. (2006). They showed that upregulation of auxin signaling led to a higher susceptibility of the plant toward *Pst*; conversely, downregulating auxin signaling led to an increase in resistance against these bacteria. The induction of auxin, as well as ABA-dependent genes, was shown to be mediated by the T3SS of the bacteria (Truman et al. 2006; de Torres-Zabala et al. 2007). Additionally, resistance was only expressed when the ABA signaling pathway was fully functional. Recent evidence points to a possible interaction between ABA- and auxin signaling, and *ABA insensitive 3* (*ABI3*) transcription factors might serve as the link between the two pathways (Nag et al. 2005).

ABA and GA

Historically, GA and ABA have been widely regarded as having primarily antagonistic effects on plant growth and development. Interestingly, this does not seem to be true for the defense-related gene thionin, which was shown to be upregulated by treatment with each of the two hormones in rice callus cultures (Yazaki et al., 2003). Microarray analysis of genes affected by either GA or ABA in rice calli showed that the upstream regions of all the genes that were regulated by both hormones had *cis*-elements for ABA and GA response. In addition, the gene coding for thionin contained numerous types of *cis*-element motives not present in additional 18 ABA- and GA-responsive genes considered by the authors. Rice calli responding to GA also expressed genes such as thaumatin-like protein, class III chitinase, *PAL*, and, again, thionin. On the basis of these cDNA microarray results, the authors proposed that ABA and GA signaling pathways have crosstalk with the pathogen-related signaling pathway in rice calli.

ABA and NO

NO has emerged as a component in the signaling of ABA-mediated stomatal closure (Neill et al. 2002). Foissner et al. (2000) showed that treatment with fungal elicitors can induce rapid NO synthesis in tobacco. In guard cells, functional NO synthesis is needed for ABA-induced stomatal closure; in contrast, ABA applied exogenously induces NO increases.

Despite the evidence linking ABA to NO signaling, NO function in plant-pathogen interactions are not restricted to guard cells and ABA mediation. NO is also implicated in other processes, such as activation of the phenylpropanoid pathway. During incompatible interactions, NO inhibits the antioxidant machinery, ensuring high levels of H_2O_2. It also affects SA accumulation and HR levels upon infection with avirulent *Ps* (Zeier et al. 2004). ABA-independent roles of NO are not discussed further in this chapter.

ABA and Brassinosteroids

Brassinosteroids (BRs) have been described to interact with other plant growth regulators in complicated ways. In the case of GA, there is an additive interaction with BRs, whereas BRs exhibit synergisms with auxins and an antagonistic interaction with ABA (Mandava, 1988). The *brassinosteroid-insensitive2* (*bin2*) mutants are insensitive only to BRs but display no altered sensitivity toward auxins, cytokinins, ET, or GAs. However, several BR mutants display enhanced sensitivity to ABA. The *bin2*, *bri1* (brassinosteroid insensitive), and the *sax1* (hypersensitive to salicylic acid and auxin) mutants all are hypersensitive to ABA (Clouse et al. 1996; Ephritikhine et al. 1999; Li et al. 2001). Therefore, some crosstalk between the BR and the ABA signaling pathways can also reasonably be expected in the case of plant defense against pathogens.

ABA in Priming

Once a plant is infected by a necrotizing pathogen or colonized by root beneficial microbes, a unique physiological state called *priming* is induced. Priming is a phenomenon that stimulates plants to express pertinent defense mechanisms. It has also been called *elicitation competency* (Graham and Graham 994; Conrath et al. 2002). Primed plants display faster or stronger (or both) activation of cellular defense responses that are induced following attack by pathogens or insects or in response to abiotic stress (Prime-A-Plant Group 2006).

Although the mechanisms underlying the priming phenomenon remain unknown, ABA has been shown to be a key factor. As described earlier, evidence suggests that ABA plays a role in secondary cell-wall and callose formation (Rezzonico et al. 1998, Hernandez-Blanco et al. 2007). Callose priming is one of the mechanisms mediated by β-amino butyric acid (BABA). Zimmerli et al. (2000) demonstrated that BABA-treated plants accumulate callose at the sites of attempted penetration by *H. arabidopsidis*, whereas this callose accumulation is not observed in water-treated plants. The callose-impaired mutant *pmr4-1* cannot express BABA-induced resistance (BABA-IR) against *A. brassicicola* and *P. cucumerina*. Therefore, callose priming is also needed to express BABA-IR against necrotrophs (Ton and Mauch-Mani 2004; Flors et al. 2008). However, ABA is also an essential signal to express priming, and accordingly *aba1* and *abi4* are not protected by BABA against *P. cucumerina* and are as susceptible as wildtype plants. The callose response in *abi4-1* is not different from that in Col-0-infected plants, and thus, ABA perception does not affect basal callose accumulation. However, *abi4-1* cannot express callose priming after BABA treatment, indicating that ABA could be involved in the enhancement of callose deposition upon infection (Ton and Mauch-Mani 2004). This hypothesis is further substantiated by experiments in which ABA-treated plants display a similar protection level and callose priming to those treated with BABA. Along the same line of evidence, the T-DNA insertion mutant *insensitive to BABA-induced sterility3* is an insertion that confers deregulation of the *ABA1* gene encoding the ABA biosynthetic enzyme zeaxantin epoxidase (Ton et al. 2005). As is expected from an ABA mutant, *ibs3* is more sensitive to NaCl. With regard to pathogen infection, *ibs3* is impaired in BABA-induced priming for enhanced papillae formation upon infection with *H. arabidopsis*. This defect in the *ABA1* gene correlates with a defect in ABA-inducible *Rab18* and *RD29* gene expression. To summarize, the *IBS3* gene is involved in potentiating signals mediated by ABA that are induced in *Arabidopsis* after BABA treatment (Ton et al. 2005).

Although callose priming seems to be controlled by ABA in *Arabidopsis*, recent findings show that control of callose deposition is different in each plant species. In fact, JA can stimulate callose deposition against *P. viticola* in grapevine more efficiently than ABA. In addition, JA-induced callose deposition is linked to a reduction in the level of infection (Hamiduzzaman et al. 2005). Therefore, ABA plays a role in callose priming in *Arabidopsis* but not in grape. Stronger evidence for callose priming by ABA was recently presented (Flors et al. 2008). The *pmr4-1* mutant is highly susceptible to *A. brassicicola* and is impaired in BABA-IR against this pathogen. However, *Arabidopsis* Col-0 plants can be further protected by BABA against this pathogen. Although basal resistance of Col-0 is already high, following BABA treatment, a strong callose accumulation surrounding penetration sites is observed. Both genotypes, Col-0 and *pmr4-1*, show a reduction of ABA accumulation between 24 and 48 hours after inoculation. This lower ABA content may result in a lower callose deposition in Col-0. Interestingly, BABA treatment of Col-0 and *pmr4-1* antagonizes the pathogen's ABA repression and causes an increase in ABA sensitization in the plant, as indicated by enhanced *ABI1* expression at 24 hpi. This hyperactive ABA signaling in BABA-treated plants results in a faster and stronger accumulation of callose in Col-0 plants that is not present in *pmr4-1*. This explains why pmr4 is impaired in BABA-IR (Flors et al. 2008). The ABA-impaired mutant *npq2* shows a *pmr4-1*-like phenotype for BABA-IR against *A. brassicicola*: it is more susceptible to the pathogen and is not protected by BABA. A possible explanation for this phenotype is that *npq2* has lost the ability to express enhanced callose priming after BABA treatment. However, SA and JA accumulation are both altered in this mutant; therefore, the resultant phenotype may be a contribution of both abnormal hormone accumulation and callose priming impairment (Flors et al. 2008).

ABA participation in priming for enhanced defenses seems to be specific to the priming agent used and the plant-pathogen interaction because vitamin B1 and BABA prime for enhanced defenses against *Pst* through the SA pathway (Ahn et al. 2007; Zimmerli et al. 2000).

During *Arabidopsis-L. maculans* interactions, both ABA and BABA can prime for enhanced resistance. This is another example of the different mechanisms that plants employ to express priming (Kaliff et al. 2007). BABA enhances ABA levels, but ABA also induces protection. Although BABA stimulates callose accumulation,

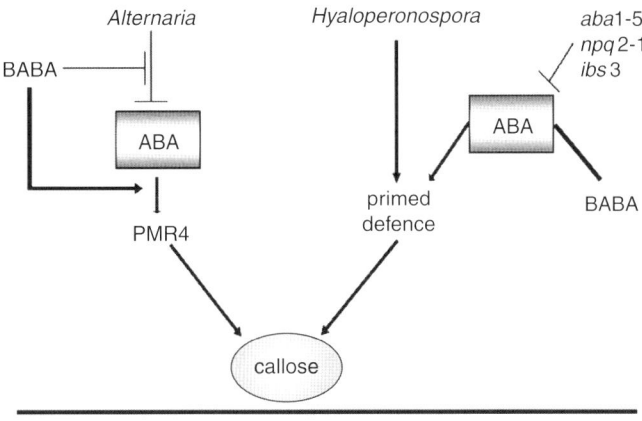

Fig. 1.3 Abscisic acid (ABA) is involved in the expression of β-amino butyric acid (BABA)-induced resistance against biotrophs (*Hyaloperonospora*) and necrotrophs (*Alternaria*) through deposition of callose at the point of penetration into the plant tissue. (Adapted from Flors et al., 2008, and Ton et al., 2005).

there are additional elements that contribute to BABA priming against *L. maculans* because *pmr4-1* is protected by BABA. Interestingly, ABA can also protect this mutant; therefore, ABA induces resistance not only through callose accumulation but through another unknown mechanism as well.

Similarly, Flors et al. (2008) have shown that ABA, but not BABA, can protect *pmr4-1* against *A. brassicicola*, although in this case, callose is also induced by ABA, probably through the induction of a callose synthase other than *PMR4* (Figure 1.3).

Although the *pmr4-1* mutant is more susceptible than the wildtype to *P. irregulare*, it is not as susceptible as ABA-impaired mutants. This demonstrates that callose deposition plays a role in defense against this pathogen, but priming for callose deposition is not the only defense mechanism regulated by ABA (Adie et al. 2007).

Interactions between Biotic and Abiotic Stress Regulation

Given the considerable data from large-scale expression profiling, it is logical to expect the existence of regulatory networks involving both biotic and abiotic stress signaling, especially in light of the importance that stress signaling seems to have in plants. Kreps and colleagues, for example, reported that roughly 30% of the transcriptome on the *Arabidopsis* GeneChip 8 K oligoarray was affected by stress treatments (Kreps et al. 2002). Interestingly, several genes induced transcriptionally by ABA-dependent osmotic stress are also part of plant defense responses to wounding and pathogen attack. Expression of peroxidase, *PR-1*, *PR-10*, and osmotin (*PR-5*) was shown to be increased by water stress, although the role of these proteins in abiotic stress has not been fully clarified (Zhu et al. 1995; Ingram and Bartels 1996).

As mentioned earlier, Anderson et al. (2004) showed that the antagonistic interactions between some components of ABA and the JA/ET signaling pathways modulate defense- and stress-responsive gene expression in response to biotic and abiotic stress. For instance, responsive to dessication 22 (*rd22*) was induced by ABA but strongly suppressed by ET. MJ and ABA both induced *VSP2* (*vegetative storage protein2*), whereas ET suppressed the expression of this gene. These and other results show a clear interaction of signaling among JA, ET, and ABA pathways concerning regulation of wound-, dehydration-, and pathogen-responsive gene expression in *Arabidopsis*.

Jasmonate and ethylene-responsive factor 3 (*JERF3*), an in vivo transcription activator in yeast, binds to an element responsive to ET/JA signaling (GCC box) and to a dehydration-responsive element (DRE) that reacts to dehydration, high salt, and low temperature. In tomato, *JERF3* is induced by ET, JA, cold, salt, and ABA. Interestingly, when *JERF3* is constitutively expressed in transgenic tobacco, the expression of *PR* genes is significantly activated (Wang et al. 2004).

In cotton (*Gossypium barbadense*), a transcription factor gene (*GbERF2*) is constitutively expressed in various tissues, but the highest level of expression is found in leaves after treatment with ET or infection with the pathogenic fungus *Verticillium daliae*. Interestingly, in contrast to the induction of *JERF3* by ABA, in this case, treatment with ABA leads to a marked downregulation of *GbERF2* transcripts. Transgenic tobacco plants constitutively expressing *GbERF2* accumulate higher levels of *PR* gene transcripts, such as *PR-1b*, *PR2*, and *PR4*, and are more resistant to *Alternaria longipes*, but not to bacterial infection by *Ps* pv. *tabaci* (Zuo et al. 2007). *ERF* transcription factors also play important roles in monocots. In wheat (*Triticum eastivum*), the ET-responsive factor 1 (*TaERF1*) is a potential phosphorylation substrate for *TaMAPK1* protein kinase. Transcription of the *TaERF1* gene is induced not only by abiotic stresses but also by infection with the pathogen *Blumeria graminis f.* sp. *tritici*. Overexpression of *TaERF1* induced the expression of stress-related genes, among them *PR* and *COR/RD* genes. Overexpressing plants showed a higher pathogen and abiotic stress tolerance (Xu et al. 2007). These results again point to the involvement of an *ERF* factor—here *TaERF1*—in multiple stress signal transduction pathways.

The ascorbate peroxidase *swAPX1* from sweet potato (*Ipomoea batatas*) was shown to be highly induced in leaves through treatment with methyl viologen, hydrogen peroxide, or ABA and by wounding or exposure to elevated temperatures. Inoculation of the plants with the bacterial pathogen *Pectobacterium chrysanthemi* also led to a strong induction of the gene (S. Y. Park et al. 2004). The authors suggested that *swAPX1* may be involved in hydrogen peroxide detoxification and thus help plants counter the oxidative stress induced by abiotic and biotic stresses.

A clear interaction between biotic and abiotic stress regulation is also obvious in the *ibs3* mutant (discussed earlier). *ibs3* plants show BABA-IR when infected with *Pst* but fail to express BABA-IR against the oomycete *H. arabidopsis*. Conversely, they are also impaired in BABA-IR against salt stress (Ton et al. 2005). These results show that mutations in *ABA1/IBS3* affect priming for ABA-dependent defenses, causing defects in BABA-induced tolerance to salt and priming for callose deposition.

Additional players in stress regulation in the form of retrotransposons have been suggested. Long terminal repeat (LTR) retrotransposons have been suggested to play an important role in genome reorganization induced by environmental challenges. Their promoters respond to various signaling pathways that regulate plant adaptation to biotic and abiotic stresses (Grandbastien 1998). In *Solanum chilense*, the promoter of the *TLC1.1* retrotransposon has two primary ET-responsive elements (PERE boxes) that are essential for stress-induced expression. ABA and SA activate this promoter through a PERE box–independent pathway, but they still require the PERE box for maximal activation. The promoter of TLC1.1 may play a role in the integration of various signal transduction pathways and thus lead to the responsiveness to multiple challenges observed for members of the *TLC1* retrotransposon family.

The possible role of *AtMYC2* was discussed in the section on crosstalk in this chapter (Anderson et al. 2004). The activation of *AtMYC* by ABA is probably not the only reason for the observed downregulation of JA-regulated defense genes, but *AtMYC* still seems to be a key player in the fine-tuning of hormone signaling in response to biotic and abiotic stress.

In addition to *AtMYC2*, *AtMYB2* has also been shown to bind *cis*-elements in *RD22*, and both transcription factors together activate *RD22* expression (Abe et al. 2003). When both *AtMYC2* and *AtMYB2* are overexpressed in *Arabidopsis* plants, they become more sensitive to ABA and more tolerant to osmotic stress compared with wildtype plants. A gene with considerable sequence similarity to *AtMYB2* is Botrytis SUSCEPTIBLE1 (*BOS1*). *bos1* mutants are more sensitive to necrotrophs and display lesser drought, salt, and oxidative stress tolerance (Mengiste et al. 2003).

Many more examples of interactions between biotic and abiotic stress pathways, as well as a possible role for ABA in this crosstalk, can be found. See Fujita et al. (2006) for an extensive discussion on this topic.

Prospects and Conclusions

As presented in this chapter, the large majority of reports describe antagonistic interactions between ABA-mediated abiotic stress signaling pathways and disease resistance. Thus, the suppression of disease signaling pathways, and, therefore, also many biotic stress responses, probably serves as a control mechanism for plants to prioritize their response to abiotic stresses over biotic stress. The application of postgenomic techniques, together with the generation and subsequent analysis of large amounts of microarray data and mutants, have unraveled a complex network of responses from the subcellular to the whole plant level. Such complex networks can only be seriously investigated with an integrated systems approach. The study of plant responses to multiple stimuli and the evolution of the corresponding signaling networks is a relatively new field of research and reminiscent of the study of neural network behavior in animals; first attempts to study such signaling networks have been undertaken (Genoud and Metraux, 1999; Pastori and Foyer, 2002). These new approaches are expected to shed more light on the sometimes controversial roles the hormone ABA plays in plant biotic stress responses.

Acknowledgments

We apologize to the colleagues whose work could not be cited here because of space restrictions. We thank Vicente Ramirez and Pablo Vera, Universidad Politécnica de Valencia, Spain, for sharing their unpublished results.

References

Abe, H., Urao, T., Ito, T., Seki, M., Shinozaki, K., and Yamaguchi-Shinozaki, K. 2003. *Arabidopsis AtMYC2 (bHLH) and AtMYB2 (MYB)* function as transcriptional activators in abscisic acid signaling. *Plant Cell* 15 (1):63–78.

Adie, B., Pérez-Pérez, J., Pérez-Pérez, M.M., Godoy, M., Sánchez-Serrano, J.J., Schmelz, E.A., and Solano, R. 2007. ABA is an essential signal for plant resistance to pathogens affecting JA biosynthesis and the activation of defenses in *Arabidopsis*. *Plant Cell* 19:1665–1681.

Agrios, G.N. 2005. *Plant Pathology*, 5th ed. London: Academic Press.

Ahn, P., Kim, S., Lee, Y.H., and Suh, S.C. 2007. Vitamin B1-induced priming is dependent on hydrogen peroxide and the NPR1 gene in *Arabidopsis*. *Plant Physiol* 143:838–848.

Allègre, M., Daire, X., Heloir, M.C., Trouvelot, S., Mercier, L., Adrian, M., and Pugin, A. 2007. Stomatal deregulation in *Plasmopara viticola*-infected grapevine leaves. *New Phytol* 173:832–840.

Anderson, J.P., Badruzsaufari, E., Schenk, P.M., Manners, J.M., Desmond, O.J., Ehlert, C., Maclean, D.J., Ebert, P.R., and Kazan, K. 2004. Antagonistic interaction between abscisic acid and jasmonate-ethylene signaling pathways modulates defense gene expression and disease resistance in *Arabidopsis*. *Plant Cell* 16:3460–3479.

Asselbergh, B., Curvers, K., Francx, S., Audenaert, K., Vuylsteke, M., Breusegem, F.V., and Hoffte, M.M. 2007. Resistance to *Botrytis cinerea* in *sitiens*, an abscisic acid-deficient tomato mutant, involves timely production of hydrogen peroxide and cell wall modifications in the epidermis. *Plant Physiol* 144:1863–1877.

Audenaert, K., De Meyer, G.B., and Hoffte, M.M. 2002. Abscisic acid determines basal susceptibility of tomato to *B. cinerea* and suppresses salicylic acid dependent signalling mechanisms. *Plant Physiol* 128:491–501.

Barth, C., Moeder, W., Klessig, D.F., and Conklin, P.L. 2004. The timing of senescence and response to pathogens is altered in the ascorbate-deficient arabidopsis mutant vitamin c-1. *Plant Physiol* 134:1784–1792

Berrocal-Lobo, M., Molina, A., and Solano, R. 2002. Constitutive expression of ETHYLENE-RESPONSE-FACTOR1 in *Arabidopsis* confers resistance to several necrotrophic fungi. *Plant J* 29:23–32.

Boyko, E.V., Smith, C.M., Thara, V.K., Bruno, J.M., and Deng, Y. 2006. The molecular basis of plant gene expression during aphid invasion: wheat Pto- and Pti-like sequences are involved in interactions between wheat and Russian wheat aphid (*Homoptera: Aphididae*). *J Economic Entomol.* 99:1430–1445.

Brown, D.M., Zeef, L.A., Ellis, J., Goodacre, R., and Turner, S.R. 2005. Identification of novel genes in *Arabidopsis* involved in secondary cell wall formation using expression profiling and reverse genetics. *Plant Cell* 17:2281–2295.

Cahill, D.M., and Ward, E.W.B. 1989. Rapid localized changes in abscisic acid concentrations in soybean in interactions with *Phytophthora megasperma* f. sp. *glycinea* or after treatment with elicitors. *Physiol Mol Plant Pathol* 35:483–493.

Chen, Z., Hong, X., Zhang, H., Wang, Y., Li, X., Zhu, J.K., and Gong, Z. 2005. Disruption of the cellulose synthase gene, AtCesA8/IRX1, enhances drought and osmotic stress tolerance in *Arabidopsis*. *Plant J* 43:273–283.

Clouse, S.D., Langford, M., and McMorris, T.C. 1996. A brassinosteroid-insensitive mutant in *Arabidopsis thaliana* exhibits multiple defects in growth and development. *Plant Physiol* 111:671–678.

Coego, A., Ramirez, V., Gil, M.J., Flors, V., Mauch-Mani, B., Vera, P. 2005. An *Arabidopsis* homeodomain transcription factor, OVEREXPRESSOR OF CATIONIC PEROXIDASE 3, mediates resistance to infection by necrotrophic pathogens. *Plant Cell* 17:2123–2137

Conrath, U., Pieterse, C.M.J., and Mauch-Mani, B. 2002. Priming in plant-pathogen interactions. *Trends Plant Sci* 7:210–216

De Paepe, A., Vuylsteke, M., Van Hummelen, P., Zabeau, M., and Van Der Straeten, D. 2004. Transcriptional profiling by cDNA-AFLP and microarray analysis reveals novel insights into the early response to ethylene in *Arabidopsis*. *Plant J* 39:537–559.

de Torres-Zabala, M., Truman, W., Bennett, M.H., Lafforgue, G., Mansfield, J.W., Egea, P.R., Bogre, L., and Grant, M. 2007. *Pseudomonas syringae* pv. *tomato* hijacks the *Arabidopsis* abscisic acid signalling pathway to cause disease. *Embo J* 26:1434–1443.

Divol, F., Vilaine, F., Thibivilliers, S., Amselem, J., Palauqui, J.C., Kusiak, C., and Dinant, S. 2005 Systemic response to aphid infestation by *Myzus persicae* in the phloem of Apium graveolens. *Plant Mol Biol* 57:517–540.

Durgbanshi, A., Arbona, V., Pozo, O., Miersch, O., Sancho, J.V., and Gomez-Cadenas, A. 2005. Simultaneous determination of multiple phytohormones from plant extracts by liquid chromatography-electrospray tandem mass spectrometry. *J Agric Food Chem* 53:8437–8442.

Ephritikhine, G., Fellner, M., Vannini, C., Lapous, D., and Barbier-Brygoo, H. 1999. The *sax1* dwarf mutant of *Arabidopsis thaliana* shows altered sensitivity of growth responses to abscisic acid, auxin, gibberellins and ethylene and is partially rescued by exogenous brassinosteroid. *Plant J* 18:303–314.

Flors, V., Ton, J., Jakab, G., and Mauch-Mani, B. 2005. Abscisic acid and callose: team players in defense against pathogens? *J Phytopathol* 153; 1–7.

Flors, V., Ton, J., van Doorn, R., Jakab, G., García-Agustín, P., and Mauch-Mani B. 2008. Interplay between JA, SA, and ABA signalling during basal and induced resistance against *Pseudomonas syringae* and *Alternaria brassicicola*. *Plant J* 54:81–92.

Foissner, I., Wendehenne, D., Langebartels, C., and Durner, J. 2000. In vivo imaging of an elicitor-induced nitric oxide burst in tobacco. *Plant J* 23:817–824.

Fujita, M., Fujita, Y., Noutoshi, Y., Takahashi, F., Narusaka, Y., Yamaguchi-Shinozaki, K., and Shinozaki, K. 2006. Crosstalk between abiotic and biotic stress responses: a current view from the points of convergence in the stress signaling networks. *Curr Opin Plant Biol* 9:436–442.

Gazzarrini, S., and McCourt, P. 2003. Crosstalk in plant hormone signalling: what *Arabidopsis* mutants are telling us. *Ann Bot (Lond)* 91:605–612.

Genoud, T., and Metraux, J.P. 1999. Crosstalk in plant cell signaling: structure and function of the genetic network. *Trends Plant Sci* 4:503–507.

Godfrey, L.D., and Holtzer, T.O. 1991. Influence of temperature and humidity on European corn-borer (*Lepidoptera*, *Pyralidae*) egg hatchability. *Environm Entomol* 20:8–14.

Graham, M.Y., and Graham, T.L. 1994. Wound-associated competency factors are required for the proximal cell responses of soybean to the *Phytophthora sojae* wall glucan elicitor. *Plant Physiol* 105; 571–578

Grandbastien, M.A. 1998. Activation of plant retrotransposons under stress conditions. *Trends Plant Sci* 3:181–187.

Grubb, C.D., and Abel, S. (2006). Glucosinolate metabolism and its control. *Trends Plant Sci* 11; 89–100.

Hamiduzzaman, M.M., Jakab, G., Barnavon, L., Neuhaus, J.-M., and Mauch-Mani, B. 2005. β-amino butyric acid-induced resistance against downy mildew in grapevine acts through the potentiation of callose formation and jasmonic acid signaling. *Mol Plant Microbe Interact* 18:819–829.

Heath, M.C. 2000. Hypersensitive response-related death. *Plant Mol Biol* 44:321–334.

Henfling, J.W.D.M., Bostock, R., and Kuc, J. 1980. Effect of abscisic acid on rishitin and lubimin accumulation and resistance to *Phytophthora infestans* and *Cladosporium cucumerinum* in potato tuber tissue slices. *Phytopathology* 70:1074–1078.

Hernandez-Blanco, C., Feng, D.X., Hu, J., Sanchez-Vallet, A., Deslandes, L., Llorente, F., Berrocal-Lobo, M., Keller, H., Barlet, X., Sanchez-Rodriguez, C., Anderson, L.K., Somerville, S., Marco, Y., and Molina, A. 2007. Impairment of cellulose synthases required for *Arabidopsis* secondary cell wall formation enhances disease resistance. *Plant Cell* 19:890–903.

Ingram, J., and Bartels, D. 1996. The molecular basis of dehydration tolerance in plants. *Ann Rev Plant Physiol Plant Mol Biol* 47:377–403.

Jakab, G., Ton, J., Flors, V., Zimmerli, L., Metraux, J.P., and Mauch-Mani, B. 2005. Enhancing *Arabidopsis* salt and drought stress tolerance by chemical priming for its abscisic acid responses. *Plant Physiol* 139:267–274.

Jones, J.D.G., and Dangl, J.L. 2006. The plant immune system. *Nature* 444:323–329.

Jung, E.H., Jung, H.W., Lee, S.C., Han, S.W., Heu, S., and Hwang, B.K. 2004. Identification of a novel pathogen-induced gene encoding a leucine-rich repeat protein expressed in phloem cells of *Capsicum annuum*. *Biochim Biophys Acta* 1676:211–222.

Kaliff, M., Staal, J., Myrenas, M., and Dixelius, C. 2007. ABA is required for *Leptosphaeria maculans* resistance via ABI1 and ABI4 dependent signalling. *Mol Plant Microbe Interact* 20; 335–345.

Klüsener, B., Young, J.J., Murata, Y., Allen, G.J., Mori, I.C., Hugouvieux, V., and Schroeder, J.I. 2002. Convergence of calcium signaling pathways of pathogenic elicitors and abscisic acid in *Arabidopsis* guard cells. *Plant Physiol* 130:2152–2163.

Koga, H., Dohi, K., and Mori, M. 2004. Abscisic acid and low temperatures suppress the whole plant-specific resistance reaction of rice plants to the infection with *Magnaporthe grisea*. *Physiol Mol Plant Pathol* 65:3–9.

Kreps, J.A., Wu, Y.J., Chang, H.S., Zhu, T., Wang, X., and Harper, J.F. 2002. Transcriptome changes for *Arabidopsis* in response to salt, osmotic, and cold stress. *Plant Physiol* 130:2129–2141.

Kunkel, B.N., and Brooks, D.M. 2002. Crosstalk between signaling pathways in pathogen defense. *Curr Opin Plant Biol* 5:325–331.

Li, J. M., Nam, K.H., Vafeados, D., and Chory, J. 2001. *BIN2*, a new brassinosteroid-insensitive locus in *Arabidopsis*. *Plant Physiol* 127:14–22.

Lorenzo, O., Chico, J.M., Sanchez-Serrano, J.J., and Solano, R. 2004. JASMONATE-INSENSITIVE1 encodes a MYC transcription factor essential to discriminate between different jasmonate-regulated defense responses in *Arabidopsis*. *Plant Cell* 16:1938–1950.

Lorenzo, O., Piqueras, R., Sanchez-Serrano, J.J., and Solano, R. 2003. *ETHYLENE RESPONSE FACTOR1* integrates signals from ethylene and jasmonate pathways in plant defense. *Plant Cell* 15:165–178.

Lucas, W.J., and Wolf, S. 1993. Plasmodesmata: the intercellular organelle of green plants. *Trends Cell Biol* 3:308–315.

Mahalingam, R., Gomez-Buitrago, A., Eckardt, N., Shah, N., Guevara-Garcia, A., Day, P., Raina, R., and Fedoroff, N.V. 2003. Characterizing the stress/defense transcriptome of *Arabidopsis*. *Genome Biol* 4:R20.

Mandava, N.B. 1988. Plant growth-promoting brassinosteroids. *Annu Rev Plant Physiol Plant Molec Biol* 39:23–52.

Mattson, W.J., and Scriber, J.M. 1987. Nutritional ecology of insect folivores of woody plants: nitrogen, water, fiber, and mineral considerations. In: Slansky, F., and Rodriguez, J.G. (eds.), *Nutritional Ecology of Insects, Mites, Spiders and Related Invertebrates*. Wiley, New York, pp. 105–146.

Mauch-Mani, B., and Mauch, F. 2005. The role of abscisic acid in plant–pathogen interactions. *Curr Opin Plant Biol* 8:409–414.

McCourt, P. 1999. Genetic analysis of hormone signaling. *Annu Rev Plant Physiol Plant Molec Biol* 50:219–243.

McDonald, K.L., and Cahill, D.M. 1999. Influence of abscisic acid and the abscisic acid biosynthesis inhibitor, norfluorazon, on interactions between Phytophthora sojae and soybean (*Glycine max*). *Eur J Plant Pathol* 105:651–658.

Melotto, M., Underwood, W., Koczan, J., Nomura, K., and He, S.Y. 2006. Plant stomata function in innate immunity against bacterial invasion. *Cell* 126; 969–980.

Mengiste, T., Chen, X., Salmeron, J., and Dietrich, R. 2003. The *BOTRYTIS SUSCEPTIBLE1* gene encodes an *R2R3MYB* transcription factor protein that is required for biotic and abiotic stress responses in *Arabidopsis*. *Plant Cell* 15:2551–2565.

Mohr, P.G., and Cahill, D.M. 2003. Abscisic acid influences the susceptibility of *Arabidopsis* thaliana to *Pseudomonas syringae* pv. *tomato* and *Peronospora parasitica*. *Funct Plant Biol* 30:461–469.

Mohr, P.G., and Cahill, D.M. 2007. Suppression by ABA of salicylic acid and lignin accumulation and the expression of multiple genes, in *Arabidopsis* infected with *Pseudomonas syringae* pv. *tomato*. *Funct Integr Genomics* 7:181–191.

Molina, A., Segura, A., and García-Olmedo, F. 1993. Lipid transfer proteins (nsLTPs) from barley and maize leaves are potent inchibitors of bacterial and fungal plant pathogens. *FEBS Lett*. 316; 119–122.

Moller, S.G., and Chua, N.H. 1999. Interactions and intersections of plant signaling pathways. *J Mol Biol* 293:219–234.

Nag, R., Maity, M.K., and DasGupta, M. 2005. Dual DNA binding property of ABA insensitive 3 like factors targeted to promoters responsive to ABA and auxin. *Plant Mol Biol* 59:821–838.

Navarro, L., Dunoyer, P., Jay, F., Arnold, B., Dharmasiri, N., Estelle, M., Voinnet, O., and Jones, J.D. 2006. A plant miRNA contributes to antibacterial resistance by repressing auxin signaling. *Science* 312:436–439.

Neil, S.T., Desikan, R., Clarke, A., and Hancok, J.T. 2002. Nitric oxide is a novel component of absicisic acid signaling in stomatal guard cells. *Plant Physiol* 128:13–16.

O'Donnell, P.J., Schmelz, E.A., Moussatche, P., Lund, S.T., Jones, J.B., and Klee, H.J. 2003. Susceptible to intolerance—a range of hormonal actions in a susceptible *Arabidopsis* pathogen response. *Plant J* 33:245–257.

Park, S.J., Huang, Y., and Ayoubi, P. 2006. Identification of expression profiles of sorghum genes in response to greenbug phloemfeeding using cDNA subtraction and microarray analysis. *Planta* 223:932–947.

Park, S.Y., Ryu, S.H., Jang, I.C., Kwon, S.Y., Kim, J.G., and Kwak, S.S. 2004. Molecular cloning of a cytosolic ascorbate peroxidase cDNA from cell cultures of sweet potato and its expression in response to stress. *Mol Genet Genomics* 271:339–346.

Pastori, G.M., and Foyer, C.H. 2002. Common components, networks, and pathways of cross-tolerance to stress. The central role of "redox" and abscisic acid-mediated controls. *Plant Physiol* 129:460–468.

Pastori, G.M., Kiddle, G., Antoni, J., Bernard, S., Veljovic-Jovanovic, S., Verrier, P.J., Noctor, G., and Foyer, C.H. 2003. Leaf vitamin C contents modulate plant defense transcripts and regulate genes that control development through hormone signaling. *Plant Cell* 15:939–951.

Perring, T.M., Holtzer, T.O., Toole, J.L., Norman, J.M., and Myers, G.L. 1984. Influences of temperature and humidity on pre-adult development of the banks grass mite (*Acari Tetranychidae*). *Environ Entomol* 13:338–343.

Persson, S., Wei, H., Milne, J., Page, G.P., and Somerville, C.R. 2005. Identification of genes required for cellulose synthesis by regression analysis of public microarray data sets. *Proc Natl Acad Sci USA* 102:8633–8638.

Prime-A-Plant Group: Conrath, U., Beckers, G.J.M., Flors, V., García-Agustín, P., Jakab, G., Mauch, F., Newman, M.A., Pieterse, C.M.J., Poinssot, B., Pozo, M.J., Pugin, A., Schaffrath, U., Ton, J., Wendehenne, D., Zimmerli, L., and Mauch-Mani, B. 2006. Priming: getting ready for battle. *Mol Plant Microbe Interact* 19:1062–1071.

Rezzonico, E., Flury, N., Meins, F., Jr., and Beffa, R. 1998. Transcriptional down-regulation by abscisic acid of pathogenesis-related beta-1,3-glucanase genes in tobacco cell cultures. *Plant Physiol* 117:585–592.

Robert-Seilaniantz, A., Navarro, L., Bari, R.J., and Jones, J.D.G. 2007. Pathological hormone imbalances. *Curr Opin Plant Biol* 10; 372–379.

Rojo, E., Leon, J., and Sanchez-Serrano, J.J. 1999. Crosstalk between wound signalling pathways determines local versus systemic gene expression in *Arabidopsis thaliana*. *Plant J* 20:135–142.

Ryals, J.A., Neuenschwander, U.H., Willits, M.H., Molina, A., Steiner, H.Y., and Hunt, M.D. 1996. Systemic acquired resistance. *Plant Cell* 8:1809–1819.

Schmelz, E.A., Engelberth, J., Alborn, H.T., O'Donnell, P., Sammons, M., Toshima, H., and Tumlinson, J.H., 3rd. 2003. Simultaneous analysis of phytohormones, phytotoxins, and volatile organic compounds in plants. *Proc Natl Acad Sci USA* 100:10552–10557.

Schwartz, S.H., Tan, B.C., Gage, D.A., Zeevart, J.A., and McCarthy, D.R. 1997. Specific oxidative cleavage of carotenoids by VP14 of maize. *Science* 276:1872–1874.

Singh, P.P., Basra, A.S., and Pannu, P.P.S. 1997. Abscisic acid is a potent inhibitor of growth and sporidial formation in *Neovossia indica* cultures: dual mode of action via loss of polyamines and cellular turgidity. *Phytoparasitica* 25:111–116.

Shinozaki, K., Yamaguchi-Shinozaki, K., and Seki, M. 2003. Regulatory network of gene expression in the drought and cold stress responses. *Curr Opin Plant Biol* 6:410–417.

Shirasu, K., Nakajima, H., Rajasekhar, V.K., Dixon, R.A., and Lamb, C. 1997. Salicylic acid potentiates a gain-control which amplifies pathogen signals for activation of defense mechanisms. *Plant Cell* 9:261–270.

Smith, C.M., and Boyko, E.V. 2007. The molecular bases of plant resistance and defense responses to aphid feeding: current status. *Entomol Exp Applicata* 122:1–16.

Somerville, C., Bauer, S., Brininstool, G., Facette, M., Hamann, T., Milne, J., Osborne, E., Paredez, A., Persson, S., Raab, T., Vorwerk, S., and Youngs, H. 2004. Toward a systems approach to understanding plant cell walls. *Science* 306:2206–2211.

Srivastava, L.M. 2002. *Plant growth and development. Hormones and environment.* Academic Press, Amsterdam.

Staxen, I., Pical, C., Montgomery, L.T., Gray, J.E., Hetherington, A.M., and McAinsh, M.R. (1999) Abscisic acid induces oscillations in guard-cell cytosolic free calcium that involve phosphoinositide-specific phospholipase C. *Proc Natl Acad Sci USA* 96:1779–1784.

Swidzinski, J.A., Sweetlove, L.J., and Leaver, C.J. 2002. A custom microarray analysis of gene expression during programmed cell death in *Arabidopsis thaliana*. *Plant J* 30:431–446.

Taylor, N.G., Howells, R.M., Huttly, A.K., Vickers, K., and Turner, S.R. 2003. Interactions among three distinct CesA proteins essential for cellulose synthesis. *Proc Natl Acad Sci USA* 100:1450–1455.

Thaler, J., and Bostock, R. 2004. Interactions between abscisic acid-mediated responses and plant resistance to pathogens and insects. *Ecology* 85:48–58.

Thomma, B.P., Penninckx, I.A., Broekaert, W.F., and Cammue, B.P. 2001. The complexity of disease signaling in *Arabidopsis*. *Curr Opin Immunol* 13:63–68.

Ton, J., Jakab, G., Toquin, V., Flors, V., Iavicoli, A., Maeder, M.N., Metraux, J.P., and Mauch-Mani, B. 2005. Dissecting the beta-aminobutyric acid-induced priming phenomenon in *Arabidopsis*. *Plant Cell* 17:987–999.

Ton, J., and Mauch-Mani, B. 2004. β-amino-butyric acid-induced resistance against necrotrophic pathogens is based on ABA-dependent priming for callose. *Plant J* 38:119–130.

Truman, W., de Zabala, M.T., and Grant, M. 2006. Type III effectors orchestrate a complex interplay between transcriptional networks to modify basal defence responses during pathogenesis and resistance. *Plant J* 46:14–33.

Unger, C.H., Kleta, S., Jandl, G., Tiedemann, A.V. 2005. Suppression of the defence-related oxidative burst in bean leaf tissue and bean suspension cells by the necrotrophic pathogen *Botrytis cinerea*. *J Phytopathol* 153:15–26.

Voelckel, C., Weisser, W.W., and Baldwin, I.T. 2004. An analysis of plant-aphid interactions by different microarray hybridization strategies. *Molec Ecol* 13:3187–3195.

Wang, H., Huang, A., Chen, Q., Zhang, Z., Zhang, H., Wu, Y., Huang, D., and Huang, R. 2004. Ectopic overexpression of tomato *JERF3* in tobacco activates downstream gene expression and enhances salt tolerance. *Plant Mol Biol* 55:183–192.

Ward, E.W., Cahill, D.M., and Bhattacharyya, M.K. 1989. Abscisic acid suppression of phenylalanine ammonia-lyase activity and mRNA, and resistance of soybeans to *Phytophthora megasperma* f.sp. *glycinea*. *Plant Physiol* 91:23–27.

White, T.C.R. 1974. A hypothesis to explain outbreaks of looper carterpilars with special reference to populations of *Slidosema suavis* in a plantation of *Pinus radiata* in New Zealand. *Oecologia* 16:279–301.

Xu, Z.S., Xia, L.Q., Chen, M., Cheng, X.G., Zhang, R.Y., Li, L.C., Zhao, Y.X., Lu, Y., Ni, Z.Y., Liu, L., Qiu, Z.G., and Ma, Y.Z. 2007. Isolation and molecular characterization of the *Triticum aestivum* L. ethylene-responsive factor 1 (TaERF1) that increases multiple stress tolerance. *Plant Mol Biol* 65:719–732.

Yadav, V., Mallappa, C., Gangappa, S.N., Bhatia, S., and Chattopadhyay, S. 2005. A basic helix-loop-helix transcription factor in *Arabidopsis*, *MYC2*, acts as a repressor of blue light-mediated photomorphogenic growth. *Plant Cell* 17:1953–1966.

Yazaki, J., Kishimoto, N., Nagata, Y., Ishikawa, M., Fujii, F., Hashimoto, A., Shimbo, K., Shimatani, Z., Kojma, K., Suzuki, K., Yamamoto, M., Honda, A., Endo, A., Yoshida, Y., Sato, Y., Takeuchi, K., Toyoshima, K., Miyamoto, C., Wu, J.Z., Sasaki, T., Sakata, K., Yamamoto, K., Iba, K., Oda, T., Otomo, Y., Murakami, K., Matsubara, K., Kawai, J., Carninci, P., Hayashizaki, Y., and Kikuchi, S. 2003. Genomics approach to abscisic acid- and gibberellin-responsive genes in rice. *DNA Res* 10:249–261.

Zeier, J., Delledonne, M., Mishina, T., Severi, E., Sonoda, M., and Lamb, C. 2004. Genetic elucidation of nitric oxide signaling in incompatible plant-pathogen interactions. *Plant Physiol* 136:2875–2886.

Zhu, B.L., Chen, T.H.H., and Li, P.H. 1995. Activation of 2 osmotin-like protein genes by abiotic stimuli and fungal pathogen in transgenic potato plants. *Plant Physiol* 108:929–937.

Zimmerli, L., Jakab, G., Metraux, J.P., and Mauch-Mani, B. 2000. Potentiation of pathogen-specific defense mechanisms in *Arabidopsis* by beta-aminobutyric acid. *Proc Natl Acad Sci USA* 97:12920–12925.

Zuo, K.J., Qin, J., Zhao, J.Y., Ling, H., Zhang, L.D., Cao, Y.F., and Tang, K.X. 2007. Over-expression *GbERF2* transcription factor in tobacco enhances brown spots disease resistance by activating expression of downstream genes. *Gene* 391:80–90.

2 Plant Mitogen-Activated Protein Kinase Cascades in Signaling Crosstalk

Fuminori Takahashi, Kazuya Ichimura, Kazuo Shinozaki, and Ken Shirasu

Introduction

MAPK cascades are a key signaling relay apparatus for responding to various extracellular stimuli in eukaryotic cells (Lewis et al. 1998; Madhani and Fink 1998; Schaeffer and Weber 1999; Widmann et al. 1999). The activation of a MAPK cascade often occurs within 1 to several minutes upon stimulation, representing one of the earliest cellular responses to environmental cues. A typical MAPK cascade consists of three protein kinases—MAPK kinase kinase (MAPKKK), MAPK kinase (MAPKK), and MAPK—that sequentially phosphorylate the corresponding downstream substrates. A phosphorylated MAPK becomes activated and then phosphorylates various proteins such as transcription factors, protein kinases, metabolic enzymes, and cytoskeletal proteins. Because of space constraints, not all literature pertaining to MAPKs is discussed in this review. For additional information, two excellent reviews on MAPK cascades in animals and yeast have been recently published (Avruch 2007; Chen and Thorner 2007).

Despite their obvious biological importance, the number of MAPK cascade components is limited. For example, budding yeast *Saccharomyces cerevisiae* only contains four MAPKKKs, four MAPKKs, and six MAPKs. It is intriguing to consider how an organism can perceive and transmit multiple environmental inputs with such a limited number of MAPK cascade components. Indeed, different stimuli often activate common MAPKKs and MAPKs, and yet a cell still has the capability of triggering distinct and proper downstream responses. Thus, to create signal specificities, various combinations of MAPK cascade components should facilitate specific responses for different stimuli. Such combinations can be brought about by specific docking interaction or with the help of scaffolding proteins (Whitmarsh et al. 1998; Elion 2001; Tanoue and Nishida 2003).

Over the past decade, a number of plant MAPK cascades have been analyzed, especially pertaining to the specific interaction and activation between MAPKKs and MAPKs. Recent investigations have confirmed major roles of defined MAPK pathways in development, cell proliferation, and hormone physiology, as well as in biotic and abiotic stress signaling (Figure 2.1). Surprisingly, it has been discovered that common MAPKK–MAPK combinations are used in response to apparently distinct stimuli. In this review, we cover the recent findings on these MAPK cascades in plant stress and developmental signaling and discuss their specificities.

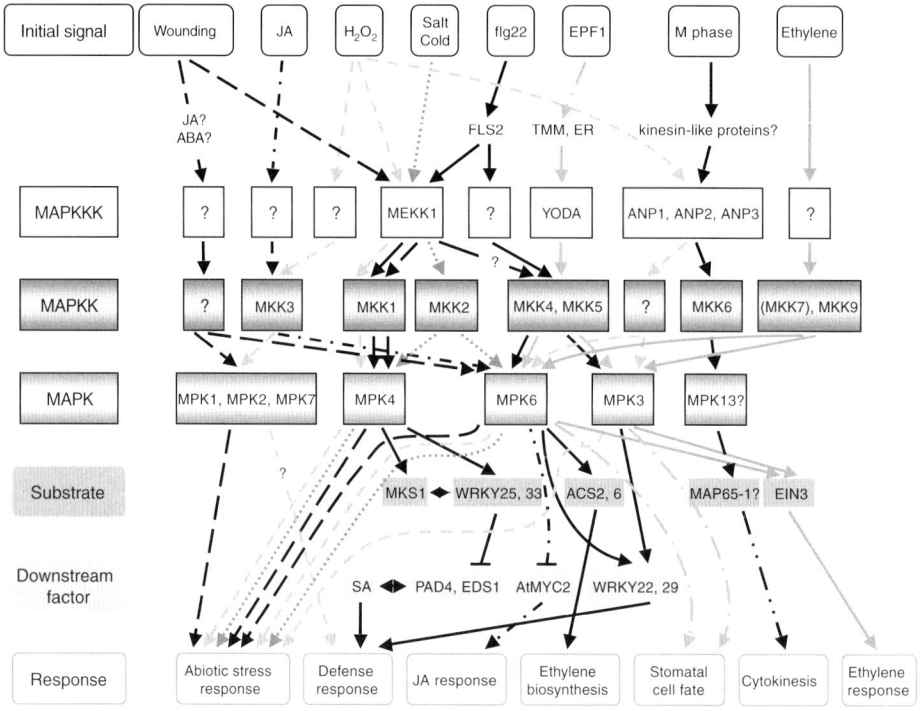

Fig. 2.1 Proposed signal networks of *Arabidopsis* mitogen-activated protein kinase (MAPK) pathway. Arrows and lines indicate signal-flow path and different signals, respectively. Question marks indicate that unknown factors or connections need to be confirmed. JA = jasmonic acid; MAPKK = MAPK kinases; MAPKKK = MAPK kinase kinases; SA = salicylic acid.

MAPK Cascade Components

MAPKs

MAPKs are the last component of MAPK cascades and are therefore likely to specify downstream components. In mammalian cells, there are at least five subfamilies of MAPKs (18 in total). These subfamilies include extracellular signal-regulated kinase (ERK), c-Jun N-terminal kinase (JNK; also called SAPK), p38 (also called RK and CSBP), ERK3/ERK4, and big MAPK (BMK or ERK5; Chang and Karin 2001; Chen et al. 2001; Kyriakis and Avruch 2001; Johnson and Lapadat 2002). In the budding yeast *S. cerevisiae* genome, six MAPKs are encoded, and these have been intensively studied for characterization of their cellular functions (Herskowitz 1995; Gustin et al. 1998; Hohmann 2002; Johnson and Lapadat 2002; O'Rourke et al. 2002). In plants, the *Arabidopsis* genome contains 20 MPK genes (Ichimura et al. 2002) and rice (*Oryza sativa*) and poplar genomes contain similar number of MPK genes (15 and 21, respectively; Hamel et al. 2006; Leu and Xue 2007). Interestingly, it appears that all the plant MPKs are highly homologous to the ERK subfamily, especially in the serine/threonine protein kinase domain. The N- and C-terminal extensions outside of the kinase domain are more divergent and thus more likely to be the determinants of their specificity.

Plant MPKs can be grouped into four distinct groups: A, B, C, and D (Table 2.1). MPKs in the groups A, B, and C contain a conserved dual phosphorylation motif TEY, which is generally phosphorylated by MAPKKs, whereas group D has a TDY motif in this same position. MPKs in group D also have an extended C-terminal region that is not found in the other groups. The group A and B MPKs contain an evolutionarily conserved common docking (CD) domain in their C-terminal extension that serves as the binding site for MAPKKs (Tanoue et al. 2000). Group C MPKs also have a modified CD domain, whereas group D MPKs lack a CD domain (Ichimura et al. 2002).

MAPKKs

Plant MAPKKs (MKKs) consist of relatively smaller gene families compared with MPKs and MAPKKKs. For instance, there are only 10 MKKs in *Arabidopsis*, whereas MPKs and MAPKKKs are encoded by 20 genes and >60 genes, respectively. Similar numbers of MPKs and MKKs are found in rice (15 MPKs and 8 MKKs) and poplar (21 MPKs and 11 MKKs) (Hamel et al. 2006). Thus, plant MKKs are likely to be activated by several distinct MAPKKKs, and in turn activates multiple MPKs. In other words, MKKs may be used as phosphorylation relay module (between MAPKKKs and MPKs), regardless extracellular stimuli. Plant MKKs have been classified into four groups: A, B, C, and D (Table 2.2; Ichimura et al. 2002). MKKs in group A, C, and D have a short N-and C-terminal extension, whereas group B MKKs possess the nuclear transport factor 2 (NTF2) domain (Steggerda and Paschal 2002) at the extended C-terminal region. Similar to animal MAPKKs, most MKKs possess the D domain in the N-terminal region, which functions as a putative MAPK docking site (Bardwell and Thorner 1996; Tanoue and Nishida 2003).

MAPKKKs

Plant MAPKKK families are constituted by extensive number of genes (>60). In addition to the protein kinase domain, they often possess a variety of other functional domains (Ichimura et al. 2002). *Arabidopsis* MAPKKKs are classified into three groups: A, B, and C. Group A consists of 21 MEKK1/STE11/BCK1-type protein kinases including MEKK1, ANPs, YDA, MAPKKKα, and MAP3Kϵ. Collectively, group B and C consist of 49 Raf-like protein kinases. Group B MAPKKKs have an extended N-terminal domain, whereas members from group C have shorter ones. Well-known members such as a negative regulator of ethylene signal transduction, CTR1 (constitutive triple response 1), and a disease signaling component, EDR1 (enhanced disease resistance 1), belong to group B. Functions of group C MAPKKKs are not well understood with exception of HT1 (high leaf temperature 1), which was shown to regulate stomatal opening under low CO_2 condition (Hashimoto et al. 2006).

MPK3 and MPK6: Promiscuous MAPKs or Crosstalk Junction?

Inarguably, the most well-studied plant MPKs are *Arabidopsis* MPK3 and MPK6, which are group A MPKs, the tobacco orthologs of which are WIPK and SIPK,

Table 2.1 Plant mitogen-activated protein kinase (MPKs) from major model plant species.

Group	TxY Motif	Arabidopsis		Rice		Alfalfa	Tobacco	Tomato	Parsley
		Gene	Gene Code	Gene	Gene Code				
A	TEY	MPK3	At3g45640		Os03g17700	SAMK	WIPK	LeMPK3	PsMPK3a
									PcMPK3b
		MPK6	At2g43790	MPK6	Os06g06090	SIMK	SIPK	LeMPK1	PcMPK6
		MPK10	At3g59790				Ntf4	LeMPK2	
B	TEY	MPK4	At4g01370	MPK4	Os10g38950		NtMPK4		PcMPK4
		MPK5	At4g11330						
		MPK11	At1g01560			MMK2			
		MPK12	At2g46070						
		MPK13	At1g07880			MMK3	Ntf6/NRK1		
C	TEY	MPK1	At1g10210						
		MPK2	At1g59580				Ntf3		
		MPK7	At2g18170	MPK7	Os06g48590				
		MPK14	At4g36450	MPK14	Os02g05480				
		MPK8	At1g18150	MPK21-1	Os05g50120				
				MPK21-2	Os01g45620				
		MPK9	At3g18040						
		MPK15	At1g73670						
		MPK16	At5g19010	MPK16	Os11g17080				
		MPK17	At2g01450	MPK17-1	Os06g49430				
				MPK17-2	Os02g04230				
D	TDY	MPK18	At1g53510			TDY1			
		MPK19	At3g14720						
		MPK20	At2g42880	MPK20-1	Os01g43910				
				MPK20-2	Os05g50560				
				MPK20-3	Os06g26340				
				MPK20-4	Os01g47530				
				MPK20-5	Os05g49140				

Accession numbers of alfalfa, tobacco, tomato, and parsley MPKs appear in this paper: SAMK (CAA57721), SIMK (Q07176), MMK2 (CAA57719), MMK3 (CAB37188), TDY1 (AAD28617), WIPK (BAA09600), SIPK (AAB58396), Ntf4 (Q40532), NtMPK4 (BAE46985), Ntf6/NRK1 (CAA58760), Ntf3 (CAA49592), LeMPK3 (AAP20421), LeMPK1 (AAP20419), LeMPK2 (AAP20420), PcMPK3a (CAA73323), PcMPK3b (AAN65181), PcMPK6 (AAN65179), PcMPK4 (AAN65180).

Table 2.2 Plant MKKs from major model plant species.

Group	Arabidopsis Gene	Arabidopsis Gene Code	Rice Gene	Rice Gene Code	Alfalfa	Tobacco	Tomato	Parsley
A	MKK1	At4g26070	MKK1	Os06g05520	PRKK1	SIPKK	tMEK1	PcMKK2
	MKK2	At4g29810						
	MKK6	At5g56580	MKK6	Os01g32660		NtMEK1/NQK1	LeMKK3	
B	MKK3	At5g40440	MKK3	Os06g27890		NPK2		
C	MKK4	At1g51660	MKK4	Os02g54600	SIMKK	NtMEK2	LeMKK2	PcMKK5
	MKK5	At3g21220	MKK5	Os06g09180				
D	MKK7	At1g18350						
	MKK8	At3g06230				NbMKK1	LeMKK4	
	MKK9	At1g73500						
	MKK10	At1g32320	MKK10-1	Os02g46760				
			MKK10-2	Os03g12390				
			MKK10-3	Os03g50550				

Accession numbers of alfalfa, tobacco, tomato, and parsley MKKs appear in this paper: PRKK1 (CAC69138), SIMKK (CAC69137), SIPKK (AAF67262), NtMEK1/NQK1 (CAC24705), NPK2 (BAA06731), NtMEK2 (AAG53979), NbMKK1 (BAE95414), tMEK1 (CAA04261), LeMKK2 (AAU04435), LeMKK3 (AAU04435), LeMKK2 (AAU04435), LeMKK4 (AAU04436), PcMKK2 (AAS21304), PcMKK5 (AAS21305).

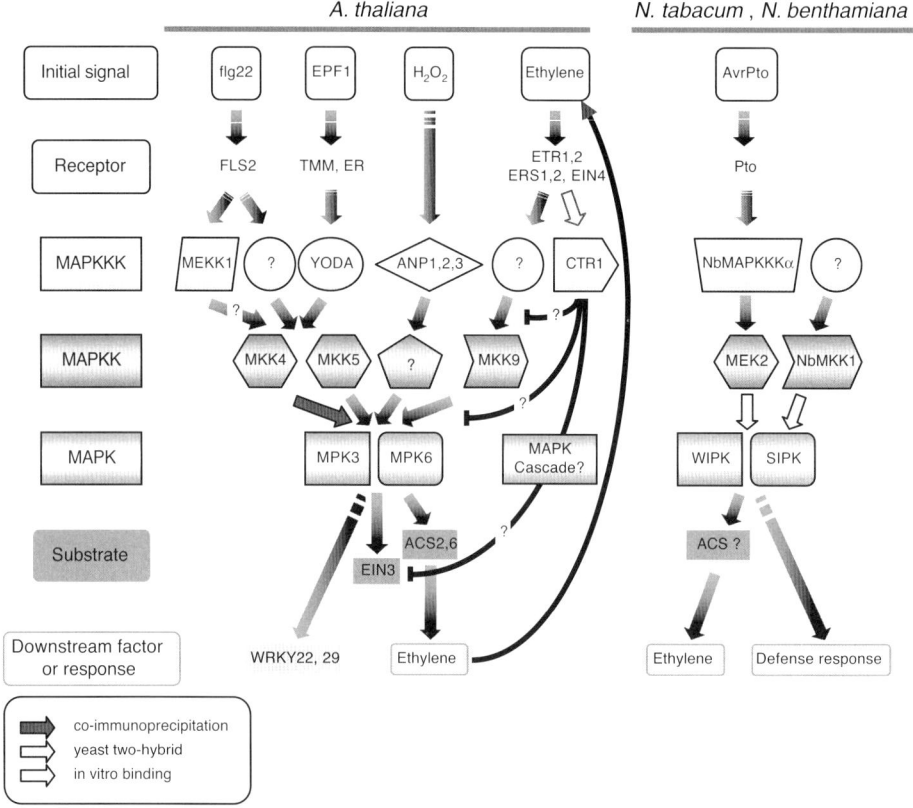

Fig. 2.2 *Arabidopsis* MPK3/MPK6 pathways and their *Nicotiana* orthologs. Shapes of mitogen-activated protein kinase (MAPKs), MAPK kinases (MAPKKs), and MAPK kinase kinases (MAPKKKs) indicate sequence similarity in order to visualize the relationship between *Arabidopsis* and *Nicotiana* species. Different kinds of arrows indicate signal-flow path and support evidences of binding or association. End-broken arrows indicate an indirect connection. Question marks indicate that unknown factors or connections need to be confirmed.

respectively (Figure 2.2; Ichimura et al. 2002; Nakagami et al. 2005; Zhang et al. 2006). These MPKs are involved in multiple responses stimuli, such as stress, pathogen determinants, jasmonic acid (JA), salicylic acid (SA), wounding, reactive oxygen species (ROS), ethylene, and even developmental cues. Here we discuss specific proposed MAPK cascades and potential mechanisms for their specificity.

In *Arabidopsis*, various stresses activate at least three MPKs, MPK3, MPK4, and MPK6, resulting in altered expression of various stress responsive genes (Figure 2.1) (Nakagami et al. 2005; Zhang et al. 2006). For MPK3 and MPK6, the direct upstream MKKs are most likely MKK4 and MKK5 and their corresponding tobacco orthologs is NtMEK2 (Figure 2.2; Yang et al. 2001). Conditional expression of the constitutive active form of MKK4 or MKK5 activates both MPK3 and MPK6 in the protoplast transient assay or dexamethazone (DEX)-inducible system in plants (Asai et al. 2002; Ren et al. 2002). Similarly, the constitutive active form of NtMEK2 also activates WIPK and SIPK in tobacco (Yang et al. 2001). Functional orthologs were also well studied in tomato and parsley, and similar results were presented (Lee et al.

2004; Pedley and Martin 2004). Because MKK4 and MKK5 double silencing lines are most likely lethal at an early developmental stage, their genetic requirement for MPK3 and MPK6 activation has not been established (Wang et al. 2007). Constitutive active forms MKK4 and MKK5 activated MPK6 in vitro (Liu and Zhang 2004). In addition, protoplast transient expression followed by co-immunoprecipitation analyses determined that MKK4 binds to MPK3 (Djamei et al. 2007). Consistently, in vitro pull-down experiments revealed that NtMEK2 binds to both WIPK and SIPK in a D domain–dependent manner (Tanoue et al. 2000; Jin et al. 2003; Tanoue and Nishida 2003).

The direct upstream component of MKK4 and MKK5 in the flagellin signaling pathway is proposed to be MEKK1 (Figure 2.2; Asai et al. 2002). This is based on the fact that overexpression of the N-terminal deleted MEKK1 activates MKK4 and MKK5 in *Arabidopsis* protoplasts. The N-terminal domain of a MAPKKK often negatively regulates its kinase activity, and thus deletion of the regulatory domain would create a constitutive active form. However, this assumption may be misleading because it is possible that some MAPKKK may use this domain as a specificity determinant factor (discussed subsequently). Indeed, in *mekk1* mutants MPK3 and MPK6 are still activated upon flagellin (flg22) or H_2O_2 treatment, suggesting that MEKK1 is not a major upstream factor for MPK3 and MPK6 (Ichimura et al. 2006; Nakagami et al. 2006; Suarez-Rodriguez et al. 2007). Thus, the signaling cascades between the flagellin receptor FLS2, a leucine-rich-repeat (LRR) receptor kinase and MKK4/MKK5-MPK3/MPK6 are still unclear. However, in *Nicotiana benthamiana*, the MAPKKK upstream of MEK2-SIPK is likely to be NbMAPKKKα. This is supported by the observation that defense responses activated by the interaction between disease resistance protein Pto and *Pseudomonas syringae* pv. *tomato* (*Pst*) effector AvrPto is compromised in NbMAPKKKα silencing plants (Figure 2.2; del Pozo et al. 2004). To determine whether this is the case for *Arabidopsis*, future functional analysis of its *Arabidopsis* MAPKKKα ortholog is still required.

Interestingly, the MAPKKK upstream of MKK4/MKK5-MPK3/MPK6 in stomatal patterning has been identified as YODA (Figure 2.2), which is originally identified as a key gene required for the partitioning of embryonic and extra-embryonic cell fates in the basal lineage (Lukowitz et al. 2004). This finding illustrates that a MAPK cascade plays a crucial role not only in responses to external stimuli but also in developmental regulation and, in this case, for the first cell fate decision in embryogenesis and stomatal patterning (Bergmann et al. 2004). Potential upstream signal of YODA is a small secretory peptide, EPF1, which controls stomatal patterning through TOO MANY MOUTH (TMM) receptor-like protein and ERECTA family LRR receptor kinases (Hara et al. 2007; Pillitteri and Torii 2007). Loss of function of *MKK4/MKK5* or *MPK3/MPK6* resulted in disrupted proper epidermal patterning and the formation of clustered stomata. However, the constitutive activation of MKK4/MKK5-MPK3/MPK6 suppresses asymmetric cell divisions and stomatal cell fate specification, resulting in a lack of stomatal differentiation. Similarly, conditional expression of an active form of NtMEK2, a tobacco ortholog for MKK4 and MKK5, in the *YODA* loss of function mutant suppressed the clustered-stomata phenotype (Wang et al. 2007), suggesting that NtMEK2 and MKK4/MKK5 are interchangeable and that the YODA-dependent stomatal patterning pathway is conserved in plants.

ANP1 as Upstream MAPKKK of MPK3 and MPK6

Another MAPKKK proposed to be upstream of MPK3 and MPK6 in oxidative stress signaling is ANP1, or its paralogs ANP2 and ANP3 (Figure 2.2). Similar to MEKK1, when the regulator domain is removed, ANP1 constitutively activates MPK3 and MPK6 in the protoplast transient overexpression system (Kovtun et al. 2000). In the same system, overexpression of full-length ANP1 preferentially enhanced H_2O_2-induced MPK3 and MPK6 activation. These data suggest that the H_2O_2 signal is mediated by the ANP-MPK3/MPK6 pathway. However, it is important to note that there have not been any observations of physical interactions among ANPs and MPK3/MPK6 or MKK4/MMK5. Transgenic tobacco plants expressing NPK1, a tobacco ANP ortholog, show multiple tolerance against freezing, heat, and salt stresses (Kovtun et al. 2000). Similarly, transgenic maize expressing NPK1 exhibits enhanced freezing and drought tolerance (Shou et al. 2004a, 2004b). In addition, ANP1, ANP2, and ANP3 also function in cell division and growth (Krysan et al. 2002). Two of the three *anp* double mutant combinations show abnormality to the developmental stages, and the triple mutant combination has a negative effect on pollen competitiveness. It is not clear, however, that MPK3 and MPK6 are involved in this pathway. The *anp* multiple mutant plants constitutively express stress-related genes. However, no phenotypes related to stress tolerance were examined in the mutants. Thus, the relationship between oxidative stress signaling and cell division/growth mediated by ANP1 remains poorly understood at this time.

MPK3 and MPK6 in Ethylene Biosynthesis and Signal Transduction

MPK3 and MPK6 are also involved in the regulation of ethylene biosynthesis and its signal transduction (Figure 2.2). The MKK4/MKK5–MPK6 pathway regulates the ethylene production through phosphorylation of 1-aminocyclopropane-1-carboxylic acid synthases (ACS2 and ACS6; Liu and Zhang 2004). The ACS proteins catalyze the rate-limiting step in ethylene biosynthesis and are the major regulatory target in stress-induced ethylene production (De Paepe and Van der Straeten 2005). In addition, constitutive expression of an active form of NtMEK2 also induce both ethylene production and an increase of ACS activity (Kim et al. 2003). These data suggest a conserved function of orthologous MAPK cascades in both *Arabidopsis* and tobacco. Biochemical and genetic analysis indicate that MPK6 predominantly phosphorylates ACS2 and ACS6 upon flg22 treatments. MPK6 phosphorylates the C-terminal noncatalytic domain of ACS6 and the phosphorylated ACS6 becomes stable by reducing protein turnover through the 26S proteasome pathway (Liu and Zhang 2004; Joo et al. 2008).

CTR1, a putative Raf-like MAPKKK, is a well-known component that negatively regulates ethylene signal transduction (Guo and Ecker 2004). Recently, Yoo et al. (2008) reported that a novel MAPK pathway, MKK9-MPK3/MPK6, positively regulates ethylene signaling and is negatively regulated by CTR1 (Figure 2.2; Yoo et al. 2008). Protoplast transient expression screening revealed that MKK7 and MKK9 can specifically activate both MPK3 and MPK6. The MKK9-MPK3/MPK6 pathway positively regulates EIN3-mediated transcription. Consistently, MPK3 and MPK6 activation is induced by the exogenous application of 1-aminocyclopropane-1-

carboxylic acid (ACC, precursor of ethylene) and this activation is abolished in *mkk9* plants. Owing to its higher steady-state expression levels, MKK9 is likely to have predominant contributions in the ACC-induced MPK3 and MPK6 activation in comparison to MKK7. The *mkk9* plants exhibit moderate insensitivity to ACC. Consistent with this observation, the expression of an active form of MKK9 resulted in constitutive ethylene responses such as a high-level expression of two ethylene marker genes *ERF1* and *ERF5*. The downstream substrate of MKK9-MPK3/MPK6 is EIN3, which has two phosphorylation sites for the MPKs, T174 and T592. MPK3 and MPK6 phosohorylate T174 in EIN3, and phosphorylated EIN3 becomes stable. However, the phosphorylation of T592, which is likely to be regulated by CTR1, creates instability for EIN3. These results suggest that CTR1 represses ethylene signaling through two mechanisms: T592 phosphorylation in EIN3 and suppression of the MKK9-MPK3/MPK6 pathway, which phosphorylates T174. Thus, the ethylene signal transduction is controlled by a finely tuned balance of positively and negatively regulating MAPK cascades.

MEKK1-MKK1/MKK2-MPK4 Pathway in Biotic and Abiotic Stress

The first complete MAPK cascade in plants, MEKK1-MKK1/MKK2-MPK4, was originally proposed based on the yeast two-hybrid analysis and the functional complementation of budding yeast mutants (Ichimura et al. 1998; Mizoguchi et al. 1998; Figure 2.3). MEKK1 specifically binds to both MKK1 and MKK2 at its C-terminal kinase domain, whereas MKK1 and MKK2 in turn bind to MPK4 in yeast (Figure 2.3; Ichimura et al. 1998; Mizoguchi et al. 1998). Intriguingly, the N-terminal domain of MEKK1 also binds to MPK4. Moreover, in vitro pull-down assay showed that MEKK1 directly interacts with group B MAPKs, including MPK4, MPK5, and MPK13 (Nakagami et al. 2006). Thus, these components are likely to form a tight complex as a phosphorylation relay apparatus. Consistent with this model, a loss of MEKK1 or MKK1 greatly reduced MPK4 activation upon flg22, a bacterial flagellin-derived pathogen associated molecular pattern (PAMP; Ichimura et al. 2006; Meszaros et al. 2006; Suarez-Rodriguez et al. 2007). MKK1 is also activated by flg22 and directly phosphorylates MPK4 (Meszaros et al. 2006). It appears that MKK2 is not required for flg22-induced MPK4 activation (Teige et al. 2004). Although MEKK1 is likely to be activated by flg22, there has yet to be a report of endogenous or transiently expressed MEKK1 activation by flg22 (Nakagami et al. 2006). Thus, the MEKK1-MKK1-MPK4 complex mediates the phosphorylation relay in PAMP signaling. In contrast, the MEKK1-MKK2-MPK4 complex appears to be involved in cold and salt stress responses. This conclusion is based on the observations that MKK2 is activated by such stress treatments and directly phosphorylates MPK4 (Teige et al. 2004). These results strongly suggest that MKK2 is the solo direct upstream component of MPK4 in cold and salt stress signaling. The interaction between MEKK1-MKK1/MKK2-MPK4 and MPK3/MPK6 is less clear. MKK1 is not likely to be direct upstream of MPK3 and MPK6 because MKK1 does not interact or phosphorylate MPK3 and MPK6. However, MKK2 can interact and phosphorylate MPK6, but not MPK3. Considering following observations that 1) flagellin treatment does not activate MKK2, 2) MEKK1 is not essential for MPK6 activation upon flagellin treatment, and 3) no alteration of MPK6 activation occurs between wt and *mkk2* by *Pst* DC3000 infection (Brader et al. 2007), it is possible that MKK2-MPK6 might be independent from flagellin signaling and may constitute a specific complex for cold and salt stresses.

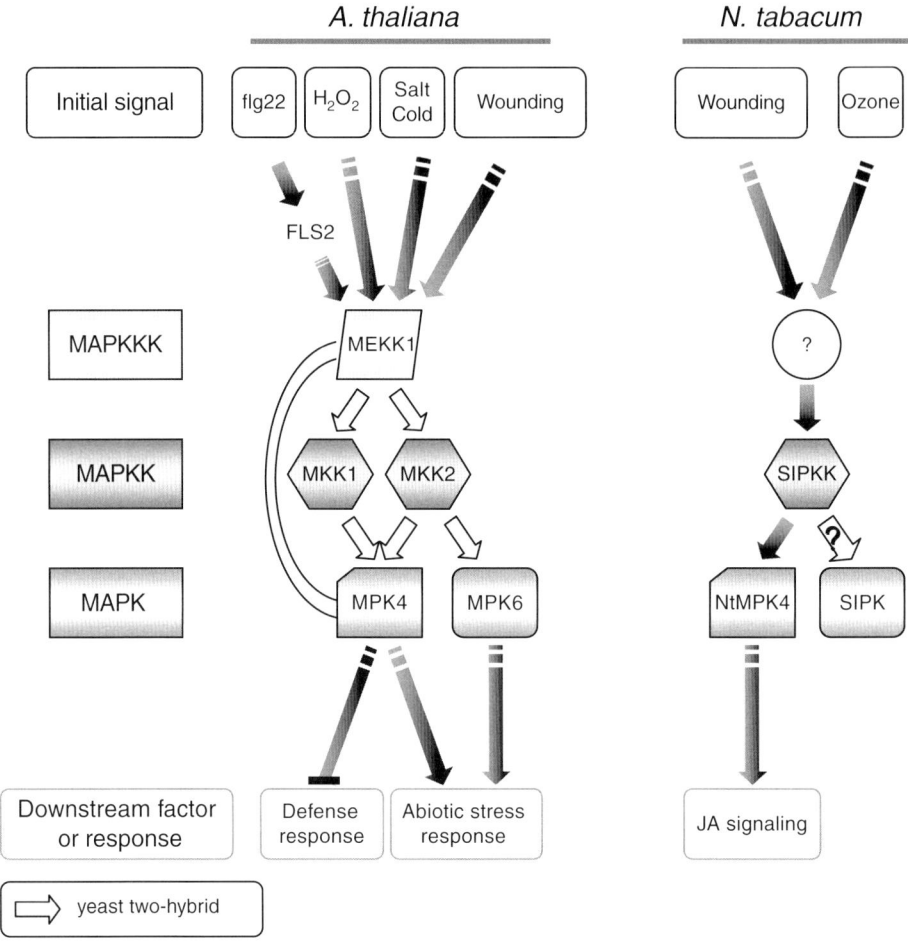

Fig. 2.3 *Arabidopsis* MEKK1-MKK1/MKK2-MPK4/MPK6 pathway and their tobacco orthologs. Shapes of mitogen-activated protein kinases (MAPKs), MAPK kinases (MAPKKs), and MAPK kinase kinases (MAPKKKs) indicate sequence similarity to visualize the relationship between *Arabidopsis* and *Nicotiana* species. Arrows indicate signal-flow path. Gray connections indicate experimental evidence of binding by yeast two-hybrid analysis. End-broken arrows indicate an indirect connection. Question marks indicate that unknown factors or connections need to be confirmed.

Interestingly, MEKK1 directly binds and phosphorylates WRKY53 transcription factor (Miao et al. 2007), a positive regulator of leaf senescence and also possibly has a role in defense response (Dong et al. 2003; Miao et al. 2004). More important, MEKK1 directly binds to the promoter region of WRKY53. These data suggest that MEKK1 directly controls WRKY53 transcription as well as its transcriptional activity. This particular function of MEKK1 seems to be independent from MKKs and MPKs because the activity can be detected in the absence of MKKs and MPKs in vitro. It is of particular interest to determine whether MKK1, MKK2, or MPK4 influence this activity.

The phenotypes of *mekk1* and *mpk4* plants appear to be very similar, also supporting the intimate functional relationship between MEKK1 and MPK4. Both *mekk1* and *mpk4* plants are dwarfed in size and SA-related defense responses such as constitutive *PR* genes expression. In addition, the JA-related defense marker gene *PDF1.2* is downregulated in both *mekk1* and *mpk4* mutants. Thus, MEKK1- MPK4 can be a negative regulatory cascade for SA-related and a positive regulator of JA-related defense responses (Petersen et al. 2000; Ichimura et al. 2006). Interestingly, however, neither *mkk1* nor *mkk2* mutants shows similar phenotypes. It is possible that in the absence of MKK1, its close homolog MKK2 may compensate the loss. One might assume that loss of *mkk1* shows an enhanced disease resistance phenotype similar to *mpk4*, a negative regulator of SA-mediated defense response. Surprisingly, however, *mkk1* plants are in fact more susceptible for *Pst* infection, whereas *mpk4* are more resistant (Meszaros et al. 2006). Consistent with this phenotype, the activation of MPK3 and MPK6 by flg22 is decreased in *mkk1* plants. These data are indicative of a positive regulation of MPK3 and MPK6 by MKK1, although it may not be a direct effect. Another interesting feature of *mekk1* and *mpk4* phenotypes is that their dwarfism is temperature-dependent (Figure 2.4; Ichimura et al. 2006; Su et al. 2007). The temperature-dependent dwarfism is often observed when immune receptors are inappropriately activated (Yang and Hua 2004). Furthermore, the dwarfism in *mekk1* is dependent on RAR1, a key HSP90 co-chaperone that is involved in stabilizing immune receptors (Takahashi et al. 2003; Ichimura et al. 2006). Thus, it is plausible that loss of MEKK1 may inappropriately activate such immune receptors and thereby results in constitutive defense responses and dwarfism.

Tobacco NtMPK4 is an ortholog for *Arabidopsis* MPK4 and is activated by wounding and ozone exposure (see Figure 2.3). In agreement with the *mpk4* mutant phenotype, loss-of-function analysis of NtMPK4 showed a similar phenotype to the *Arabidopsis mpk4* mutant (Gomi et al. 2005). SIPKK, which was originally isolated as a SIPK-interacting MKK (Liu et al. 2000), appears to be an upstream MKK for NtMPK4. This conclusion is supported by the observation that its activity is enhanced by incubation with SIPKK in vitro and by the overexpression of SIPKK in planta (Gomi et al. 2005). It is likely that SIPKK is an ortholog of MKK2 because of NtMPK4 activation and SIPK binding, although no SIPK activation by SIPKK has been reported to date.

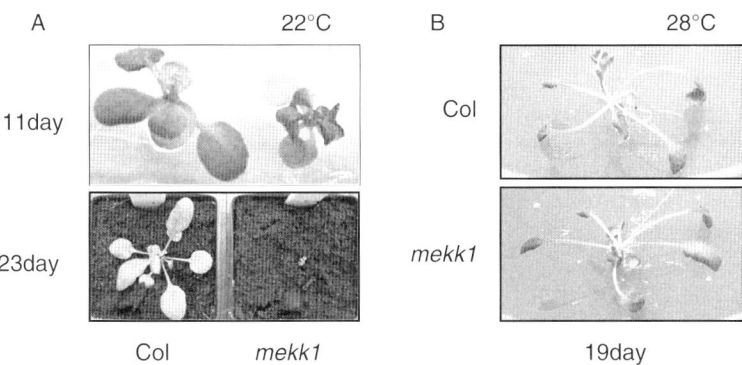

Fig. 2.4 Dwarf phenotype of *mekk1* mutant and its suppression by high temperature. (A) *mekk1* dwarf phenotype at 22°C after 11 or 23 days germination. (B) Phenotype suppression by high temperature (28°C) after 19 days germination.

Arabidopsis MKK3-MPK6 Pathway in JA Signaling

MKK3 is another candidate for a direct upstream component of MPK6, especially in JA signal transduction. MKK3 can activate MPK6 both in vitro and in planta, although direct physical interaction has not yet been observed (see Figure 2.1). JA treatments induce rapid activation of MPK6 in an MKK3-dependent manner, and both *mkk3* and *mpk6* mutants show the reduced root growth sensitivity to JA (Takahashi et al. 2007). In addition, upon JA treatment, *mkk3* and *mpk6* plants exhibit reduced and enhanced expression levels of *PDF1.2* and *VSP2*, useful markers for ET/JA and JA pathways, respectively (Wasternack 2007). Consistently, expression levels of *PDF1.2* and *VSP2* are reversed in plants overexpressing *MKK3* or *MPK6*. Genetic analysis by producing double mutant *mkk3/atmyc2* shows that AtMYC2, the key transcriptional factor of JA pathway, is hypostatic to MKK3 (Takahashi et al. 2007). JA-hypersensitive phenotype of *mkk3* is suppressed by loss of function of *AtMYC2*, resulting in JA insensitive phenotype as *atmyc2* single mutant. These results suggest that MKK3-MPK6 cascade controls the JA signaling pathway by negatively regulating AtMYC2. Therefore, it is suggested that *AtMYC2* function is controlled by a finely turned balance of positive and negative pathways. This report also indicates that at least six MKKs (MKK2, MKK3, MKK4, MKK5, MKK7, and MKK9) activate MPK6, which is activated by various stress stimuli. However, distinct scaffold proteins may play a key role in each signaling pathway.

Arabidopsis MKK3 and the Group C MPKs (MPK1, MPK2, MPK7, and MPK14) in Wounding and Pathogen Infection

Arabidopsis MKK3 is likely to be directly upstream of the group C MPK members. Dóczi et al. showed that MPK1, MPK2, MPK7, and MPK14 are downstream MPK for MKK3 by using yeast two-hybrid analysis, co-immunoprecipitation, and protein kinase assays (Figure 2.1; Dóczi et al. 2007). A protoplast transient expression assay revealed that MKK3 enhanced MPK7 activation by H_2O_2. The activation and specificity of the group C MPKs by MKK3 are clear, although genetic requirement of *MKK3* has not been shown. Interestingly, *MKK3* promoter-GUS lines showed strong induction by infection with the *Pst* DC3000. Moreover, *PR1* promoter-GUS expression was strongly enhanced by coexpression of MKK3-MPK7. Resistance to *Pst* DC3000 was reduced or enhanced by loss-of-function or gain-of-function of MKK3, respectively. These results strongly suggest that the MKK3-group C MPK pathway plays a role in defense signaling and resistance to bacterial pathogens.

MPK1 and MPK2 are also activated by wounding (Figure 2.1; Ortiz-Masia et al. 2007). In contrast to the rapid activation of MPK4 and MPK6 by wounding (Ichimura et al. 2000), MPK1 and MPK2 are activated only after 10 minutes and reach a maximum approximately 2 hours after wounding. This relatively slower activation profile is probably due to the requirement of de novo synthesized protein(s), which may be mediated by the induction of JA, abscisic acid (ABA), or ROS. Exogenous application of JA, ABA and H_2O_2 induced activation of both MPK1 and MPK2 at earlier time points than wounding (Ortiz-Masia et al. 2007). Further study of the activation mechanism of

MPK1 and MPK2 by wounding may provide a new discovery in relation to the signal transduction of wounding responses.

Tobacco NPK1-NtMEK1-NTF6 Cascade in Cell Plate Formation and Defense Responses

Another well-studied complete MAPK cascade is tobacco NPK1-NtMEK1 (also known as NQK1)-NTF6 (NRK1), which is essential for formation of the cell plate during cell division (Soyano et al. 2003; Figure 2.5). A MAPKKK NPK1 is an ortholog of

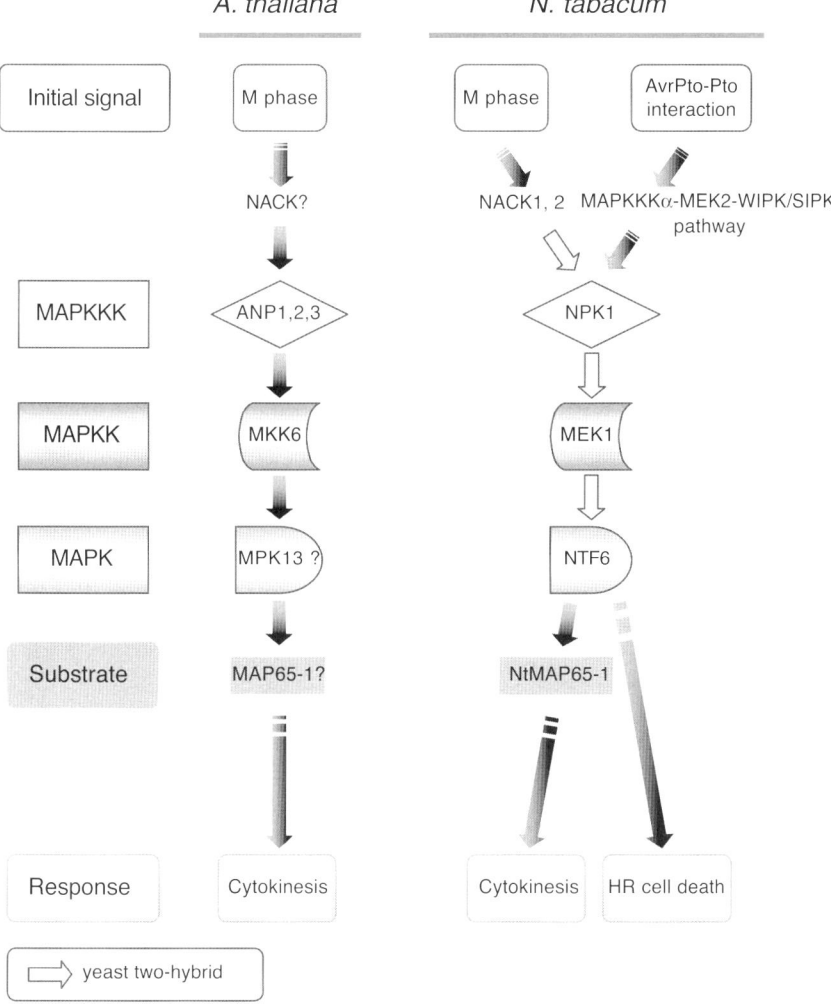

Fig. 2.5 Tobacco NPK1-MEK1-NTF6 (NACK-PQR) pathway and their *Arabidopsis* orthologs. Shapes of mitogen-activated protein kinase (MAPK) cascade components indicate sequence similarity to visualize the relationship between *Arabidopsis* and *Nicotiana* species. Arrows indicate signal-flow path. Gray connections indicate experimental evidences of binding by yeast two-hybrid analysis. End-broken arrows indicate an indirect connection. Question marks indicate that unknown factors or connections need to be confirmed.

Arabidopsis ANP1, ANP2, and ANP3, which were discussed earlier. At the late M phase in the cell cycle, NACK1, a kinesin-like protein, accumulates and forms a complex with NPK1, a MAPKKK. The complex formation between NACK1 and NPK1 induces the activation of NPK1, which interacts and phosphorylates NtMEK1. In turn, NtMEK1 also interacts and phosphorylates NTF6. Both the overexpression of catalytically inactive NPK1 or silencing of NPK1 resulted in the generation of multinucleate guard cells (Nishihama et al. 2001; Jin et al. 2002; Soyano et al. 2003). Similarly, *anp2/anp3* double mutant, also exhibits multinucleate cell phenotype (Krysan et al. 2002). In addition, the overexpression of the catalytically inactive NtMEK1 in tobacco or a knockout mutant of its *Arabidospsis* ortholog *MKK6* also shows multinucleate cells (Soyano et al. 2003). The substrate for NTF6 is NtMAP65-1, a microtubule (MT)-associated protein (MAP) that was identified as a NTF6-phosphorylated protein by using column chromatography (Sasabe et al. 2006). Phosphorylation of NtMAP65-1 by NTF6 downregulates its MT-bundling activity in vitro. This result suggests that the phosphorylation of NtMAP65-1 by NTF6 reduces its microtubule bundling activity in vivo, which enhances the destabilization and turnover of microtubules. This in turn facilitates phragmoplast expansion during cytokinesis. Because MKK6 activates MPK13 (the ortholog of NTF6) in vitro and in yeast (Melikant et al. 2004), the *Arabidopsis* MKK6-MPK13 cascade is likely to have same function as their orthologs in tobacco.

Interestingly, silencing of *NPK1* in *N. benthamiana* resulted in compromised disease resistance against tobacco mosaic virus and hypersensitive responses induced by several resistance genes such as *N*, *Bs2*, and *Rx* (Jin et al. 2002). In addition, silencing of *NPK1*, *NtMEK1*, or *NTF6* suppresses cell death induced by overexpression of a cell death inducer such as a constitutive active form of tomato *MAPKKKα* or *MEK2*. These results suggest that the NPK1-NtMEK1-NTF6 pathway also positively regulates cell death induced by another MAPK cascade (del Pozo et al. 2004). Thus, the NPK1-NtMEK1-NTF6 cascade, and its probable *Arabidopsis* orthologous pathway, ANP-MKK6-MPK13, have at least two functions in the regulation of cell division and disease resistance–related responses.

Protein Phosphatases Negatively Regulate MAPK Cascades

MAPK cascades are often negatively regulated by the dephosphorylation by various protein phosphatases (Schweighofer et al. 2004; Farkas et al. 2007). The *Arabidopsis* genome contains 122 genes encoding putative protein phosphatases, which are classified into five distinct classes; protein phosphatase 2C (PP2C), dual specificity protein phosphatases (DSP; also called MAPK phosphatase [MKP]), serine/threonine protein phosphatases (STPP), protein tyrosine phosphatase (PTP), and low molecular weight (LMW) PTP (Kerk et al. 2002; Schweighofer et al. 2004). PTPs and DSPs are known to be negative regulators of MAPKs or MAPKKs in yeasts and in animals, respectively (Martin et al. 2005; Tamura et al. 2006; Owens and Keyse 2007).

In *Arabidopsis*, a PP2C-type phophatase, AP2C1, is a negative regulator of MPK4 and MPK6 in wounding and pathogen responses. AP2C1 specifically binds to both MPK4 and MPK6 and reduces their kinase activity upon wounding. Plants overexpressing AP2C1 produce less ethylene and JA and are more sensitive to *Botrytis cinerea*. These results suggest that the phosphorylation and dephosphorylation status on MPK6 and MPK4 controlled by AP2C1 are key regulatory mechanisms in JA, wounding, and pathogen signaling pathways (Schweighofer et al. 2007).

Arabidopsis MKP1 is another phosphatase that appears to be a negative regulator of MPK3, MPK4, and MPK6. MKP1 was originally identified from a mutant that was hypersensitive to genotoxic stress (Ulm et al. 2001). With the yeast two-hybrid analysis, MKP1 strongly binds to MPK6 and weakly to MPK3 and MPK4. Genotoxic stresses such as UV-C treatment induce MPK6 activation, which is enhanced in the *mkp1* mutant and reduced in the MKP1 overexpressor. Interestingly, *mkp1* plants exhibits higher salt tolerance (Ulm et al. 2002). These results indicate the negative regulation of MPK6 by MKP1 and an occurrence of signal crosstalk at least between genotoxic and salt stresses. Lee and Ellis (2007) also found MKP2 as a negative regulator of MPK3 and MPK6 by the systematic silencing of five *Arabidopsis* MKP genes. The *mkp2* plants show prolonged ozone-induced activation of MPK3 and MPK6 and are hypersensitive to ozone treatments. When these results are taken into consideration along with MKP2's ability to inactivate MPK3 and MPK6 in vitro, it is likely that MKP2 functions as a second negative regulator of MPK3 and MPK6. Interestingly, NtMKP1 a tobacco ortholog of MKP1, was isolated as a calmodulin (CaM) binding protein (Yamakawa et al. 2004). No NtMKP1 activation occurs by CaM binding; instead, the phosphatase activity of NtMKP1 is strongly increased by the binding of SIPK. Overexpression of NtMKP1 in *N. benthamiana* leaf compromises both cell death and SIPK activation induced by constitutive active potato MEK1 (Katou et al. 2005). In the near future, it is expected that additional protein phosphatases that regulate MAPK cascade components will be isolated and their fine-tuning and crosstalk mechanisms revealed.

Conclusions

We have presented a number of MAPK cascades that control diverse signaling pathways and may serve as potential crosstalk points within the cell. It is intriguing that distinct external stimuli activate common MAPK cascade components. How do these common components control different cellular responses? In yeasts or mammals, scaffolding proteins or interactions within cascade components play an important role in determining combinations of MAPK, MAPKK, and MAPKKK for signaling specificity (Figure 2.6; Dard and Peter 2006). However, in plants, such examples are still rare. Most of the presented studies built MAPK cascade models using in vivo or in vitro activation assays using protein–protein pair combinations rather than threesome interaction of a complete MAPK cascade. The only example for the scaffolding function is the case for MEKK1 and its alfalfa homolog (Ichimura et al. 1998; Nakagami et al. 2004). MEKK1 interacts with the direct downstream MKK1 and MKK2 at the C-terminal kinase domain and their downstream MPK4 (and other group B MPKs) at the N-terminal regulatory domain. Thus, it is likely that MEKK1 holds MKK1/MKK2 and MPK4 for specific phosphorylation relays. This hypothetical scenario is similar to well-known MAPK cascades in yeasts and animals where many scaffolding proteins are known (Figure 2.6). In addition, plant MAPKKK might also specify downstream of MPK, such as transcriptional factors, as seen in the case of MEKK1, which directly binds to WRKY53. Whether MEKK1 actually is a unique example in plants remains unknown at this time. However, as we understand interaction specificities and functions of other MAPKKKs,

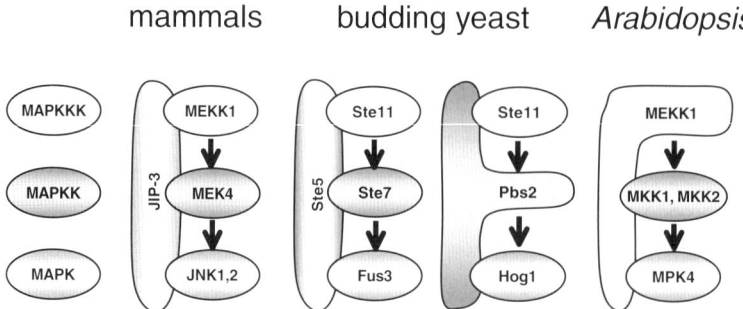

Fig. 2.6 Mitogen-activated protein kinase (MAPK) cascade complex formation by scaffolding in mammal, budding yeast, and plants. In mammalian cells, numerous scaffold proteins have been identified. As an example, JIP-3 protein (Verhey et al. 2001) and components of the c-Jun N-terminal kinase (JNK) signaling pathway are schematically presented. In budding yeast *Saccharomyces cerevisiae*, two scaffold proteins, Ste5 and Pbs2, are identified. Ste5 functions as a scaffold protein and forms a complex with the mating pathway components, Ste11, Ste7, and Fus3. Pbs2 itself is a MAPK kinase (MAPKK) and also functions as a scaffold by tethering its pathway components. In the case of yeast, Ste11 is involved in both the mating and the Hog pathways; however, no crosstalk occurs between these pathways because of the presence of scaffold proteins. MAPKKK = MAPK kinase kinase.

we may find a similar case like MEKK1. For example, it is of particular interest to determine whether MPK3 and MPK6 interact with such scaffolding proteins as they seem to be exceptionally promiscuous. We would also like to know how these complexes are regulated by specific protein phosphatases. MKP1 and AP2C1 were the first two to be isolated. However, it is necessary to isolate more phosphatases that control MAPK cascades for detailed analyses. In addition, we know only a few substrates of MPKs. It is likely that many more exist and remain to be identified. After these substrates have been identified, we may then be in a better position to understand the temporal and spatial mechanisms governing how each MPK specifically finds its substrate.

Acknowledgment

This work is supported by Grant-in-Aid for Scientific Research Nos. 19770045 (to K.I.) and KAKENHI 19678001 (to K.S.). We thank Hirofumi Nakagami for comments on the chapter.

References

Asai, T., Tena, G., Plotnikova, J., Willmann, M.R., Chiu, W.L., Gomez-Gomez, L., Boller, T., Ausubel, F.M., and Sheen, J. (2002). MAP kinase signalling cascade in *Arabidopsis* innate immunity. *Nature* 415:977–983.

Avruch, J. (2007). MAP kinase pathways: the first twenty years. *Biochim Biophys Acta* 1773:1150–1160.

Bardwell, L., and Thorner, J. (1996). A conserved motif at the amino termini of MEKs might mediate high-affinity interaction with the cognate MAPKs. *Trends Biochem Sci* 21:373–374.

Bergmann, D.C., Lukowitz, W., and Somerville, C.R. (2004). Stomatal development and pattern controlled by a MAPKK kinase. *Science* 304:1494–1497.

Brader, G., Djamei, A., Teige, M., Palva, E.T., and Hirt, H. (2007). The MAP kinase kinase MKK2 affects disease resistance in *Arabidopsis*. *Mol Plant Microbe Interact* 20:589–596.

Chang, L., and Karin, M. (2001). Mammalian MAP kinase signalling cascades. *Nature* 410:37–40.

Chen, R.E., and Thorner, J. (2007). Function and regulation in MAPK signaling pathways: lessons learned from the yeast *Saccharomyces cerevisiae*. *Biochim Biophys Acta* 1773:1311–1140.

Chen, Z., Gibson, T.B., Robinson, F., Silvestro, L., Pearson, G., Xu, B., Wright, A., Vanderbilt, C., and Cobb, M.H. (2001). MAP kinases. *Chem Rev* 101:2449–2476.

Dard, N., and Peter, M. (2006). Scaffold proteins in MAP kinase signaling: more than simple passive activating platforms. *Bioessays* 28:146–156.

De Paepe, A., and Van der Straeten, D. (2005). Ethylene biosynthesis and signaling: an overview. *Vitam Horm* 72:399–430.

del Pozo, O., Pedley, K.F., and Martin, G.B. (2004). MAPKKKα is a positive regulator of cell death associated with both plant immunity and disease. *Embo J* 23:3072–3082.

Djamei, A., Pitzschke, A., Nakagami, H., Rajh, I., and Hirt, H. (2007). Trojan horse strategy in Agrobacterium transformation: abusing MAPK defense signaling. *Science* 318:453–456.

Dóczi, R., Brader, G., Pettko-Szandtner, A., Rajh, I., Djamei, A., Pitzschke, A., Teige, M., and Hirt, H. (2007). The *Arabidopsis* mitogen-activated protein kinase kinase MKK3 is upstream of group C mitogen-activated protein kinases and participates in pathogen signaling. *Plant Cell* 19:3266–3279.

Dong, J., Chen, C., and Chen, Z. (2003). Expression profiles of the *Arabidopsis* WRKY gene superfamily during plant defense response. *Plant Mol Biol* 51:21–37.

Elion, E.A. (2001). The Ste5p scaffold. *J Cell Sci* 114:3967–3978.

Farkas, I., Dombradi, V., Miskei, M., Szabados, L., and Koncz, C. (2007). *Arabidopsis* PPP family of serine/threonine phosphatases. *Trends Plant Sci* 12:169–176.

Gomi, K., Ogawa, D., Katou, S., Kamada, H., Nakajima, N., Saji, H., Soyano, T., Sasabe, M., Machida, Y., Mitsuhara, I., Ohashi, Y., and Seo, S. (2005). A mitogen-activated protein kinase NtMPK4 activated by SIPKK is required for jasmonic acid signaling and involved in ozone tolerance via stomatal movement in tobacco. *Plant Cell Physiol* 46:1902–1914.

Guo, H., and Ecker, J.R. (2004). The ethylene signaling pathway: new insights. *Curr Opin Plant Biol* 7:40–49.

Gustin, M.C., Albertyn, J., Alexander, M., and Davenport, K. (1998). MAP kinase pathways in the yeast *Saccharomyces cerevisiae*. *Microbiol Mol Biol Rev* 62:1264–1300.

Hamel, L.P., Nicole, M.C., Sritubtim, S., Morency, M.J., Ellis, M., Ehlting, J., Beaudoin, N., Barbazuk, B., Klessig, D., Lee, J., Martin, G., Mundy, J., Ohashi, Y., Scheel, D., Sheen, J., Xing, T., Zhang, S., Seguin, A., and Ellis, B.E. (2006). Ancient signals: comparative genomics of plant MAPK and MAPKK gene families. *Trends Plant Sci* 11:192–198.

Hara, K., Kajita, R., Torii, U., Bergmann, D.C., and Kakimoto, T. (2007). The secretory peptide gene EPF1 enforces the stomatal one-cell-spacing rule. *Genes Dev* 21:1720–1725.

Hashimoto, M., Negi, J., Young, J., Israelsson, M., Schroeder, J.I., and Iba, K. (2006). *Arabidopsis* HT1 kinase controls stomatal movements in response to CO_2. *Nat Cell Biol* 8:391–397.

Herskowitz, I. (1995). MAP kinase pathways in yeast: for mating and more. *Cell* 80:187–197.

Hohmann, S. (2002). Osmotic adaptation in yeast—control of the yeast osmolyte system. *Int Rev Cytol* 215:149–187.

Ichimura, K., Casais, C., Peck, S.C., Shinozaki, K., and Shirasu, K. (2006). MEKK1 is required for MPK4 activation and regulates tissue specific and temperature dependent cell death in *Arabidopsis*. *J Biol Chem* 281:36969–36976.

Ichimura, K., Mizoguchi, T., Irie, K., Morris, P., Giraudat, J., Matsumoto, K., and Shinozaki, K. (1998). Isolation of ATMEKK1 (a MAP kinase kinase kinase)-interacting proteins and analysis of a MAP kinase cascade in *Arabidopsis*. *Biochem Biophys Res Commun* 253:532–543.

Ichimura, K., Mizoguchi, T., Yoshida, R., Yuasa, T., and Shinozaki, K. (2000). Various abiotic stresses rapidly activate *Arabidopsis* MAP kinases ATMPK4 and ATMPK6. *Plant J* 24:655–665.

Ichimura, K., Shinozaki, K., Tena, G., Sheen, J., Henry, Y., Champion, A., Kreis, M., Zhang, S., Hirt, H., Wilson, C., Heberle-Bors, E., Ellis, B.E., Morris, P.C., Innes, R.W., Ecker, J.R., Scheel, D., Klessig, D.F., Machida, Y., Mundy, J., Ohashi, Y., and Walker, J.C. (2002). Mitogen-activated protein kinase cascades in plants: a new nomenclature. *Trends Plant Sci* 7:301–308.

Jin, H.L., Axtell, M.J., Dahlbeck, D., Ekwenna, O., Zhang, S., Staskawicz, B., and Baker, B. (2002). NPK1, an MEKK1-like mitogen-activated protein kinase kinase kinase, regulates innate immunity and development in plants. *Dev Cell* 3:291–297.

Jin, H.L., Liu, Y.D., Yang, K.Y., Kim, C.Y., Baker, B., and Zhang, S. (2003). Function of a mitogen-activated protein kinase pathway in N gene-mediated resistance in tobacco. *Plant J* 33:719–731.

Johnson, G.L., and Lapadat, R. (2002). Mitogen-activated protein kinase pathways mediated by ERK, JNK, and p38 protein kinases. *Science* 298:1911–1912.

Joo, S., Liu, Y., Lueth, A., and Zhang, S. (2008). MAPK phosphorylation-induced stabilization of ACS6 protein is mediated by the non-catalytic C-terminal domain, which also contains the cis-determinant for rapid degradation by the 26S proteasome pathway. *Plant J* 54:129–140.

Katou, S., Karita, E., Yamakawa, H., Seo, S., Mitsuhara, I., Kuchitsu, K., and Ohashi, Y. (2005). Catalytic activation of the plant MAPK phosphatase NtMKP1 by its physiological substrate salicylic acid-induced protein kinase but not by calmodulins. *J Biol Chem* 280:39569–39581.

Kerk, D., Bulgrien, J., Smith, D.W., Barsam, B., Veretnik, S., and Gribskov, M. (2002). The complement of protein phosphatase catalytic subunits encoded in the genome of *Arabidopsis*. *Plant Physiol* 129:908–925.

Kim, C.Y., Liu, Y.D., Thorne, E.T., Yang, H., Fukushige, H., Gassmann, W., Hildebrand, D., Sharp, R.E., and Zhang, S. (2003). Activation of a stress-responsive mitogen-activated protein kinase cascade induces the biosynthesis of ethylene in plants. *Plant Cell* 15:2707–2718.

Kovtun, Y., Chiu, W.L., Tena, G., and Sheen, J. (2000). Functional analysis of oxidative stress-activated mitogen-activated protein kinase cascade in plants. *Proc Natl Acad Sci USA* 97:2940–2945.

Krysan, P.J., Jester, P.J., Gottwald, J.R., and Sussman, M.R. (2002). An *Arabidopsis* mitogen-activated protein kinase kinase kinase gene family encodes essential positive regulators of cytokinesis. *Plant Cell* 14:1109–1120.

Kyriakis, J.M., and Avruch, J. (2001). Mammalian mitogen-activated protein kinase signal transduction pathways activated by stress and inflammation. *Physiol Rev* 81:807–869.

Lee, J., Rudd, J.J., Macioszek, V.K., and Scheel, D. (2004). Dynamic changes in the localization of MAPK cascade components controlling pathogenesis-related (PR) gene expression during innate immunity in parsley. *J Biol Chem* 279:22440–22448.

Lee, J.S., and Ellis, B.E. (2007). *Arabidopsis* MAPK phosphatase 2 (MKP2) positively regulates oxidative stress tolerance and inactivates the MPK3 and MPK6 MAPKs. *J Biol Chem* 282:25020–25029.

Lewis, T.S., Shapiro, P.S., and Ahn, N.G. (1998). Signal transduction through MAP kinase cascades. *Adv Cancer Res* 74:49–139.

Liu, Q., and Xue, Q. (2007). Computational identification and phylogenetic analysis of the MAPK gene family in *Oryza sativa*. *Plant Physiol Biochem* 45:6–14.

Liu, Y., and Zhang, S. (2004). Phosphorylation of 1-aminocyclopropane-1-carboxylic acid synthase by MPK6, a stress-responsive mitogen-activated protein kinase, induces ethylene biosynthesis in *Arabidopsis*. *Plant Cell* 16:3386–3399.

Liu, Y., Zhang, S., and Klessig, D.F. (2000). Molecular cloning and characterization of a tobacco MAP kinase kinase that interacts with SIPK. *Mol Plant Microbe Interact* 13:118–124.

Lukowitz, W., Roeder, A., Parmenter, D., and Somerville, C. (2004). A MAPKK kinase gene regulates extra-embryonic cell fate in *Arabidopsis*. *Cell* 116:109–119.

Madhani, H.D., and Fink, G.R. (1998). The riddle of MAP kinase signaling specificity. *Trends Genet* 14:151–155.

Martin, H., Flandez, M., Nombela, C., and Molina, M. (2005). Protein phosphatases in MAPK signalling: we keep learning from yeast. *Mol Microbiol* 58:6–16.

Melikant, B., Giuliani, C., Halbmayer-Watzina, S., Limmongkon, A., Heberle-Bors, E., and Wilson, C. (2004). The *Arabidopsis thaliana* MEK AtMKK6 activates the MAP kinase AtMPK13. *FEBS Lett* 576:5–8.

Meszaros, T., Helfer, A., Hatzimasoura, E., Magyar, Z., Serazetinova, L., Rios, G., Bardoczy, V., Teige, M., Koncz, C., Peck, S., and Bogre, L. (2006). The *Arabidopsis* MAP kinase kinase MKK1 participates in defence responses to the bacterial elicitor flagellin. *Plant J* 48:485–498.

Miao, Y., Laun, T.M., Smykowski, A., and Zentgraf, U. (2007). *Arabidopsis* MEKK1 can take a short cut: it can directly interact with senescence-related WRKY53 transcription factor on the protein level and can bind to its promoter. *Plant Mol Biol* 65:63–76.

Miao, Y., Laun, T., Zimmermann, P., and Zentgraf, U. (2004). Targets of the WRKY53 transcription factor and its role during leaf senescence in *Arabidopsis*. *Plant Mol Biol* 55:853–867.

Mizoguchi, T., Ichimura, K., Irie, K., Morris, P., Giraudat, J., Matsumoto, K., and Shinozaki, K. (1998). Identification of a possible MAP kinase cascade in *Arabidopsis thaliana* based on pairwise yeast two-hybrid analysis and functional complementation tests of yeast mutants. *FEBS Lett* 437:56–60.

Nakagami, H., Kiegerl, S., and Hirt, H. (2004). OMTK1, a novel MAPKKK, channels oxidative stress signaling through direct MAPK interaction. *J Biol Chem* 279:26959–26966.

Nakagami, H., Pitzschke, A., and Hirt, H. (2005). Emerging MAP kinase pathways in plant stress signalling. *Trends Plant Sci* 10:339–346.

Nakagami, H., Soukupova, H., Schikora, A., Zarsky, V., and Hirt, H. (2006). A mitogen-activated protein kinase kinase kinase mediates reactive oxygen species homeostasis in *Arabidopsis*. *J Biol Chem* 281:38697–38704.

Nishihama, R., Ishikawa, M., Araki, S., Soyano, T., Asada, T., and Machida, Y. (2001). The NPK1 mitogen-activated protein kinase kinase kinase is a regulator of cell-plate formation in plant cytokinesis. *Genes Dev* 15:352–363.

O'Rourke, S.M., Herskowitz, I., and O'Shea, E.K. (2002). Yeast go the whole HOG for the hyperosmotic response. *Trends Genet* 18:405–412.

Ortiz-Masia, D., Perez-Amador, M.A., Carbonell, J., and Marcote, M.J. (2007). Diverse stress signals activate the C1 subgroup MAP kinases of *Arabidopsis*. *FEBS Lett* 581:1834–1840.

Owens, D.M., and Keyse, S.M. (2007). Differential regulation of MAP kinase signalling by dual-specificity protein phosphatases. *Oncogene* 26:3203–3213.

Pedley, K.F., and Martin, G.B. (2004). Identification of MAPKs and their possible MAPK kinase activators involved in the Pto-mediated defense response of tomato. *J Biol Chem* 279:49229–49235.

Petersen, M., Brodersen, P., Naested, H., Andreasson, E., Lindhart, U., Johansen, B., Nielsen, H.B., Lacy, M., Austin, M.J., Parker, J.E., Sharma, S.B., Klessig, D.F., Martienssen, R., Mattsson, O., Jensen, A.B., and Mundy, J. (2000). *Arabidopsis* MAP kinase 4 negatively regulates systemic acquired resistance. *Cell* 103:1111–1120.

Pillitteri, L.J., and Torii, K.U. (2007). Breaking the silence: three bHLH proteins direct cell-fate decisions during stomatal development. *Bioessays* 29:861–870.

Ren, D., Yang, H., and Zhang, S. (2002). Cell death mediated by MAPK is associated with hydrogen peroxide production in *Arabidopsis*. *J Biol Chem* 277:559–565.

Sasabe, M., Soyano, T., Takahashi, Y., Sonobe, S., Igarashi, H., Itoh, T.J., Hidaka, M., and Machida, Y. (2006). Phosphorylation of NtMAP65-1 by a MAP kinase down-regulates its activity of microtubule bundling and stimulates progression of cytokinesis of tobacco cells. *Genes Dev* 20:1004–1014.

Schaeffer, H.J., and Weber, M.J. (1999). Mitogen-activated protein kinases: specific messages from ubiquitous messengers. *Mol Cell Biol* 19:2435–2444.

Schweighofer, A., Hirt, H., and Meskiene, I. (2004). Plant PP2C phosphatases: emerging functions in stress signaling. *Trends Plant Sci* 9:236–243.

Schweighofer, A., Kazanaviciute, V., Scheikl, E., Teige, M., Doczi, R., Hirt, H., Schwanninger, M., Kant, M., Schuurink, R., Mauch, F., Buchala, A., Cardinale, F., and Meskiene, I. (2007). The PP2C-type phosphatase AP2C1, which negatively regulates MPK4 and MPK6, modulates innate immunity, jasmonic acid, and ethylene levels in *Arabidopsis*. *Plant Cell* 19:2213–2224.

Shou, H., Bordallo, P., Fan, B., Yeakley, J.M., Bibikova, M., Sheen, J., and Wang, K. (2004a). Expression of an active tobacco mitogen-activated protein kinase kinase kinase enhances freezing tolerance in transgenic maize. *Proc Natl Acad Sci USA* 101:3298–3303.

Shou, H., Bordallo, P., and Wang, K. (2004b). Expression of the *Nicotiana* protein kinase (NPK1) enhanced drought tolerance in transgenic maize. *J Exp Bot* 55:1013–109.

Soyano, T., Nishihama, R., Morikiyo, K., Ishikawa, M., and Machida, Y. (2003). NQK1/NtMEK1 is a MAPKK that acts in the NPK1 MAPKKK-mediated MAPK cascade and is required for plant cytokinesis. *Genes Dev* 17:1055–1067.

Steggerda, S.M., and Paschal, B.M. (2002). Regulation of nuclear import and export by the GTPase Ran. *Int Rev Cytol* 217:41–91.

Su, S.H., Suarez-Rodriguez, M.C., and Krysan, P. (2007). Genetic interaction and phenotypic analysis of the *Arabidopsis* MAP kinase pathway mutations mekk1 and mpk4 suggests signaling pathway complexity. *FEBS Lett* 581:3171–3177.

Suarez-Rodriguez, M.C., Adams-Phillips, L., Liu, Y., Wang, H., Su, S.H., Jester, P.J., Zhang, S., Bent, A.F., and Krysan, P.J. (2007). MEKK1 is required for flg22-induced MPK4 activation in *Arabidopsis* plants. *Plant Physiol* 143:661–669.

Takahashi, A., Casais, C., Ichimura, K., and Shirasu, K. (2003). HSP90 interacts with RAR1 and SGT1, and is essential for RPS2-mediated resistance in *Arabidopsis*. *Proc Natl Acad Sci USA* 100:11777–11782.

Takahashi, F., Yoshida, R., Ichimura, K., Mizoguchi, T., Seo, S., Yonezawa, M., Maruyama, K., Yamaguchi-Shinozaki, K., and Shinozaki, K. (2007). The mitogen-activated protein kinase cascade MKK3-MPK6 is an important part of the jasmonate signal transduction pathway in *Arabidopsis*. *Plant Cell* 19:805–818.

Tamura, S., Toriumi, S., Saito, J., Awano, K., Kudo, T.A., and Kobayashi, T. (2006). PP2C family members play key roles in regulation of cell survival and apoptosis. *Cancer Sci* 97:563–567.

Tanoue, T., Adachi, M., Moriguchi, T., and Nishida, E. (2000). A conserved docking motif in MAP kinases common to substrates, activators and regulators. *Nat Cell Biol* 2:110–116.

Tanoue, T., and Nishida, E. (2003). Molecular recognitions in the MAP kinase cascades. *Cell Signal* 15:455–462.

Teige, M., Scheikl, E., Eulgem, T., Doczi, R., Ichimura, K., Shinozaki, K., Dangl, J.L., and Hirt, H. (2004). The MKK2 pathway mediates cold and salt stress signaling in *Arabidopsis*. *Mol Cell* 15:141–152.

Ulm, R., Ichimura, K., Mizoguchi, T., Peck, S.C., Zhu, T., Wang, X., Shinozaki, K., and Paszkowski, J. (2002). Distinct regulation of salinity and genotoxic stress responses by *Arabidopsis* MAP kinase phosphatase 1. *Embo J* 21:6483–6493.

Ulm, R., Revenkova, E., di Sansebastiano, G.P., Bechtold, N., and Paszkowski, J. (2001). Mitogen-activated protein kinase phosphatase is required for genotoxic stress relief in *Arabidopsis*. *Genes Dev* 15:699–709.

Verhey, K.J., Meyer, D., Deehan, R., Blenis, J., Schnapp, B.J., Rapoport, T.A., and Margolis, B. (2001). Cargo of kinesin identified as JIP scaffolding proteins and associated signaling molecules. *J Cell Biol* 152:959–970.

Wang, H., Ngwenyama, N., Liu, Y., Walker, J.C., and Zhang, S. (2007). Stomatal development and patterning are regulated by environmentally responsive mitogen-activated protein kinases in *Arabidopsis*. *Plant Cell* 19:63–73.

Wasternack, C. (2007). Jasmonates: an update on biosynthesis, signal transduction and action in plant stress response, growth and development. *Ann Botany* 100:681–697.

Whitmarsh, A.J., Cavanagh, J., Tournier, C., Yasuda, J., and Davis, R.J. (1998). A mammalian scaffold complex that selectively mediates MAP kinase activation. *Science* 281:1671–1674.

Widmann, C., Gibson, S., Jarpe, M.B., and Johnson, G.L. (1999). Mitogen-activated protein kinase: conservation of a three-kinase module from yeast to human. *Physiol Rev* 79:143–180.

Yamakawa, H., Katou, S., Seo, S., Mitsuhara, I., Kamada, H., and Ohashi, Y. (2004). Plant MAPK phosphatase interacts with calmodulins. *J Biol Chem* 279:928–936.

Yang, K.Y., Liu, Y., and Zhang, S. (2001). Activation of a mitogen-activated protein kinase pathway is involved in disease resistance in tobacco. *Proc Natl Acad Sci USA* 98:741–746.

Yang, S., and Hua, J. (2004). A haplotype-specific Resistance gene regulated by BONZAI1 mediates temperature-dependent growth control in *Arabidopsis*. *Plant Cell* 16:1060–1071.

Yoo, S.D., Cho, Y.H., Tena, G., Xiong, Y., and Sheen, J. (2008). Dual control of nuclear EIN3 by bifurcate MAPK cascades in C2H4 signalling. *Nature* 451:789–795.

Zhang, T., Liu, Y., Yang, T., Zhang, L., Xu, S., Xue, L., and An, L. (2006). Diverse signals converge at MAPK cascades in plant. *Plant Physiol Biochem* 44:274–283.

3 Transcription Factors Involved in the Crosstalk between Abiotic and Biotic Stress-Signaling Networks

Yasunari Fujita,* Miki Fujita,* Kazuko Yamaguchi-Shinozaki, and Kazuo Shinozaki

Introduction

Plants have evolved a variety of intricate mechanisms to cope with biotic and abiotic stresses (Fujita 2006). Transcription factors are a group of master proteins that regulate gene expression by recognizing and binding to *cis*-elements in the promoter regions upstream of the target genes. As target gene regulators, transcription factors are involved in myriad biological processes, such as growth, development, cell cycle progression, metabolism, and responses to environmental stimuli. Although numerous studies have demonstrated that transcription factors play key roles in biotic or abiotic stress responses and tolerance in plants, there have been a small number of reports on transcription factors that function in the crosstalk between biotic and abiotic stress signaling pathways. However, recent studies have revealed that several transcription factors appear to be key players involved in the crosstalk between these stress signaling pathways. Increasing evidence suggests that these stress signaling pathways are modulated by multiple hormone signaling pathways regulated by abscisic acid (ABA), jasmonic acid (JA), ethylene (ET), and salicylic acid (SA), as well as reactive oxygen species (ROS) signaling pathways (Fujita 2006). In this chapter, we focus on transcriptional regulation of gene expression in the crosstalk between biotic and abiotic stress signaling pathways, with particular emphasis on the role of transcription factors and *cis*-acting elements in stress-inducible promoters. Various transcription factors involved in crosstalk are discussed according to family.

Zinc Finger Transcription Factors in ROS Signaling

In *Arabidopsis*, three Cys2/His2-type zinc finger transcription factors with the Ethylene-Responsive element binding Factor (ERF)-associated amphiphilic repression (EAR) domain, *RESPONSIVE TO HIGH LIGHT 41 (RHL41)/ZAT12, SALT TOLERANCE ZINC FINGER (STZ)/ZAT10*, and *ZAT7*, are involved in various stress responses via ROS signaling. The tight regulation of the steady-state levels of ROS, such as hydrogen peroxide (H_2O_2), superoxide (O_2^-), singlet oxygen ($^1O_2$), and hydroxyl radicals, is implicated in multiple cellular processes in plants (Apel and Hirt 2004). ROS are thought to play a dual role as toxic byproducts of aerobic metabolism and as signaling molecules in development, growth, disease resistance, and stress-response pathways (Apel and Hirt 2004). Various lines of evidence suggest that nicotinamide adenine dinucleotide phosphate (NADPH)-dependent respiratory burst oxidase homolog (rboh)

---

*These authors contributed equally.

genes are required for ROS generation, leading to ABA-induced stomatal closure and hypersensitive cell death in response to avirulent pathogen infection (Torres et al. 2002; Kwak et al. 2003; Torres and Dangl 2005). ROS-scavenging enzymes, including superoxide dismutase (SOD), ascorbate peroxidase (APX), and catalase (CAT), and the antioxidants ascorbic acid and glutathione, are considered to detoxify the cytotoxic effects of ROS under various stress conditions (Apel and Hilt 2004; Mittler et al. 2004). Microarray analyses of plants subjected to various stress treatments have demonstrated the induction of a large set of genes that encode ROS-scavenging enzymes under these conditions (Schenk et al. 2000; Seki et al. 2002; Mittler et al. 2004).

RHL41/ZAT12, which was originally identified as an acclimatization response protein (Iida et al. 2000), likely regulates the ROS-scavenging mechanism of multiple stress responses. The disruption of *RHL41/ZAT12*, which responds strongly to multiple stresses including wounding and pathogen infection as well as abiotic stresses (Seki et al. 2002; Zimmermann et al. 2004; Davletova et al. 2005b; Vogel et al. 2005), suppresses the expression of the defense enzyme cytosolic *APX1* gene (induced by H_2O_2) and increases the level of H_2O_2-induced protein oxidation (Rizhsky et al. 2004). The constitutive expression of *RHL41/ZAT12* upregulates oxidative- and light stress–responsive genes and enhances tolerance to high light, freezing, and oxidative stresses (Iida et al. 2000; Rizhsky et al. 2004; Davletova et al. 2005b; Vogel et al. 2005). The expression of *RHL41/ZAT12* is regulated by a redox-sensitive transcription factor, *HEAT SHOCK FACTOR21* (*HSF21*), which likely functions as an initial sensor for H_2O_2 that accumulates in response to stress (Davletova et al. 2005a). These findings, along with expression analyses of *apx1* and *rbohD* knockout mutants (Davletova et al. 2005a), suggest that *RHL41/ZAT12* may play a key role in the H_2O_2-mediated crosstalk between biotic and abiotic stresses. In *RHL41/ZAT12* homologs, *STZ/ZAT10* has similar expression profiles in response to stress (Mittler 2006). The constitutive expression of *STZ/ZAT10*, which is thought to be a downstream target gene of *DREB1A* (Maruyama et al. 2004), results in growth retardation and enhanced tolerance to various abiotic stresses such as drought, heat, osmotic stress, and high salinity (Sakamoto et al. 2004; Mittler 2006). The overexpression of *STZ/ZAT10* results in the increased expression of antioxidant genes, improved ROS detoxification, and enhanced tolerance to light stress and exogenous H_2O_2 (Mittler 2006; Rossel et al. 2007). Taken together with the results of loss- and reduction-of-function analyses of *STZ/ZAT10*, these findings suggest that *STZ/ZAT10* is involved in abiotic stress tolerance by modulating ROS signaling.

ZAT7 is also a key player in the response to abiotic stresses such as high salinity and cold (Ciftci-Yilmaz et al. 2007). Overexpression of *ZAT7* in transgenic *Arabidopsis* plants leads to suppressed growth and enhanced tolerance to salinity stress. Further mutational analyses have demonstrated that the EAR motif is directly involved in tolerance but not growth retardation. Moreover, analysis of *ZAT7* using RNAi lines also supports the view that *ZAT7* negatively regulates a repressor of the salt-defense response. The EAR motif of *ZAT7* physically interacts with WRKY70 and HASTY, a protein involved in miRNA transport. The constitutive expression of either ZAT7, WRKY70, or HASTY is observed in *apx1* knockout mutants, suggesting that *ZAT7* expression may also be associated with H_2O_2 stress. Moreover, WRKY70 plays an important role at the point of convergence between JA and SA signaling in defense pathways and modulates abiotic stress and senescence (Li et al. 2004, 2006; AbuQamar et al. 2006; Ülker et al. 2007). Taken together, these results suggest that these proteins may also function in mediating the crosstalk between biotic and abiotic stresses. This is consistent with the upregulation of the *WRKY70* and *ZAT7* genes in leaves infected with *Botrytis cinerea* and the observation that *WRKY70* and *ZAT7* have similar expression patterns in *coronatine insensitive1* (*coi1*) mutant plants and plants expressing the *nahG* gene, which inhibits the accumulation of SA and is not susceptible to *Botrytis* (Veronese et al. 2004). The expression of the *WRKY70* and *ZAT7* genes are severely reduced in *nahG* plants but induced in *coi1* plants compared with wildtype plants (AbuQamar et al. 2006). These findings indicate that the expression of *ZAT7* and *WRKY70* is regulated by both JA and SA, implying that the *ZAT7* gene may play a role in the biotic stress response. A large-scale transcriptome analysis also showed that in addition to the *ZAT7* gene, *RHL41/ZAT10* and *STZ/ZAT12* were induced in response to *Botrytis* infection (AbuQamar et al. 2006). These observations support the view that the *RHL41/ZAT12* and *STZ/ZAT10* genes may also help transmit pathogen signals, although the T-DNA insertion mutant alleles of *RHL41/ZAT12* and *STZ/ZAT10* showed no detectable effect on the resistance phenotype (AbuQamar et al. 2006). Overall, these findings suggest that ZAT-mediated ROS signaling may mediate crosstalk between biotic and abiotic stress-responsive gene expression networks.

MYC and MYB Transcription Factors in JA and ABA Signaling Pathways

The basic helix–loop–helix (bHLH) transcription factor ATMYC2 plays multiple roles in ABA, JA, ET, SA, and light-signaling pathways in *Arabidopsis* (Figure 3.1). ATMYC2 is synonymized with RD22BP1 (Abe et al. 1997), R-homologous *Arabidopsis* protein-1 (RAP-1; de Pater et al. 1997), AtbHLH006 (Heim et al. 2003), Z-box binding factor (ZBF1; Yadav et al. 2005), and MYC2 (Dombrecht et al. 2007). In addition, genetic analysis of *jasmonate insensitive1* (*jin1/jai1*) mutants revealed that *JIN1/JAI1* encodes *ATMYC2* (Lorenzo et al. 2004). DNA binding experiments showed that AtMYC2 preferentially binds to the G-box-like MYC recognition sequences 5'-CACATG-3' (Abe et al. 1997), 5'-CACNTG-3' (de Pater et al. 1997), 5'-(T/C)ACGTG-3' (Yadav et al. 2005), and 5'-CACGTG-3' (Dombrecht et al. 2007), clearly indicating that the ATMYC2 core binding sequence is CACNTG.

ATMYC2 was initially identified as a transcriptional activator involved in the ABA-mediated drought stress signaling pathway (Abe et al. 2003). *ATMYC2* knockout mutants are less sensitive to ABA and display reduced ABA-responsive gene expression (Abe et al. 2003). Moreover, ATMYC2 upregulates the expression of JA-mediated wounding-responsive genes such as *VSP* and *LOX* and represses the expression of JA/ET-mediated pathogen-defense genes such as *PR4/HEL* and *PDF1.2* (Anderson et al. 2004; Boter et al. 2004; Lorenzo et al. 2004). Several lines of evidence suggest that the negative regulation of *PDF1.2* expression by ATMYC2 is most likely mediated by the suppression of *ERF1*, a transcriptional activator of *PDF1.2* (Dombrecht et al. 2007). Disruption of ATMYC2 results in increased JA- or ET-regulated defense gene expression and enhanced resistance to necrotrophic fungal pathogens such as *B. cinerea*, *Fusarium oxysporum*, and *Plectosphaerella cucumerina* (Anderson et al. 2004; Lorenzo et al. 2004). However, even in *atmyc2/jin1* plants, exogenous ABA has an inhibitory effect on the JA-regulated expression of defense genes, suggesting that ATMYC2 is not involved in the antagonistic effect of ABA on the JA–ET defense pathway and that it is not the only point of convergence (Anderson et al. 2004).

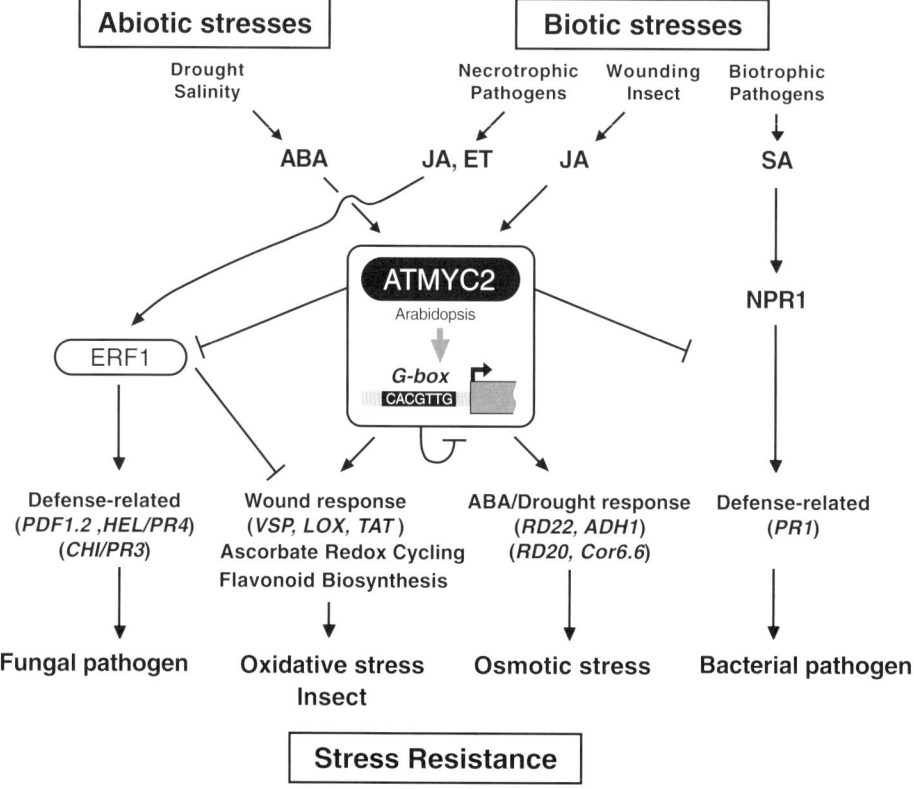

Fig. 3.1 Model describing the role of ATMYC2 in the crosstalk between biotic and abiotic stresses. For explanation of gene symbols, see text.

Indeed, genetic analyses using JA- and ABA-signaling mutants supports the view that ABA likely regulates ATMYC2 expression by activating the COI1-dependent JA-signaling pathway (Lorenzo et al. 2004). Thus, the antagonistic interactions between ABA and the JA–ET signaling pathways appear to regulate defense- and stress-responsive gene expression in response to biotic and abiotic stresses (Anderson et al. 2004; Lorenzo et al. 2004). The SA signaling pathway is also relevant to ATMYC2 (Laurie-Berry et al. 2006). The *atmyc2/jin1* mutants show increased resistant to the biotrophic bacterial pathogen *Pseudomonas syringae* as well as the necrotrophic fungal pathogens (Nickstadt et al. 2004; Laurie-Berry et al. 2006). This enhanced disease resistance is correlated with increased expression of *PATHOGENESIS-RELATED1* (*PR1*) gene and depends on the accumulation of SA (Laurie-Berry et al. 2006). Moreover, JIN1/ATMYC2 is implicated in the normal symptom development of *P. syringae* through an SA-independent mechanism (Laurie-Berry et al. 2006). Collectively, ATMYC2 is also a key player in defense responses against biotrophic pathogens through the crosstalk between the JA and SA hormone-signaling pathways.

ATMYC2 positively regulates JA-mediated resistance to insect herbivory and tolerance to oxidative stress, possibly by enhancing flavonoid biosynthesis and ascorbate redox cycling, and negatively regulates Trp metabolism, leading to JA-dependent synthesis of defensive compounds such as indole glucosinolates (Dombrecht et al. 2007). JA-dependent *ATMYC2* expression appears to be negatively regulated by the JA-activated MKK3–MPK6 pathway, whereas an unidentified signaling pathway may be implicated in the JA-mediated positive regulation of *ATMYC2* expression (Takahashi et al. 2007). Taken together with the observation that ATMYC2 can negatively regulate its own expression, the negative and positive regulation of *ATMYC2* expression might be an important mechanism underlying the JA-signaling pathways (Dombrecht et al. 2007; Takahashi et al. 2007).

Both *ATMYC2* and the *R2R3MYB* family transcription factor *ATMYB2* bind *cis*-elements in the dehydration-inducible *RD22* gene and cooperatively activate its expression (Abe et al. 2003). Interestingly, transgenic *Arabidopsis* plants overexpressing both *ATMYC2* and *ATMYB2* are more sensitive to ABA, and enhance osmotic stress tolerance (Abe 2003). In addition, *Botrytis* infection induces the expression of *BOS1*, which also belongs to the R2R3MYB family, through a JA-mediated defense signaling pathway (Mengiste et al. 2003). BOS1 deficiency results in increased sensitivity to necrotrophic pathogens and impaired tolerance to drought, salinity, and oxidative stresses. In plant defense responses, ROS are involved in cell death (Lamb and Dixon 1997), which appears to promote pathogen growth in necrotrophic pathogens (Govrin and Levine 2000). Accordingly, the increased sensitivity to ROS in *bos1* is consistent with its enhanced susceptibility to necrotrophic pathogens. BOS1 mediates both biotic and abiotic stress signaling through ROS. Therefore, these *R2R3MYB* transcription factors might serve as important mediators of complex activities spanning multiple stress signaling pathways.

NAC Transcription Factors in ABA, JA, and Senescence Signaling Pathways

Plant-specific NAC family transcription factors contain N-terminal NAC DNA-binding domains and form a plant-specific gene family with diverse functions (Olsen et al. 2005). To date, four NAC family transcription factors—*OsNAC6*, *ARABIDOPSIS TRANSCRIPTION ACTIVATION FACTOR 1 (ATAF1)*, *ATAF2*, and *RESPONSIVE TO DEHYDRATION 26 (RD26)*, which belong to the ATAF subfamily (Ooka et al. 2003; Fujita et al. 2004)—are viable candidate molecules that potentially regulate aspects of both biotic and abiotic signaling (Figure 3.2). In rice, *OsNAC6* is involved in both response and tolerance to biotic and abiotic stresses (Nakashima et al. 2007). *OsNAC6* mRNA transcript is upregulated by cold, drought, high salinity, and exogenous ABA, as well as wounding and pathogen infection (Ohnishi 2005; Nakashima et al. 2007). The constitutive expression of *OsNAC6* in transgenic rice activates a subset of stress-related genes and displays enhanced tolerance to drought and high-salt stresses and slightly increased resistance to blast diseases (Nakashima et al. 2007). *Arabidopsis ATAF2*, one of the closest orthologs of rice *OsNAC6*, functions as a repressor of pathogenesis-related protein genes (Delessert et al. 2005). Overexpression of *ATAF2*, which responds to both biotic and abiotic stresses such as wounding and exogenous SA and JA in addition to high-salinity stress, results in the repression of pathogenesis-related proteins and elevated susceptibility to soil-borne fungal pathogens. A close homolog of *ATAF2*, *ATAF1*, is also involved in both biotic and abiotic stress responses including dehydration and exogenous ABA treatments (Lu et al. 2007) as well as wounding and pathogen

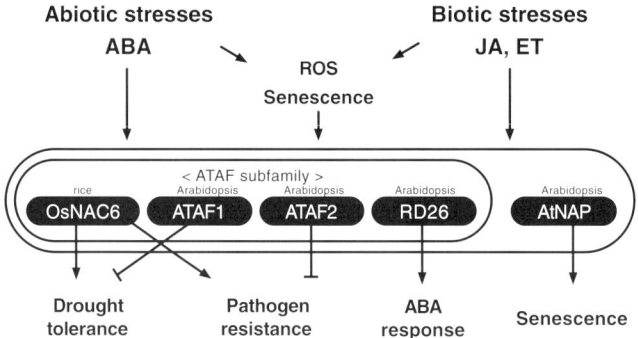

Fig. 3.2 NAC genes involved in the crosstalk between biotic and abiotic stresses.

infection (Schenk et al. 2000; Collinge and Boller 2001). *Arabidopsis* ATAF1 T-DNA insertion lines activate the expression of dehydration-responsive genes and exhibit enhanced drought tolerance, indicating that ATAF1 is likely a transcriptional repressor of dehydration-responsive genes (Lu et al. 2007). Another *Arabidopsis* ATAF homolog is *RD26*, which is involved in the ABA-dependent stress signaling pathway, as demonstrated by both gain-of- and reduction-of-function mutants of RD26 (Fujita et al. 2004). The expression of *RD26* is induced by JA, H_2O_2, and pathogen infections as well as by drought, high salinity, and ABA treatment (Fujita et al. 2004; Zimmermann et al. 2004). Large-scale transcriptome analysis with the both types of mutants revealed that RD26-regulated genes are involved in the detoxification of ROS, defense, and senescence in an ABA-dependent manner (Fujita et al. 2004). These findings suggest that RD26 may function at the node of convergence among the ABA-, pathogen defense-, and senescence-signaling pathways (Fujita et al. 2004). Leaf senescence is induced by both biotic and abiotic stresses such as pathogen infection, drought, extreme temperature, and oxidative stress (Buchanan-Wollaston 2003; Gepstein 2003). Recently, AtNAP, which belongs to the closest NAC subfamily of the ATAF subfamily, has been shown to be involved in senescence (Guo and Gan 2006). In addition, all ATAF subfamily NAC genes, including *ATAF1*, *ATAF2*, and *RD26*, are upregulated during senescence in *Arabidopsis* leaves (Guo et al. 2004) as well as by abiotic stresses such as drought, high salinity, and ABA treatment (Fujita et al. 2004). Moreover, the suppression of drought-induced senescence in transgenic tobacco plants expressing the isopentenyltransferase gene results in drought tolerance, driven by a stress- and maturation-induced promoter (Rivero et al. 2007). Taken together, these findings support the notion that leaf senescence may be closely related to NAC-mediated stress responses.

ERF/AP2 Family Transcription Factors

DREB/CBF Subfamily Transcription Factors in Abiotic and Pathogen Signaling

DEHYDRATION-RESPONSIVE ELEMENT BINDING PROTEINS/C-REPEAT BINDING FACTORS (DREB/CBFs) and ETHYLENE-RESPONSIVE ELEMENT

BINDING FACTORS (ERFs) are two major subfamilies of the plant-specific ERF/APETALA2 (AP2) family transcription factors that play multiple roles in biotic and abiotic stress signaling pathways (Sakuma et al. 2002; Nakano et al. 2006). DREB1/CBFs and DREB2s interact with the *cis*-acting elements dehydration-responsive elements/C-repeat (DRE/CRT), which is involved in the expression of many stress-inducible genes in *Arabidopsis* (Maruyama et al. 2004; Agarwal et al. 2006; Yamaguchi-Shinozaki and Shinozaki 2006; Shinozaki et al. 2007). *DREB1A/CBF3*, *DREB1B/CBF1*, and *DREB1C/CBF2* mainly activate downstream genes implicated in the cold stress response, whereas *DREB2A* regulates genes in response to drought, salt, and high temperature stresses (Sakuma 2006a; Yamaguchi-Shinozaki and Shinozaki 2006). The *DREB1A/CBF3*, *DREB1B/CBF1*, and *DREB1C/CBF2* genes are upregulated by cold (Liu et al. 1998), whereas the *DREB1D/CBF4* gene is induced by osmotic stress (Haake et al. 2002) and the *DREB1E/DDF2* and *DREB1F/DDF1* genes are induced by high salinity stress (Magome 2004). The *DREB2A* and *DREB2B* genes are strongly upregulated by drought, high salinity, and high temperature stresses (Liu et al. 1998; Nakashima et al. 2000; Sakuma et al. 2002, 2006a, 2006b). The transgenic expression of *DREB1B/CBF1* in *Arabidopsis* leads to enhanced freezing tolerance (Jaglo-Ottosen 1998). *DREB1A/CBF3* overexpressors also display enhanced tolerance to abiotic stresses such as drought, high salinity, and freezing (Gilmour 2000; Kasuga et al. 1999; Liu 1998). Moreover, overexpression of a constitutively active form of DREB2A significantly increases tolerance to drought and high temperature but only slightly increases tolerance to freezing in *Arabidopsis* (Sakuma et al. 2006a, 2006b). Therefore, the DREB transcription factors regulate many stress-inducible genes and play prominent roles in various abiotic stress signaling pathways in *Arabidopsis*. In addition to *Arabidopsis* plants, DREB transcription factors that are highly conserved in both dicot and monocot plants have been extensively studied in a broad range of higher plants such as rice, tobacco, wheat, *Brassica napus*, barley, maize, soybean, and cherry (Agarwal et al. 2006; Ito et al. 2006; Qin et al. 2007; Yamaguchi-Shinozaki and Shinozaki 2006).

Although the involvement of DREB transcription factors in various abiotic stress signaling pathways has been extensively studied, few studies have shown a link between these and biotic stress signaling pathways. Constitutive or conditional expression of *ACTIVATED RESISTANCE1* (*ADR1*) gene, which encodes a coiled-coil (CC)-nucleotide-binding site (NBS) leucine-rich repeat (LRR) protein, confer not only increased resistance to a broad spectrum of virulent pathogens (Grant et al. 2003) but also enhanced drought tolerance in *Arabidopsis* (Chini et al. 2004). Transgenic plants overexpressing *ADR1* upregulate *DREB2A* and other dehydration-responsive genes as well as SA-dependent defense genes and exhibit increased sensitivity to heat and salinity (Chini et al. 2004). Furthermore, *DREB2A* expression may be regulated by a SA-dependent but NON-EXPRESSOR OF PATHOGENESIS-RELATED GENES1 (NPR1)-independent signaling pathway in activation-tagged mutant *adr1* (Chini et al. 2004). Moreover, ROS levels are higher in *adr1* mutants than in wildtype plants (Grant et al. 2003), and *DREB2A* transcript accumulation is upregulated by ROS such as H_2O_2 (Desikan et al. 2001; Sakuma et al. 2006b). Because SA appears to amplify ROS during the expression of disease resistance (Shirasu et al. 1997; Grant et al. 2000), these findings imply that *DREB2A* expression is induced by ROS accumulation amplified by SA (Chini et al. 2004). This suggests that *DREB2A* expression is under redox control (Chini et al. 2004). Thus, because *adr1*-mediated drought tolerance appears to be controlled by SA and ABA rather than JA and ET (Chini et al. 2004), DREB2A may play a role in the crosstalk between biotic and abiotic stress tolerance under complex signaling pathways of redox and plant hormones such as SA and ABA. However, although *DREB2A* transcript accumulation is induced by the exogenous application of benzothiadiazole (BTH), a functional analog of SA but not SA itself (Gorlach et al. 1996; Chini et al. 2004; Sakuma et al. 2006b), the exogenous application of BTH alone is insufficient to establish considerable drought tolerance (Chini et al. 2004). In addition, DREB2A has been reported to function in ABA-independent osmotic signaling pathway, although the target genes of DERB2A contain ABA-inducible genes and are partially overlapped with those of AREB1, which is a transcriptional activator of the ABRE-dependent ABA/osmotic signaling pathway (Fujita et al. 2005; Sakuma et al. 2006b). Thus, these observations may reflect part of an elaborate network of biotic and abiotic stress signaling pathways.

TINY, an *Arabidopsis* transcription factor belonging to a subfamily closely related to the DREB subfamily, may function in both DRE- and ethylene responsive element (ERE)–dependent signaling pathways (Sun et al. 2008). It binds to *cis*-elements including DRE and ERE with similar affinity and activates the expression of reporter genes driven by either of these two elements in tobacco cells. Transgenic *Arabidopsis* plants overexpressing TINY, whose expression is strongly induced by drought, cold, and ET, exhibit increased expression of both DRE- and ERE-regulated genes. In transgenic plants, the expression of DRE-containing genes is upregulated by exogenous ET treatment, whereas the expression of ERE-containing genes is also upregulated by cold stress, suggesting that TINY might act as a key regulator at the point of convergence between abiotic and biotic stress gene expression by linking to DRE- and ERE-dependent signaling pathways.

ERF Subfamily Transcription Factors in Pathogen and Salinity Signaling

Although no reports have implicated ERF subfamily transcription factors in the crosstalk between biotic and abiotic stress responses in *Arabidopsis*, tobacco and pepper ERF transcription factors have been reported to play a role in the crosstalk. TOBACCO STRESS-INDUCED-1 (TSI1) is a tobacco ERF subfamily transcription factor that plays an important role in both biotic and abiotic stress signaling pathways (Park et al. 2001). *TSI1* cDNA is an ortholog of a salt-induced transcript cDNA (Park et al. 2001) but is transiently activated not only by salt treatment but also by ethephon, SA, methyl jasmonate, and wounding treatments. In contrast, both drought and ABA treatments did not induce its transcription (Park et al. 2001). More interestingly, electrophoretic mobility shift assays demonstrated that the TSI1 protein can bind to both *cis*-acting elements GCC box and DRE/CRT motif, although the binding activity to GCC is stronger (Park et al. 2001). The proteins of ERFs, such as PTIs and AtEBP, bind specifically to the GCC box containing the core GCCGCC sequence, which exists in the promoter regions of many ET-inducible genes that encode PR proteins (Ohme-Takagi and Shinshi 1995). In contrast, DREB/CBF proteins bind specifically to the DRE/CRT sequence containing core sequence A/GCCGAC, which is present in the promoter regions of many abiotic stress-inducible genes (Yamaguchi-Shinozaki and Shinozaki 2006). Both GCC box and the DRE/CRT motif contain CCGNC as a common core sequence. These findings appear to be correlated with the observation that overexpression of the *TSI1* in transgenic tobacco plants activates the expression of multiple PR genes and increases resistance to biotrophic pathogen invasion and tolerance to osmotic stress (Park et al. 2001). In addition, ectopic expression of tobacco *TSI1* in transgenic hot pepper plants also triggers constitutive expression of several PR genes and enhances resistance to a wide spectrum of pathogens, including viruses, a bacterium, and an oomycete (Shin et al. 2002).

TSI1-INTERACTING PROTEIN1 (TSIP1), a DnaJ type zinc finger protein, interacts with TSI1 in vitro and in yeast (Ham et al. 2006). The Cys-rich zinc finger motif (CXXCXGXG) found in some members of the DnaJ class of molecular chaperones (Kelley 1998) is both required and sufficient for the interaction (Ham et al. 2006). In addition, TSIP1 colocalizes and coimmunoprecipitates with TSI1 in plant cells after treatment with SA (Ham et al. 2006). The expression of *TSIP1* is similar to that of *TSI1* upon exposure of plants to the same stimuli (Ham et al. 2006). Note that transgenic tobacco plants simultaneously overexpressing *TSIP1* and *TSI1* confer stronger salt tolerance and pathogen resistance than transgenic plants overexpressing either *TSI1* or *TSIP1* alone (Ham et al. 2006). Consistent with this, gain- and reduction-of-function analyses demonstrate that TSIP1 and TSI1 functionally cooperate to activate the expression of a subset of stress-inducible target genes (Ham et al. 2006). These findings corroborate evidence that TSIP1 enhances the TSI1-mediated transcriptional activation of GCC- or DRE/CRT-containing promoters through a direct interaction with TSI1, suggesting that TSI1 and TSIP1 cooperatively regulate biotic and abiotic stress-inducible genes in plants (Ham et al. 2006).

The pepper ERF transcription factors CaERFLP1 and CaPF1 can bind both GCC and DRE/CRT *cis*-elements in vitro (Lee et al. 2004; Yi et al. 2004). The expression of *CaERFLP1* is induced by exogenous ET, pathogen infection, mechanical wounding, and high-salinity stress, whereas the expression of *CaPF1* is upregulated by methyl jasmonate, ethephon, and cold stress. Overexpression of *CaERFLP1* in transgenic tobacco plants enhances tolerance to salt stress and increases resistance to biotrophic pathogen infection, whereas transgenic *Arabidopsis* plants overexpressing *CaPF1* display enhanced tolerance to freezing and increased resistance to biotrophic pathogen infection. Transgenic plants overexpressing *CaERFLP1* or *CaPF1* express GCC- and DRE/CRT-containing genes. These findings suggest that these two pepper ERF transcription factors might play a role in the crosstalk between biotic and abiotic stress signaling pathways by regulating GCC- and DRE/CRT-dependent gene expression.

Transcription Factors Involved in the Biosynthesis and Regulation of Cuticular Components

The cuticular layer, consisting of cutin and waxes, represents a primary barrier that minimizes water and solute loss and protects aerial organs in plants from damage caused by various abiotic and biotic stresses (Jeffree and Christopher 1996; Nawrath 2006). Although several transcription factors involved in the biosynthesis or regulation of wax and cutin components have been identified, few known transcription factors function as the convergence points between biotic and abiotic stress signaling pathways. Accumulating evidence, however, suggests that transcription factors are

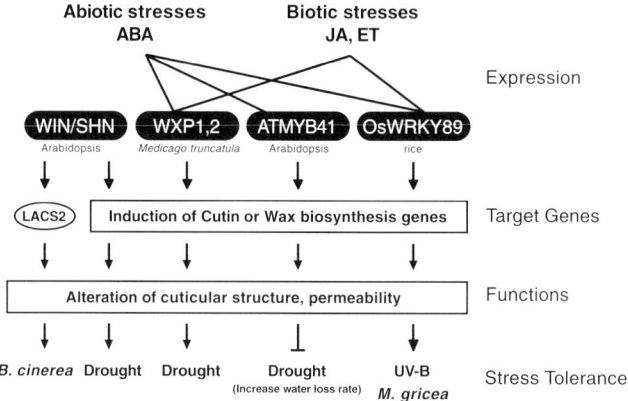

Fig. 3.3 Factors involved in the cuticular-mediated crosstalk between abiotic and biotic stresses.

directly or indirectly involved in the tolerance response to both biotic or abiotic stresses (Figure 3.3).

In *Arabidopsis*, WAX INDUCER1/SHINE1 (WIN1/SHN1), a transcription factor in the ERF/AP2 family, has recently been shown to activate the expression of genes associated with wax and cutin biosynthesis and controls cuticular lipid production (Aharoni et al. 2004; Broun et al. 2004; Kannangara et al. 2007). Transgenic *Arabidopsis* plants overexpressing the *WIN1/SHN1* display increased cuticle permeability, reduced stomatal density, and enhanced tolerance to drought (Aharoni et al. 2004). However, although a direct link between the *WIN1/SHN1* gene and biotic stress tolerance has not been reported, the disruption of one of the direct target genes of *WIN1/SHN1*, *LONG-CHAIN ACYL-CoA SYNTHETASE2* (*LACS2*), results in increased resistance to *Botrytis cinerea* and *Sclerotinia sclerotiorum* (Bessire et al. 2007; Kannangara et al. 2007). Genetic analysis of the *botrytis resistant1* (*bre1*) mutant of *Arabidopsis* revealed that *BRE1* is allelic to *LACS2* (Schnurr et al. 2004; Bessire et al. 2007). LACS2 is required for the biosynthesis of cutin polyester and implicated in the alteration of cuticular permeability (Bessire et al. 2007). Moreover, *bre1/lacs2* does not exhibit an altered resistance or susceptibility to a broad range of pathogens. These data suggest that increased cuticle permeability allows the induction of antifungal compounds to arrest invasion of specific necrotrophic pathogens. This further suggests that a permeable cuticle plays an important role by altering the perception of elicitors rather than having the direct effect of a major barrier that prevents pathogen invasions (Bessire et al. 2007). This view has been corroborated by recent studies on chitinase-expressing plants and the *bodyguard* (*bdg*) mutant, which creates defects in cuticle structure and integrity (Kurdyukov et al. 2006; Chassot et al. 2007).

WAX PRODUCTION1 (WXP1) and WAX PRODUCTION2 (WXP2) of the model legume *Medicago truncatula* are also transcription factors in the ERF/AP2 family but are distinctly different from other well-characterized members related to abiotic stress or wax/cutin accumulation in the family such as WIN1/SHN1, DREB1/CBFs, and DREB2s (Zhang et al. 2005, 2007). Overexpression of the *WXP1* gene in transgenic alfalfa (*Medicago sativa*) leads to increased cuticular wax accumulation and enhanced drought tolerance (Zhang et al. 2005). The constitutive expression of the *WXP1* or

WXP2 gene in transgenic *Arabidopsis* results in increased wax deposition and enhanced drought tolerance (Zhang et al. 2007). However, *WXP1*-expressing plants are more freeze-tolerant, whereas *WXP2*-expressing plants shows more freeze-sensitive phenotypes, implying that wax quantity may not be directly linked to freeze tolerance in plants (Zhang et al. 2007). Although the transcriptional induction of the *WIN1/SHN1* gene in response to environmental stress has not been reported, like *DREB/CBF* genes, the expression of *WXP1* and *WXP2* genes are upregulated by several biotic and abiotic stresses such as fungal pathogen, drought, cold, and ABA (Zhang et al. 2005, 2007). However, whether the *WXP1* and *WXP2* genes are involved in tolerance to biotic stress remains unknown.

Another family of transcription factors is involved in cuticular deposition and stress responses. The *Arabidopsis AtMYB41* gene encodes an R2R3-MYB transcription factor in which the transcript is induced to a high level by drought, salt, and ABA treatment, although it is not detectable in any organ or developmental stage (thus far analyzed) under nonstressed conditions (Cominelli et al. 2008). Because *AtMYB41* regulates wax accumulation and cell expansion when overexpressed in *Arabidopsis*, it may play an important role in cell wall modification, cuticle synthesis, and deposition, especially in response to osmotic stress (Cominelli et al. 2008). The plant-specific WRKY family of transcription factors interact with the W-box *cis*-elements found in the promoters of many plant defense-related genes (Dong et al. 2003; Ülker et al. 2004). In rice plants, overexpression of *OsWRKY89*, a member of the rice WRKY family of transcription factors, confers ultraviolet B (UVB) tolerance and enhances resistance to the rice blast fungus and white-backed planthopper (Wang et al. 2007). The *OsWRKY89* gene is inducible by methyl jasmonate, ABA, UVB, and wounding and is also involved in wax loading, lignification, and accumulation of cell wall-bound phenolic compounds and SA (Wang et al. 2007). These data suggest that OsWRKY89 is a key regulator of the link between biotic stresses and UVB radiation (Wang et al. 2007). Thus, transcription factors associated with cuticular deposition may be directly or indirectly involved in the tolerance response to biotic and abiotic stresses. Note that cuticular layers likely play pivotal roles not only as major barriers in plants that protect aerial organs from damage but also in biotic and abiotic signal transduction pathways themselves. Furthermore, because several transcription factors involved in the regulation of cuticular composition are also implicated in transcriptional regulation in response to environmental stresses (e.g., WXP1, AtMYB41, and OsWRKY89), cuticular composition is likely modulated flexibly by specific environmental cues. Recent studies therefore support the view that the cuticular layer itself is the crucial point of convergence between biotic and abiotic stress response signaling pathways as a first defense against environmental stresses.

Conclusions

The acclimation of plants to various biotic and abiotic stresses in their natural environment is orchestrated by a complex network of regulatory genes and signal molecules (Fujita 2006). Recent research has shown that plant transcription factors play key roles in the response to abiotic and biotic stresses through ROS and multiple hormone-signaling pathways. Moreover, transcription factors associated with cuticular deposition also appear to play crucial roles in the crosstalk. Nevertheless, our current understanding of transcription factors in crosstalk between these pathways is limited. Have we already identified most of the players implicated in crosstalk? Alternatively, do these represent just a fraction of the major players? When combined, the accumulating data on large-scale phenome, transcriptome, and metabolome analyses in plants will help clarify the transcriptional regulatory networks in response to biotic and abiotic stresses in plants. Much work, however, is still needed to clarify the impact of transcription factors on the biotic and abiotic stress signaling pathways and discover transcription factors implicated in the crosstalk.

References

Abe, H., Urao, T., Ito, T., Seki, M., Shinozaki, K., and Yamaguchi-Shinozaki, K. 2003. Arabidopsis AtMYC2 (bHLH) and AtMYB2 (MYB) function as transcriptional activators in abscisic acid signaling. *Plant Cell* 15:63–78.

Abe, H., Yamaguchi-Shinozaki, K., Urao, T., Iwasaki, T., Hosokawa, D., and Shinozaki, K. 1997. Role of arabidopsis MYC and MYB homologs in drought- and abscisic acid-regulated gene expression. *Plant Cell* 9:1859–1868.

AbuQamar, S., Chen, X., Dhawan, R., Bluhm, B., Salmeron, J., Lam, S., Dietrich, R.A., and Mengiste, T. 2006. Expression profiling and mutant analysis reveals complex regulatory networks involved in *Arabidopsis* response to *Botrytis* infection. *Plant J* 48:28–44.

Agarwal, P.K., Agarwal, P., Reddy, M.K., and Sopory, S.K. 2006. Role of DREB transcription factors in abiotic and biotic stress tolerance in plants. *Plant Cell Rep* 25:1263–1274.

Aharoni, A., Dixit, S., Jetter, R., Thoenes, E., Van Arkel, G., and Pereira, A. 2004. The SHINE clade of AP2 domain transcription factors activates wax biosynthesis, alters cuticle properties, and confers drought tolerance when overexpressed in *Arabidopsis*. *Plant Cell* 16:2463–2480.

Anderson, J.P., Badruzsaufari, E., Schenk, P.M., Manners, J.M., Desmond, O.J., Ehlert, C., Maclean, D.J., Ebert, P.R., and Kazan, K. 2004. Antagonistic interaction between abscisic acid and jasmonate-ethylene signaling pathways modulates defense gene expression and disease resistance in *Arabidopsis*. *Plant Cell* 16:3460–3479.

Apel, K., and Hirt, H. 2004. Reactive oxygen species: metabolism, oxidative stress, and signal transduction. *Annu Rev Plant Biol* 55:373–399.

Bessire, M., Chassot, C., Jacquat, A.C., Humphry, M., Borel, S., Petetot, J.M., Metraux, J.P., and Nawrath, C. 2007. A permeable cuticle in *Arabidopsis* leads to a strong resistance to *Botrytis cinerea*. *Embo J* 26:2158–2168.

Boter, M., Ruiz-Rivero, O., Abdeen, A., and Prat, S. 2004. Conserved MYC transcription factors play a key role in jasmonate signaling both in tomato and *Arabidopsis*. *Genes Dev* 18:1577–1591.

Broun, P., Poindexter, P., Osborne, E., Jiang, C.Z., and Riechmann, J.L. 2004. WIN1, a transcriptional activator of epidermal wax accumulation in *Arabidopsis*. *Proc Natl Acad Sci USA* 101:4706–4711.

Buchanan-Wollaston, V., Earl, S., Harrison, E., Mathas, E., Navabpour, S., Page, T., and Pink, D. 2003. The molecular analysis of leaf senescence—a genomics approach. *Plant Biotechnol J* 1:3–22.

Chassot, C., Nawrath, C., and Metraux, J.P. 2007. Cuticular defects lead to full immunity to a major plant pathogen. *Plant J* 49:972–980.

Chini, A., Grant, J.J., Seki, M., Shinozaki, K., and Loake, G.J. 2004. Drought tolerance established by enhanced expression of the CC-NBS-LRR gene, ADR1, requires salicylic acid, EDS1 and ABI1. *Plant J* 38:810–822.

Ciftci-Yilmaz, S., Morsy, M.R., Song, L., Coutu, A., Krizek, B.A., Lewis, M.W., Warren, D., Cushman, J., Connolly, E.L., and Mittler, R. 2007. The EAR-motif of the Cys2/His2-type zinc finger protein Zat7 plays a key role in the defense response of *Arabidopsis* to salinity stress. *J Biol Chem* 282:9260–9268.

Collinge, M., and Boller, T. 2001. Differential induction of two potato genes, Stprx2 and StNAC, in response to infection by *Phytophthora infestans* and to wounding. *Plant Mol Biol* 46:521–529.

Cominelli, E., Sala, T., Calvi, D., Gusmaroli, G., and Tonelli, C. 2008. Over-expression of the *Arabidopsis* AtMYB41 gene alters cell expansion and leaf surface permeability. *Plant J* 53:53–64.

Davletova, S., Rizhsky, L., Liang, H., Shengqiang, Z., Oliver, D.J., Coutu, J., Shulaev, V., Schlauch, K., and Mittler, R. 2005a. Cytosolic ascorbate peroxidase 1 is a central component of the reactive oxygen gene network of *Arabidopsis*. *Plant Cell* 17:268–281.

Davletova, S., Schlauch, K., Coutu, J., and Mittler, R. 2005b. The zinc-finger protein Zat12 plays a central role in reactive oxygen and abiotic stress signaling in *Arabidopsis*. *Plant Physiol* 139:847–856.

De Pater, S, Pham, K., Memelink, J., and Kijne, J. 1997. RAP-1 is an *Arabidopsis* MYC-like R protein homologue, that binds to G-box sequence motifs. *Plant Mol Biol* 34:169–174.

Delessert, C., Kazan, K., Wilson, I.W., Van Der Straeten, D., Manners, J., Dennis, E.S., and Dolferus, R. 2005. The transcription factor ATAF2 represses the expression of pathogenesis-related genes in *Arabidopsis*. *Plant J* 43:745–757.

Desikan, R., Mackerness, S.A.H., Hancock, J.T., and Neill, S.J. 2001. Regulation of the *Arabidopsis* transcriptome by oxidative stress. *Plant Physiol* 127:159–172.

Dombrecht, B., Xue, G.P., Sprague, S.J., Kirkegaard, J.A., Ross, J.J., Reid, J.B., Fitt, G.P., Sewelam, N., Schenk, P.M., Manners, J.M., and Kazan, K. 2007. MYC2 differentially modulates diverse jasmonate-dependent functions in *Arabidopsis*. *Plant Cell* 19:2225–2245.

Dong, J., Chen, C., and Chen, Z. 2003. Expression profiles of the *Arabidopsis* WRKY gene superfamily during plant defense response. *Plant Mol Biol* 51:21–37.

Fujita, M., Fujita, Y., Maruyama, K., Seki, M., Hiratsu, K., Ohme-Takagi, M., Tran, L.S., Yamaguchi-Shinozaki, K., and Shinozaki, K. 2004. A dehydration-induced NAC protein, RD26, is involved in a novel ABA-dependent stress-signaling pathway. *Plant J* 39:863–876.

Fujita, M., Fujita, Y., Noutoshi, Y., Takahashi, F., Narusaka, Y., Yamaguchi-Shinozaki, K., and Shinozaki, K. 2006. Crosstalk between abiotic and biotic stress responses: a current view from the points of convergence in the stress signaling networks. *Curr Opin Plant Biol* 9:436–442.

Fujita, Y., Fujita, M., Satoh, R., Maruyama, K., Parvez, M.M., Seki, M., Hiratsu, K., Ohme-Takagi, M., Shinozaki, K., and Yamaguchi-Shinozaki, K. 2005. AREB1 Is a transcription activator of novel ABRE-dependent ABA signaling that enhances drought stress tolerance in *Arabidopsis*. *Plant Cell* 17:3470–3488.

Gepstein, S., Sabehi, G., Carp, M.J., Hajouj, T., Nesher, M.F., Yariv, I., Dor, C., and Bassani, M. 2003. Large-scale identification of leaf senescence-associated genes. *Plant J* 36:629–642.

Gilmour, D. 2000. Risk for the new or expectant mother working in the perioperative environment. *Br J Perioper Nurs* 10:306–310.

Gorlach, J., Volrath, S., Knauf-Beiter, G., Hengy, G., Beckhove, U., Kogel, K.H., Oostendorp, M., Staub, T., Ward, E., Kessmann, H., and Ryals, J. 1996. Benzothiadiazole, a novel class of inducers of systemic acquired resistance, activates gene expression and disease resistance in wheat. *Plant Cell* 8:629–643.

Govrin, E.M., and Levine, A. 2000. The hypersensitive response facilitates plant infection by the necrotrophic pathogen *Botrytis cinerea*. *Curr Biol* 10:751–757.

Grant, J.J., Chini, A., Basu, D., and Loake, G.J. 2003. Targeted activation tagging of the Arabidopsis NBS-LRR gene, ADR1, conveys resistance to virulent pathogens. *Mol Plant Microbe Interact* 16:669–680.

Grant, J.J., Yun, B.W., and Loake, G.J. 2000. Oxidative burst and cognate redox signalling reported by luciferase imaging: identification of a signal network that functions independently of ethylene, SA and Me-JA but is dependent on MAPKK activity. *Plant J* 24:569–582.

Guo, Y., Cai, Z., and Gan, S. 2004. Transcriptome of *Arabidopsis* leaf senescence. *Plant, Cell Environ* 27:521–549.

Guo, Y., and Gan, S. 2006. AtNAP, a NAC family transcription factor, has an important role in leaf senescence. *Plant J* 46:601–612.

Haake, V., Cook, D., Riechmann, J.L., Pineda, O., Thomashow, M.F., and Zhang, J.Z. 2002. Transcription factor CBF4 is a regulator of drought adaptation in *Arabidopsis*. *Plant Physiol* 130:639–648.

Ham, B.K., Park, J.M., Lee, S.B., Kim, M.J., Lee, I.J., Kim, K.J., Kwon, C.S., and Paek, K.H. 2006. Tobacco Tsip1, a DnaJ-type Zn finger protein, is recruited to and potentiates Tsi1-mediated transcriptional activation. *Plant Cell* 18:2005–2020.

Heim, M.A., Jakoby, M., Werber, M., Martin, C., Weisshaar, B., and Bailey, P.C. 2003. The basic helix-loop-helix transcription factor family in plants: a genome-wide study of protein structure and functional diversity. *Mol Biol Evol* 20:735–747.

Iida, A., Kazuoka, T., Torikai, S., Kikuchi, H., and Oeda, K. 2000. A zinc finger protein RHL41 mediates the light acclimatization response in *Arabidopsis*. *Plant J* 24:191–203.

Ito, Y., Katsura, K., Maruyama, K., Taji, T., Kobayashi, M., Seki, M., Shinozaki, K., and Yamaguchi-Shinozaki, K. 2006. Functional analysis of rice DREB1/CBF-type transcription factors involved in cold-responsive gene expression in transgenic rice. *Plant Cell Physiol* 47:141–153.

Jaglo-Ottosen, K.R., Gilmour, S.J., Zarka, D.G., Schabenberger, O., and Thomashow, M.F. 1998. *Arabidopsis* CBF1 overexpression induces COR genes and enhances freezing tolerance. *Science* 280:104–106.

Jeffree, Christopher E. 1996. Structure and ontogeny of plant cuticles. In: G. Kerstiens (ed.), *Plant Cuticles: An Integrated Functional Approach*. Oxford, UK: BIOS Scientific Publishers, pp. 33–82.

Kannangara, R., Branigan, C., Liu, Y., Penfield, T., Rao, V., Mouille, G., Hofte, H., Pauly, M., Riechmann, J.L., and Broun, P. 2007. The transcription factor WIN1/SHN1 regulates cutin biosynthesis in *Arabidopsis thaliana*. *Plant Cell* 19:1278–1294.

Kasuga, M., Liu, Q., Miura, S., Yamaguchi-Shinozaki, K., and Shinozaki, K. 1999. Improving plant drought, salt, and freezing tolerance by gene transfer of a single stress-inducible transcription factor. *Nat Biotechnol* 17:287–291.

Kelley, W.L. 1998. The J-domain family and the recruitment of chaperone power. *Trends Biochem Sci* 23:222–227.

Kurdyukov, S., Faust, A., Nawrath, C., Bar, S., Voisin, D., Efremova, N., Franke, R., Schreiber, L., Saedler, H., Metraux, J.P., and Yephremov, A. 2006. The epidermis-specific extracellular BODYGUARD controls cuticle development and morphogenesis in *Arabidopsis*. *Plant Cell* 18:321–339.

Kwak, J.M., Mori, I.C., Pei, Z.M., Leonhardt, N., Torres, M.A., Dangl, J.L., Bloom, R.E., Bodde, S., Jones, J.D., and Schroeder, J.I. 2003. NADPH oxidase AtrbohD and AtrbohF genes function in ROS-dependent ABA signaling in *Arabidopsis*. *Embo J* 22:2623–2633.

Lamb, C., and Dixon, R.A. 1997. The oxidative burst in plant disease resistance. *Annu Rev Plant Physiol Plant Mol Biol* 48:251–275.

Laurie-Berry, N., Joardar, V., Street, I.H., and Kunkel, B.N. 2006. The *Arabidopsis thaliana* JASMONATE INSENSITIVE 1 gene is required for suppression of salicylic acid-dependent defenses during infection by *Pseudomonas syringae*. *Mol Plant Microbe Interact* 19:789–800.

Lee, J.H., Hong, J.P., Oh, S.K., Lee, S., Choi, D., and Kim, W.T. 2004. The ethylene-responsive factor like protein 1 (CaERFLP1) of hot pepper (*Capsicum annuum L.*) interacts in vitro with both GCC and DRE/CRT sequences with different binding affinities: possible biological roles of CaERFLP1 in response to pathogen infection and high salinity conditions in transgenic tobacco plants. *Plant Mol Biol* 55:61–81.

Li, J., Brader, G., Kariola, T., and Palva, E.T. 2006. WRKY70 modulates the selection of signaling pathways in plant defense. *Plant J* 46:477–491.

Li, J., Brader, G., and Palva, E.T. 2004. The WRKY70 transcription factor: a node of convergence for jasmonate-mediated and salicylate-mediated signals in plant defense. *Plant Cell* 16:319–331.

Liu, Q., Kasuga, M., Sakuma, Y., Abe, H., Miura, S., Yamaguchi-Shinozaki, K., and Shinozaki, K. 1998. Two transcription factors, DREB1 and DREB2, with an EREBP/AP2 DNA binding domain separate two cellular signal transduction pathways in drought- and low-temperature-responsive gene expression, respectively, in *Arabidopsis*. *Plant Cell* 10:1391–1406.

Lorenzo, O., Chico, J.M., Sanchez-Serrano, J.J., and Solano, R. 2004. JASMONATE-INSENSITIVE1 encodes a MYC transcription factor essential to discriminate between different jasmonate-regulated defense responses in *Arabidopsis*. *Plant Cell* 16:1938–1950.

Lu, P.L., Chen, N.Z., An, R., Su, Z., Qi, B.S., Ren, F., Chen, J., and Wang, X.C. 2007. A novel drought-inducible gene, ATAF1, encodes a NAC family protein that negatively regulates the expression of stress-responsive genes in *Arabidopsis*. *Plant Mol Biol* 63:289–305.

Magome, H., Yamaguchi, S., Hanada, A., Kamiya, Y., and Oda, K. 2004. dwarf and delayed-flowering 1, a novel Arabidopsis mutant deficient in gibberellin biosynthesis because of overexpression of a putative AP2 transcription factor. *Plant J* 37:720–729.

Maruyama, K., Sakuma, Y., Kasuga, M., Ito, Y., Seki, M., Goda, H., Shimada, Y., Yoshida, S., Shinozaki, K., and Yamaguchi-Shinozaki, K. 2004. Identification of cold-inducible downstream genes of the Arabidopsis DREB1A/CBF3 transcriptional factor using two microarray systems. *Plant J* 38:982–993.

Mengiste, T., Chen, X., Salmeron, J., and Dietrich, R. 2003. The BOTRYTIS SUSCEPTIBLE1 gene encodes an R2R3MYB transcription factor protein that is required for biotic and abiotic stress responses in *Arabidopsis*. *Plant Cell* 15:2551–2565.

Mittler, R. 2006. Abiotic stress, the field environment and stress combination. *Trends Plant Sci* 11:15–19.

Mittler, R., Vanderauwera, S., Gollery, M., and Van Breusegem, F. 2004. Reactive oxygen gene network of plants. *Trends Plant Sci* 9:490–498.

Nakano, T., Suzuki, K., Fujimura, T., and Shinshi, H. 2006. Genome-wide analysis of the ERF gene family in *Arabidopsis* and rice. *Plant Physiol* 140:411–432.

Nakashima, K., Shinwari, Z.K., Sakuma, Y., Seki, M., Miura, S., Shinozaki, K., and Yamaguchi-Shinozaki, K. 2000. Organization and expression of two *Arabidopsis* DREB2 genes encoding DRE-binding proteins involved in dehydration- and high-salinity-responsive gene expression. *Plant Mol Biol* 42:657–665.

Nakashima, K., Tran, L.S., Van Nguyen, D., Fujita, M., Maruyama, K., Todaka, D., Ito, Y., Hayashi, N., Shinozaki, K., and Yamaguchi-Shinozaki, K. 2007. Functional analysis of a NAC-type transcription factor OsNAC6 involved in abiotic and biotic stress-responsive gene expression in rice. *Plant J* 51:617–630.

Nawrath, C. 2006. Unraveling the complex network of cuticular structure and function. *Curr Opin Plant Biol* 9:281–287.

Nickstadt, A., Thomma, B.P.H.J., Feussner, I.V.O., Kangasjarvi, J., Zeier, J., Loeffler, C., Scheel, D., and Berger, S. 2004. The jasmonate-insensitive mutant jin1 shows increased resistance to biotrophic as well as necrotrophic pathogens. *Mol Plant Pathol* 5:425–434.

Ohme-Takagi, M., and Shinshi, H. 1995. Ethylene-inducible DNA binding proteins that interact with an ethylene-responsive element. *Plant Cell* 7:173–182.

Ohnishi, T., Sugahara, S., Yamada, T., Kikuchi, K., Yoshiba, Y., Hirano, H.Y., and Tsutsumi, N. 2005. OsNAC6, a member of the NAC gene family, is induced by various stresses in rice. *Genes Genet Syst* 80:135–139.

Olsen, A.N., Ernst, H.A., Leggio, L.L., and Skriver, K. 2005. NAC transcription factors: structurally distinct, functionally diverse. *Trends Plant Sci* 10:79–87.

Ooka, H., Satoh, K., Doi, K., Nagata, T., Otomo, Y., Murakami, K., Matsubara, K., Osato, N., Kawai, J., Carninci, P., Hayashizaki, Y., Suzuki, K., Kojima, K., Takahara, Y., Yamamoto, K., and Kikuchi, S. 2003. Comprehensive analysis of NAC family genes in *Oryza sativa* and *Arabidopsis thaliana*. *DNA Res* 10:239–247.

Park, J.M., Park, C.J., Lee, S.B., Ham, B.K., Shin, R., and Paek, K.H. 2001. Overexpression of the tobacco Tsi1 gene encoding an EREBP/AP2-type transcription factor enhances resistance against pathogen attack and osmotic stress in tobacco. *Plant Cell* 13:1035–1046.

Qin, F., Kakimoto, M., Sakuma, Y., Maruyama, K., Osakabe, Y., Tran, L.S., Shinozaki, K. and Yamaguchi-Shinozaki, K. 2007. Regulation and functional analysis of ZmDREB2A in response to drought and heat stresses in *Zea mays* L. *Plant J* 50:54–69.

Rivero, R. M., Kojima, M., Gepstein, A., Sakakibara, H., Mittler, R., Gepstein, S., and Blumwald, E. 2007. Delayed leaf senescence induces extreme drought tolerance in a flowering plant. *Proc Natl Acad Sci USA* 104:19631–19636.

Rizhsky, L., Davletova, S., Liang, H., and Mittler, R. 2004. The zinc finger protein Zat12 is required for cytosolic ascorbate peroxidase 1 expression during oxidative stress in *Arabidopsis*. *J Biol Chem* 279:11736–11743.

Rossel, J.B., Wilson, P.B., Hussain, D., Woo, N.S., Gordon, M.J., Mewett, O.P., Howell, K.A., Whelan, J., Kazan, K., and Pogson, B.J. 2007. Systemic and intracellular responses to photooxidative stress in *Arabidopsis*. *Plant Cell* 19:4091–4110.

Sakamoto, H., Maruyama, K., Sakuma, Y., Meshi, T., Iwabuchi, M., Shinozaki, K., and Yamaguchi-Shinozaki, K. 2004. *Arabidopsis* Cys2/His2-type zinc-finger proteins function as transcription repressors under drought, cold, and high-salinity stress conditions. *Plant Physiol* 136:2734–2746.

Sakuma, Y., Liu, Q., Dubouzet, J.G., Abe, H., Shinozaki, K., and Yamaguchi-Shinozaki, K. 2002. DNA-binding specificity of the ERF/AP2 domain of *Arabidopsis* DREBs, transcription factors involved in dehydration- and cold-inducible gene expression. *Biochem Biophys Res Commun* 290:998–1009.

Sakuma, Y., Maruyama, K., Osakabe, Y., Qin, F., Seki, M., Shinozaki, K., and Yamaguchi-Shinozaki, K. 2006a. Functional analysis of an *Arabidopsis* transcription factor, DREB2A, involved in drought-responsive gene expression. *Plant Cell* 18:1292–1309.

Sakuma, Y., Maruyama, K., Qin, F., Osakabe, Y., Shinozaki, K., and Yamaguchi-Shinozaki, K. 2006b. Dual function of an *Arabidopsis* transcription factor DREB2A in water-stress-responsive and heat-stress-responsive gene expression. *Proc Natl Acad Sci USA* 103:18822–18827.

Schenk, P.M., Kazan, K., Wilson, I., Anderson, J.P., Richmond, T., Somerville, S.C., and Manners, J.M. 2000. Coordinated plant defense responses in *Arabidopsis* revealed by microarray analysis. *PNAS* 97:11655–11660.

Schnurr, J., Shockey, J., and Browse, J. 2004. The acyl-CoA synthetase encoded by LACS2 is essential for normal cuticle development in *Arabidopsis*. *Plant Cell* 16:629–642.

Seki, M., Narusaka, M., Ishida, J., Nanjo, T., Fujita, M., Oono, Y., Kamiya, A., Nakajima, M., Enju, A., Sakurai, T., Satou, M., Akiyama, K., Taji, T., Yamaguchi-Shinozaki, K., Carninci, P., Kawai, J., Hayashizaki, Y., and Shinozaki, K. 2002. Monitoring the expression profiles of 7000 *Arabidopsis* genes under drought, cold and high-salinity stresses using a full-length cDNA microarray. *Plant J* 31:279–292.

Shin, R., Park, J.M., An, J.M., and Paek, K.H. 2002. Ectopic expression of Tsi1 in transgenic hot pepper plants enhances host resistance to viral, bacterial, and oomycete pathogens. *Mol Plant Microbe Interact* 15:983–989.

Shinozaki, K., and Yamaguchi-Shinozaki, K. 2007. Gene networks involved in drought stress response and tolerance. *J Exp Bot* 58:221–227.

Shirasu, K., Nakajima, H., Rajasekhar, V.K., Dixon, R.A., and Lamb, C. 1997. Salicylic acid potentiates an agonist-dependent gain control that amplifies pathogen signals in the activation of defense mechanisms. *Plant Cell* 9:261–270.

Sun, S., Yu, J.P., Chen, F., Zhao, T.J., Fang, X.H., Li, Y.Q., and Sui, S.F. 2008. TINY, a DREB-like transcription factor connecting the DRE- and ERE-mediated signaling pathways in *Arabidopsis*. *J Biol Chem* 283:6261–6271.

Takahashi, F., Yoshida, R., Ichimura, K., Mizoguchi, T., Seo, S., Yonezawa, M., Maruyama, K., Yamaguchi-Shinozaki, K., and Shinozaki, K. 2007. The mitogen-activated protein kinase cascade MKK3-MPK6 is an important part of the jasmonate signal transduction pathway in *Arabidopsis*. *Plant Cell* 19:805–818.

Torres, M.A., and Dangl, J.L. 2005. Functions of the respiratory burst oxidase in biotic interactions, abiotic stress and development. *Curr Opin Plant Biol* 8:397–403.

Torres, M.A., Dangl, J.L., and Jones, J.D. 2002. *Arabidopsis* gp91phox homologues AtrbohD and AtrbohF are required for accumulation of reactive oxygen intermediates in the plant defense response. *Proc Natl Acad Sci USA* 99:517–522.

Ulker, B., Shahid Mukhtar, M., and Somssich, I.E. 2007. The WRKY70 transcription factor of *Arabidopsis* influences both the plant senescence and defense signaling pathways. *Planta* 226:125–137.

Ülker, B., and Somssich, I.E. 2004. WRKY transcription factors: from DNA binding towards biological function. *Curr Opin Plant Biol* 7:491–498.

Veronese, P., Chen, X., Bluhm, B., Salmeron, J., Dietrich, R., and Mengiste, T. 2004. The BOS loci of *Arabidopsis* are required for resistance to *Botrytis cinerea* infection. *Plant J* 40:558–574.

Vogel, J.T., Zarka, D.G., Van Buskirk, H.A., Fowler, S.G. and Thomashow, M.F. 2005. Roles of the CBF2 and ZAT12 transcription factors in configuring the low temperature transcriptome of *Arabidopsis*. *Plant J* 41:195–211.

Wang, H., Hao, J., Chen, X., Hao, Z., Wang, X., Lou, Y., Peng, Y., and Guo, Z. 2007. Overexpression of rice WRKY89 enhances ultraviolet B tolerance and disease resistance in rice plants. *Plant Mol Biol* 65:799–815.

Yadav, V., Mallappa, C., Gangappa, S.N., Bhatia, S., and Chattopadhyay, S. 2005. A basic helix-loop-helix transcription factor in *Arabidopsis*, MYC2, acts as a repressor of blue light-mediated photomorphogenic growth. *Plant Cell* 17:1953–1966.

Yamaguchi-Shinozaki, K., and Shinozaki, K. 2006. Transcriptional regulatory networks in cellular responses and tolerance to dehydration and cold stresses. *Annu Rev Plant Biol* 57:781–803.

Yi, S.Y., Kim, J.H., Joung, Y.H., Lee, S., Kim, W.T., Yu, S.H., and Choi, D. 2004. The pepper transcription factor CaPF1 confers pathogen and freezing tolerance in *Arabidopsis*. *Plant Physiol* 136:2862–2874.

Zhang, J.Y., Broeckling, C.D., Blancaflor, E.B., Sledge, M.K., Sumner, L.W., and Wang, Z.Y. 2005. Overexpression of WXP1, a putative *Medicago truncatula* AP2 domain-containing transcription factor gene, increases cuticular wax accumulation and enhances drought tolerance in transgenic alfalfa (*Medicago sativa*). *Plant J* 42:689–707.

Zhang, J.Y., Broeckling, C.D., Sumner, L.W., and Wang, Z.Y. 2007. Heterologous expression of two *Medicago truncatula* putative ERF transcription factor genes, WXP1 and WXP2, in *Arabidopsis* led to increased leaf wax accumulation and improved drought tolerance, but differential response in freezing tolerance. *Plant Mol Biol* 64:265–278.

Zimmermann, P., Hirsch-Hoffmann, M., Hennig, L., and Gruissem, W. 2004. GENEVESTIGATOR. *Arabidopsis* microarray database and analysis toolbox. *Plant Physiol* 136:2621–2632.

4   Crosstalk in Ca^{2+} Signaling Pathways

Martin R. McAinsh and Julian I. Schroeder

Introduction

The past 20 years has witnessed an explosion in the number of plant-Ca^{2+}-related publications from only a handful per year in the years preceding 1990 to approaching 800 in 2007. The technical and conceptual advances made over this period, many of which have been novel to plant systems, have significantly advanced our understanding of the role of Ca^{2+}-based signaling systems in plants, and the Ca^{2+} ion is now firmly established as a ubiquitous intracellular second messenger in plant cells (Rudd and Franklin-Tong 2001; Sanders et al. 2002; Hetherington and Brownlee 2004). An increase in the cytosolic concentration of free Ca^{2+} ($[Ca^{2+}]_{cyt}$) has been shown to be an important component in the signal transduction pathways by which a diverse array of environmental and developmental stimuli are coupled to their respective end response including both biotic and abiotic signals. For example, there is clear evidence that Ca^{2+} has a central signaling role in plant responses to osmotic, salt, and drought signals (Knight et al. 1991; Ranf et al. 2008), oxidative stress (Price et al. 1994; McAinsh et al. 1996; Pei et al. 2000; Evans et al. 2005), cold (Knight et al. 1991; Ranf et al. 2008), gaseous pollutants (McAinsh et al. 2002; Evans et al. 2005), light (Shacklock et al. 1992; Dodd et al. 2007), plant hormones (McAinsh et al. 1990; Allen et al. 2001), pathogens (elicitors; Knight et al. 1991; Lecourieux et al. 2002), and nodulation (Nod factors; Ehrhardt et al. 1996; Peiter et al. 2007). However, the very ubiquity of this second messenger raises important questions regarding the level of interaction or crosstalk that takes place between different Ca^{2+}-mediated signaling pathways and mechanism(s) by which specificity is controlled in Ca^{2+}-based signaling systems (McAinsh and Hetherington 1998; Evans et al. 2001; Ng and McAinsh 2003). Specifically, cells must possess the intracellular machinery to allow them to discriminate between Ca^{2+} elevations elicited by the different stress signals—for example, Ca^{2+} elevation caused by salt and cold, or even different levels of the same signal, each of which induces a specific set of responses.

The mechanisms by which Ca^{2+} signals are generated in plant cells involve the coordinated action of plasma membrane and endomembrane Ca^{2+}-permeable channels, Ca^{2+}-ATPases and H^+/Ca^{2+} exchangers, and Ca^{2+}-binding proteins, enabling the generation of complex patterns of elevations in $[Ca^{2+}]_{cyt}$ (Sanders et al. 2002; Hetherington and Brownlee 2004). These have been referred to by some authors as a stimulus-specific 'Ca^{2+} signature' (Knight et al. 1998; McAinsh and Hetherington 1998; Ng and McAinsh 2003). Therefore, the potential for plant cells to generate changes in $[Ca^{2+}]_{cyt}$ that encode information about the nature and strength of a stimuli, in a similar manner to that observed in animal systems (Berridge et al. 2000, 2003; Berridge 2007), is clear (McAinsh and Hetherington 1998; Evans et al. 2001; Harper 2001; Rudd and Franklin-Tong 2001; Schroeder et al. 2001; Sanders et al. 2002; Ng and McAinsh 2003; Hetherington and Brownlee 2004; Oldroyd and Downie 2004).

Ca^{2+} signals with specific spatio-temporal dynamics have been observed in response to a number of stimuli in a range of cell types, and it has been shown that alterations to these results in a change to the downstream response, lending support to this proposal (McAinsh et al. 1995; Allen et al. 2000, 2001; Miwa et al. 2006).

If indeed signaling information is encoded by the differing spatiotemporal dynamics of stimulus-induced changes in $[Ca^{2+}]_{cyt}$, it is essential that plants can also sense these changes and decode the encrypted signaling information. Several models have been proposed in animal cells to explain how signaling information encrypted in transients, spikes, and oscillations in $[Ca^{2+}]_{cyt}$ might be decoded based largely on Ca^{2+}-dependent protein kinases (Berridge et al. 2000, 2003; Berridge 2007). In plants, downstream effects are frequently elicited through protein modification by phosphorylation, and kinases have been implicated as being central in transducing Ca^{2+} signals in plants (Luan et al. 2002; Harper et al. 2004). *Arabidopsis* has 1,019 protein kinases of which 67 have been implicated in Ca^{2+} signaling (Hrabak et al. 2003; Harper et al. 2004). In this chapter, we consider the mechanisms by which Ca^{2+} signals are generated in plants, the evidence for stimulus-specific changes in plant $[Ca^{2+}]_{cyt}$ and specificity in plant Ca^{2+} signals, and how the encrypted signaling information may be decoded, highlighting the potential for crosstalk between Ca^{2+} signaling pathways.

Ca^{2+} Signals in Abiotic and Biotic Stress Responses

Stimulus-induced changes in plant $[Ca^{2+}]_{cyt}$ have been reported in a number of cell types in response to a range of abiotic and biotic stimuli (see Rudd and Franklin-Tong 2001; Oldroyd and Downie 2006; Lecourieux et al. 2006). Therefore, we focus on a number of key examples to illustrate the variety and physiological role of these Ca^{2+} signals.

Whole-Plant Abiotic Stress Responses

The recombinant aequorin technique has been extensively used to demonstrate the role of Ca^{2+} signals in plant abiotic stress responses (Knight et al. 1991). Changes in whole-plant $[Ca^{2+}]_{cyt}$ have been reported in response to a variety of abiotic environmental stress signals (see Rudd and Franklin-Tong 2001; Sanders et al. 2002; Israelsson et al. 2006) including drought and salt stress (Knight et al. 1997; Kiegle et al. 2000; Ranf et al. 2008), cold (Knight et al. 1991, 1996; Dodd et al. 2006; Ranf et al. 2008), hypo-osmotic shock (Knight et al. 1997, 1998; Keigle et al. 2000), mechanical stimulation (Knight et al. 1992; Haley et al. 1995), and ozone pollution (Clayton et al. 1999; Evans et al. 2005). The dynamics of the Ca^{2+} signals varies markedly among different abiotic stresses and cell types, providing the potential to encode signaling information about both the nature and strength of the stress in terms of differences in the kinetics, both the spatial and temporal dynamics, of the changes in $[Ca^{2+}]_{cyt}$.

For example, in *Arabidopsis* cold-shock induces a rapid, transient increase in whole-plant $[Ca^{2+}]_{cyt}$ of 1–2 µM of less than 30 seconds in duration (Knight et al. 1991; Keigle et al. 2000), whereas drought and salt stress result in a biphasic increase in $[Ca^{2+}]_{cyt}$ consisting of a rapid Ca^{2+} transient followed by a sustained elevation lasting hours

(Knight et al. 1997; Keigle et al. 2000). Interestingly, slow cooling also results in a biphasic increase in $[Ca^{2+}]_{cyt}$, although the second $[Ca^{2+}]_{cyt}$ elevation is much shorter, lasting minutes before $[Ca^{2+}]_{cyt}$ returns to resting levels (Knight et al. 1996). However, when these responses are examined at the tissue level using aequorin targeted to distinct cell types in the root, including the pericycle, endodermis, and elongation zone, the picture is subtlety different (Keigle et al. 2000). The spatial dynamic of the $[Ca^{2+}]_{cyt}$ changes varies among tissues. In addition, although all cell types still exhibit a characteristic $[Ca^{2+}]_{cyt}$ response to all three stresses, the responses of different cell types to the same stress, and also the same cell type to different stress, is very diverse. In particular, the magnitude of the cold-shock response is lowest in the elongation zone (0.5 ± 0.03 µM) and highest in the pericycle (0.9 ± 0.04 µM). Furthermore, the second phase of the drought and salt stress responses is most pronounced in the endodermis and pericycle and is characterized by prolonged oscillations in $[Ca^{2+}]_{cyt}$ that are not seen in another stress–cell type combination. Interestingly, Dodd et al. (2006) subsequently added an additional layer of complexity to this story. They show that cold-shock-induced changes in both whole-plant and guard cell $[Ca^{2+}]_{cyt}$ are modulated by the time of day such that the observed $[Ca^{2+}]_{cyt}$ transients were significantly higher during the middle of the photoperiod than at the beginning or end. In addition, the previously described plateaus in Ca^{2+}, which are recorded as an average response from a population of cells, were modeled as resulting from out-of-phase Ca^{2+} oscillations in individual cells (Dodd et al. 2006). $[Ca^{2+}]_{cyt}$ also participates in red light signaling, and there is an interaction between phytochrome signaling and low temperature signaling (Shacklock et al. 1992; Benedict et al. 2006). Therefore, this observation highlights the potential for crosstalk between abiotic stress signaling (e.g., low temperature responses) and other signaling pathways such as those involved in light responses.

There is an increasing body of evidence that signaling information may be encoded in different parameters of the kinetics of abiotic stress-induced Ca^{2+} signals (see Rudd and Franklin-Tong 2001; Sanders et al. 2002). This potential is clearly illustrated in the signaling pathway by which plants perceive and respond to the pollutant gas ozone. An increase in $[Ca^{2+}]_{cyt}$ is one of the earliest characterized response to ozone which takes place within seconds of exposure (Clayton et al. 1999; Evans et al. 2005). *Arabidopsis* exhibits a biphasic increase in $[Ca^{2+}]_{cyt}$ in the cotyledons and aerial parts of the plant consisting of a rapid Ca^{2+} transient that peaks within minutes of ozone exposure, and which returns to resting basal levels within 7 to 10 minutes, followed by a sustained elevation lasting for up to 60 minutes (Evans et al. 2005). In contrast, the $[Ca^{2+}]_{cyt}$ increase in the roots is monophasic. In both cases, there is a positive correlation between the magnitude of the two phases of the ozone $[Ca^{2+}]_{cyt}$ response and ozone concentration exposure (Clayton et al. 1999; Evans et al. 2005). Interestingly, the two phases appear to encode information about various facets of the Ca^{2+} signal. The first phase is sensitive to the rate of change in the ozone concentration and is only observed when plants are exposed to a rapid and steady increase in ozone (Evans et al. 2005). However, the second phase, which can be abolished by treatment with La^{3+}, suggesting a role for Ca^{2+} influx in its generation, is essential for the subsequent induction of the gene encoding the antioxidant enzyme glutathione-S-transferase (Clayton et al. 1999). This indicates that the second phase of the ozone Ca^{2+} response encodes the signaling information required for the induction of this antioxidant defence pathways.

Pollen Tube Growth

The role of $[Ca^{2+}]_{cyt}$ in pollen germination and pollen tube growth is well documented (see Rudd and Franklin-Tong 2001; Holdaway-Clarke and Hepler 2003; Dumas and Gaude 2006). Growing pollen tubes exhibit a sharp, tip-focused $[Ca^{2+}]_{cyt}$ gradient including those of lily, *Nicotiana sylvestris,* and *Tradescantia virginiana* (Obermeyer and Weisenseel 1991; Rathore et al. 1991; Miller et al. 1992; Malho et al. 1994; Pierson et al. 1994, 1996; Franklin-Tong et al. 1997). For example, in the lily, *Lilium longiflorum,* the $[Ca^{2+}]_{cyt}$ in the apical few micrometers of the tip of the growing pollen tube is 3 to 5 µM and returns to resting levels (approximately 100 nM) within 20 µm from the tip apex generating a sharp gradient (Obermeyer and Weisenseel 1991; Rathore et al. 1991; Miller et al. 1992; Pierson et al. 1994). The high point of the gradient in the immediate vicinity of the tip apex appears to be derived, at least in part, from the influx of extracellular Ca^{2+} through stretch-responsive channels activated by deformations in the nascent tip wall (Pierson et al. 1994, 1996; Malho et al. 1995; Holdaway-Clarke et al. 1997). However, although Ca^{2+}-permeable channels are likely to contribute to the gradient formation and control, the specific molecular mechanisms of this process remain largely unknown (Dutta and Robinson 2004). A recent study has shown that loss of function mutations in the *Arabidopsis* cyclic nucleotide gated channel homolog *AtCNGC18* disrupt pollen tube growth and that the AtCNGC protein is Ca^{2+} permeable and is targeted to the tip region of *Arabidopsis* pollen tubes (Frietsch et al. 2007). Further research should uncover additional Ca^{2+} channel and transporter genes that together mediate this response.

Several pieces of evidence point toward the importance of the pollen tube $[Ca^{2+}]_{cyt}$ gradient in the control and orientation of apical growth. Dissipation of the $[Ca^{2+}]_{cyt}$ gradient by the injection of Ca^{2+} buffers such as BAPTA (1,2-bis(o-aminophenoxy)ethane-N,N,N',N'-tetraacetic acid), mild thermal shock, or the application of Ca^{2+} channel blockers inhibits pollen tube growth (Rathore et al. 1991; Pierson et al. 1994; Li et al. 1996). Interestingly, studies in poppy, *Papaver rhoeas,* undergoing a self-incompatibility response also reveal a disruption of the pollen tube $[Ca^{2+}]_{cyt}$ gradient and inhibition of pollen tube growth (Franklin-Tong et al. 1993, 1995, 1997; Pierson et al. 1994). Self-incompatibility induction triggers a transient increase in $[Ca^{2+}]_{cyt}$ reaching 1 µM in the first minute and returning to resting levels within 10 to 12 minutes. This involves the influx of extracellular Ca^{2+} at the shank of the pollen tube and an increase in $[Ca^{2+}]_{cyt}$ in the form of a Ca^{2+} wave, followed by the cessation of pollen tube growth within 1 to 2 minutes (Franklin-Tong et al. 1997, 2002). In contrast, changing the position of the gradient within the pollen tube apex by the localized photolysis of caged-Ca^{2+} results in a change to the direction of tip growth (Malho et al. 1994, 1995; Malho and Trewavas 1996). These data highlight the importance of the $[Ca^{2+}]_{cyt}$ gradient in the control of pollen tube growth.

Oscillations in $[Ca^{2+}]_{cyt}$ are also observed at the tip of growing pollen tubes with a magnitude of between 750 nm and 3 µM with a period of 23 to 57 seconds (Pierson et al. 1996; Holdaway-Clarke et al. 1997; Messerli and Robinson 1997). Interestingly, there is a close correlation between the timing of the peaks of the $[Ca^{2+}]_{cyt}$ oscillations and the oscillations that are characteristic of pollen tube growth. However, the peaks in the oscillations of growth rate precede the peaks of the oscillations in $[Ca^{2+}]_{cyt}$ by 4 to 11 seconds, suggesting a complex relationship (Holdaway-Clarke et al. 1997; Messerli et al. 2000). In addition, although $[Ca^{2+}]_{cyt}$ oscillations are typical of growing pollen tubes, there are a limited number of reports of $[Ca^{2+}]_{cyt}$ oscillations in lily pollen tubes in the absence of growth (Messerli and Robinson 2003).

Stomatal Guard Cells

A change in $[Ca^{2+}]_{cyt}$ is a key hub in the signaling network by which stomatal guard cells respond to external stimuli (Hetherington and Woodward 2003; Li et al. 2006; Israelsson et al. 2006). Stimulus-induced changes in guard cell $[Ca^{2+}]_{cyt}$ were first quantified in *Commelina communis* in response to the plant hormone abscisic acid (ABA; McAinsh et al. 1990). Subsequently, changes in guard cell $[Ca^{2+}]_{cyt}$ are well documented in a number of species using a range of reporter systems and in response to a diverse array of abiotic and biotic stimuli—for example, plant hormones (McAinsh et al. 1990; Schroeder and Hagiwara 1990; Irving et al. 1992; Allen et al. 1999; Staxen et al. 1999), external Ca^{2+} (McAinsh et al. 1995; Allen et al. 1999; Han et al. 2003; Tang et al. 2007), H_2O_2 (McAinsh et al. 1996; Pei et al. 2000), cold (Allen et al. 2000), CO_2 (Webb et al. 1996; Young et al. 2006), hyperpolarization (Grabov and Blatt 1998; Staxen et al. 1999; Allen et al. 2000, 2001), inositol trisphosphate (Gilroy et al. 1990), inositol hexakisphosphate (Lemtiri-Chlieh et al. 2003), cyclic adenosine diphosphoribose (ADP)-ribose (Leckie et al. 1999), sphingosine-1-phosphate (Ng et al. 2001), pathogenic elicitors (Klusener et al. 2002), and time of day (Dodd et al. 2006). This highlights the potential for crosstalk within the guard cell signaling network and therefore the necessity for specificity in guard cell Ca^{2+} signaling. A change in guard cell turgor, and hence stomatal aperture, is associated with stimulus-induced changes in guard-cell $[Ca^{2+}]_{cyt}$, although the significance of stomata aperture changes observed in the absence of a change in $[Ca^{2+}]_{cyt}$ has been hotly debated, culminating in the accepted view that guard cell signaling networks comprise both Ca^{2+}-dependent and Ca^{2+}-independent signaling pathways (Ward et al. 1995; Hetherington and Brownlee 2004; Israelsson et al. 2006). However, the relative contributions of these proposed parallel pathways to ABA responses have not yet been investigated, and further research is direly needed.

The kinetics of the stimulus-induced changes in guard cell $[Ca^{2+}]_{cyt}$ vary markedly and, interestingly, frequently take the form of oscillations in $[Ca^{2+}]_{cyt}$ (see McAinsh 2007). Initial studies of external Ca^{2+}- (McAinsh et al. 1995) and ABA (Staxen et al. 1999) -induced oscillations in guard cell $[Ca^{2+}]_{cyt}$ in *C. communis* provide important clues to the role of oscillations in guard cell Ca^{2+} signaling. A direct correlation exists among the pattern of $[Ca^{2+}]_{cyt}$ oscillations, the strength of the external stimulus, and the resultant steady-state stomatal aperture. External Ca^{2+} at 0.1 mM induces symmetrical oscillations (amplitude, 300–560 nM; mean period 8.3 minutes) whereas 1.0 mM external Ca^{2+} induces oscillations that are asymmetrical in character (amplitude, 400–850 nM; mean period 13.6 minutes) resulting in approximately a 12.7% and 41.6% reduction in steady state stomatal aperture, respectively (McAinsh et al. 1995). A similar relationship is also observed between the pattern of ABA-induced oscillations in guard cell $[Ca^{2+}]_{cyt}$, the concentration of ABA and magnitude of the ABA-induced stomatal closure (Staxen et al. 1999). In both cases, the continued presence of the stimulus is required to maintain the oscillations in $[Ca^{2+}]_{cyt}$ and it is possible to switch reversibly among the various oscillatory patterns by changing the strength of the stimulus within the range that induced oscillations (McAinsh et al. 1995; Staxen et al. 1999). This suggests that, as in animal cells, oscillations in guard cell $[Ca^{2+}]_{cyt}$ encode important signaling information about the nature and strength of a stimuli (Berridge et al. 2000, 2003; Berridge 2007). The mechanism(s) by which guard cell signaling information is encrypted in guard cell Ca^{2+} signals has been the subject of a series of detailed studies that are discussed in subsequent sections (Allen et al. 1999, 2000, 2001; Young et al. 2006).

Root Hairs and Nodulation Signals

The role of Nod factors, bacterial lipochitooligosaccharides, in the formation of nitrogen-fixing nodules in the roots of leguminous plants is well documented (Geurts and Bisseling 2002; Oldroyd and Downie 2004). An increase in $[Ca^{2+}]_{cyt}$ appears as an early event in the Nod-factor signaling pathway in hairs of susceptible roots. An influx of Ca^{2+} is observed within 1 minute of root hairs being exposed to Nod factors (Cardenas et al. 1998; Felle et al. 1998; Shaw and Long 2003), resulting in elevations in $[Ca^{2+}]_{cyt}$. These elevations take the form of an initial rapid increase in $[Ca^{2+}]_{cyt}$, leading to a sustained increase in $[Ca^{2+}]_{cyt}$ up to a plateau lasting 2 to 4 minutes (Gehring et al. 1997). Approximately 10 to 20 minutes later, this initial elevation is followed by a series of repetitive increases or "Ca^{2+} spikes" in the nuclear region (Shaw and Long 2003). Nod-factor-induced Ca^{2+} spikes have been observed in a range of leguminous species including *Medicago sativa* (Ehrhardt et al. 1996), *Medicago truncatula* (Wais et al. 2000; Miwa et al. 2006), *Pisium sativum* (Walker et al. 2000), *Phaseolus vulgaris* (Cardenas et al. 1998), and *Lotus japonicus* (Harris et al. 2003), suggesting that they are an important component of the Nod-factor signaling pathway. Additional support for this suggestion comes from the observation that non-nodulating mutants fail to exhibit Nod-factor-induced Ca^{2+} spikes (Ehrhardt et al. 1996; Wais et al. 2000; Walker et al. 2000; Harris et al. 2003; Miwa et al. 2006). Interestingly, Miwa et al. (2006) have shown that in *M. truncatual* the frequency of the Ca^{2+} spikes varies depending on the location of the root hair cell on the root, producing a gradient of Ca^{2+} spiking frequency along the root, with younger hairs having a longer average period between Ca^{2+} spikes than older hairs. Further, they showed that the zone in which early nodulation genes such as *ENOD11* are induced correlates with a region of the root where Ca^{2+} spiking occurs with a period of approximately 100 seconds between spikes. This suggests that the kinetics of the Ca^{2+} spikes may encode information in the Ca^{2+} signal although artificial manipulation of the Nod-factor-induced Ca^{2+}-spikes, indicating that it is the number of spikes, together with additional as yet undefined inputs, rather than spike frequency that are important in coupling Nod factor signals to the induction of early nodulation genes (Miwa et al. 2006).

Plant Defense Responses

An influx of extracellular Ca^{2+} is an early event in the induction of plant defense responses to pathogen attack and changes in $[Ca^{2+}]_{cyt}$ are well documented in response to pathogenic elicitors and in natural fungus–plant interactions (Nurnberger et al. 1994; see Rudd and Franklin-Tong 2001; Lecourieux et al. 2006). The oxidative burst, which is one of the earliest and best characterized components of defense signaling in plants, is a direct response to the increases in $[Ca^{2+}]_{cyt}$ (Blume et al. 2000) and results from the Ca^{2+}-dependent activation of a reduced nicotinamide adenine dinucleotide phosphate (NADPH) oxidase (see Nurnberger and Scheel 2001). Interestingly, the H_2O_2 produced has the potential to stimulate further Ca^{2+} influx through H_2O_2-stimulated Ca^{2+} channels (Pei et al. 2000; Kwak et al. 2003) so that Ca^{2+} appears to function both upstream and downstream of pathogen/stimulus-induced increases in reactive oxygen species during defense and stress signaling.

Initial studies of elicitor-induced changes in $[Ca^{2+}]_{cyt}$ reported transient increases in $[Ca^{2+}]_{cyt}$, with a duration of <1 minute, in whole seedlings of tobacco in response to a crude yeast elicitor preparation and *Gliocladium deliquescens* elicitor (Knight et al. 1991). Interestingly, the kinetics of these elicitor-induced $[Ca^{2+}]_{cyt}$ transients were markedly different from those induced by either cold or wind, providing further evidence for stimulus-specific Ca^{2+} signals. Elicitor-induced transient increases in $[Ca^{2+}]_{cyt}$ have also been observed in *Arabidopsis* seedlings in response to oligogalacturonic acid (Hu et al. 2004) and in tobacco and soybean cells in response to and lipopolysaccharides from *Burkholderia cepacia* (Poinssot et al. 2003) and micorrhizal fungi (Navazio et al. 2007), respectively. Subsequently, it has been shown that elicitor-induced changes in $[Ca^{2+}]_{cyt}$ are frequently biphasic and that the kinetics of the resultant Ca^{2+} signals differs between elicitors. Lecourieux et al. (2002) have shown that treatment of tobacco cells with oligosaccharidic elicitors induces a rapid, transient increase in $[Ca^{2+}]_{cyt}$ characterized by two distinct peaks before returning to resting levels of $[Ca^{2+}]_{cyt}$ within 15 to 20 minutes. Two transient peaks in $[Ca^{2+}]_{cyt}$ are also observed in soybean cells treated with a *Phytophthora sojae* β-glucan elicitor and chitin (Mithofer et al. 1999). In contrast, tobacco cells treated with cryptogein, a proteinaceous elicitor derived from *Phytophthora cryptogea*, exhibit a transient increase in $[Ca^{2+}]_{cyt}$ within 6 minutes of the addition of the elicitor that peaks at 2.4 μM followed immediately by second increase in $[Ca^{2+}]_{cyt}$ that peaks at 750 nM and that remains elevated for up to 2.5 hours (Lecourieux et al. 2002). Parsley cells treated with the Pep-13 elicitor from *Phytophora sojae* (Nurnberger et al. 1994) exhibit a rapid, large, transient increase in $[Ca^{2+}]_{cyt}$ that peaks at 1 μM within 1 to 2 minutes after the addition of the elicitor followed by sustained increase in $[Ca^{2+}]_{cyt}$ to 300 nM and that remains elevated above resting levels for at least 30 minutes (Blume et al. 2000). A similar biphasic and sustained increases in $[Ca^{2+}]_{cyt}$ is observed in grapevine cells in response to the BcPG1 elicitor from *Botrytis cinerea* (Vandelle et al. 2006). Interestingly, this sustained second phase observed in parsley cells has been shown to be required to stimulate phytoalexin synthesis and the subsequent defense response to Pep-13 (Blume et al. 2000) and the Pp-elicitor from *Phytopthora parasitica* (Fellbrich et al. 2000), suggesting that important signaling information required for the induction of plant defense responses is encoded in the elicitor-induced Ca^{2+} signal, possibly through differences in the kinetics of the Ca^{2+} signal. This is further illustrated by the observation that in cowpea, increases in $[Ca^{2+}]_{cyt}$ in response to infection by the rust fungus *Uromyces vignae* are only observed in resistant plants; infection fails to elicit an increase in $[Ca^{2+}]_{cyt}$ in susceptible plants (Xu and Heath 1998). Similarly, *Arabidopsis* possessing the *RPM1* resistance gene exhibits a biphasic increases in $[Ca^{2+}]_{cyt}$ leading to the hypersensitive response following treatment with the bacterial pathogen *Pseudomonas syringae* pv. *tomato* harboring the *avrRpm1* avirulence gene before the onset of the hypersensitive response, whereas a loss of function *Arabidopsis* R-gene mutant fails to exhibit an increase in $[Ca^{2+}]_{cyt}$ or mount a hypersensitive response (Grant et al. 2000). These data confirm the suggestion that the infection-derived Ca^{2+} signal encodes important information required for induction of plant defense responses. In addition, differences in the spatial dynamics of elicitor-induced Ca^{2+} signals have also been reported. For example, proteinaceous elicitors that lead to necrosis induce a pronounced and sustained increase in nuclear free Ca^{2+} in tobacco cells, whereas nonnecrotic elicitors have little effect on nuclear free Ca^{2+} (Lecourieux et al. 2005). These data suggest the potential of both the spatial and temporal dynamics of infection-derived Ca^{2+} signals to encode signaling information required for the induction of plant defense responses.

Generation of Ca^{2+} Signals

The generation of complex Ca^{2+} signals, in terms of both the spatial and temporal dynamics of the changes in $[Ca^{2+}]_{cyt}$, requires the coordinated functioning of Ca^{2+} influx channels in the plasma membrane and endomembranes that release Ca^{2+} into the cytosol and Ca^{2+} efflux transporters that restore $[Ca^{2+}]_{cyt}$ to resting levels by removing Ca^{2+} from the cell or into internal stores (Sanders et al. 2002; Hetherington and Brownlee 2004; Figure 4.1).

Plasma Membrane Ca^{2+} Influx Channels

Research has suggested that multiple pathways mediating stress- or stimulus-induced Ca^{2+} influx exist in plants. Defined properties of plant Ca^{2+} channel classes have been characterized in patch-clamp and electrophysiological studies. Stress- and ABA-activated plasma membrane Ca^{2+} influx have been characterized in response to pathogen elicitors (Zimmermann et al. 1997) and ABA (Schroeder and Hagiwara 1990; Hamilton et al. 2000; MacRobbie 2000; Pei et al. 2000).

Early studies showed that plant Ca^{2+} channels are often not strictly Ca^{2+} selective but rather cation permeable with a physiologically relevant relative permeability to Ca^{2+} ions (Schroeder and Hagiwara, 1990; Thuleau et al. 1994a; Zimmermann et al. 1997; Pei et al. 2000; Very and Davies 2000). Indeed, to date, no Ca^{2+} channel has, to our knowledge, been characterized in the plasma membrane of plant cells that have a strong selectivity for Ca^{2+} over other cations, such as Mg^{2+} or Na^+ (Demidchik et al. 2002). Similarly, in animal systems many receptor-linked Ca^{2+} permeable channels are also permeable to other physiological cations. A marked exception are the voltage-dependent Ca^{2+} channels in animal cells, which are highly Ca^{2+} selective when Ca^{2+} ions are present in the extracellular solution (Tsien and Tsien 1990).

The main classes of plasma membrane Ca^{2+} channels that have been characterized in patch clamp studies of plant cells are as follows:

1. Ca^{2+} channels that show enhanced opening at hyperpolarized membrane potentials (Gelli and Blumwald 1997; Hamilton et al. 2000; Pei et al. 2000; Very and Davies

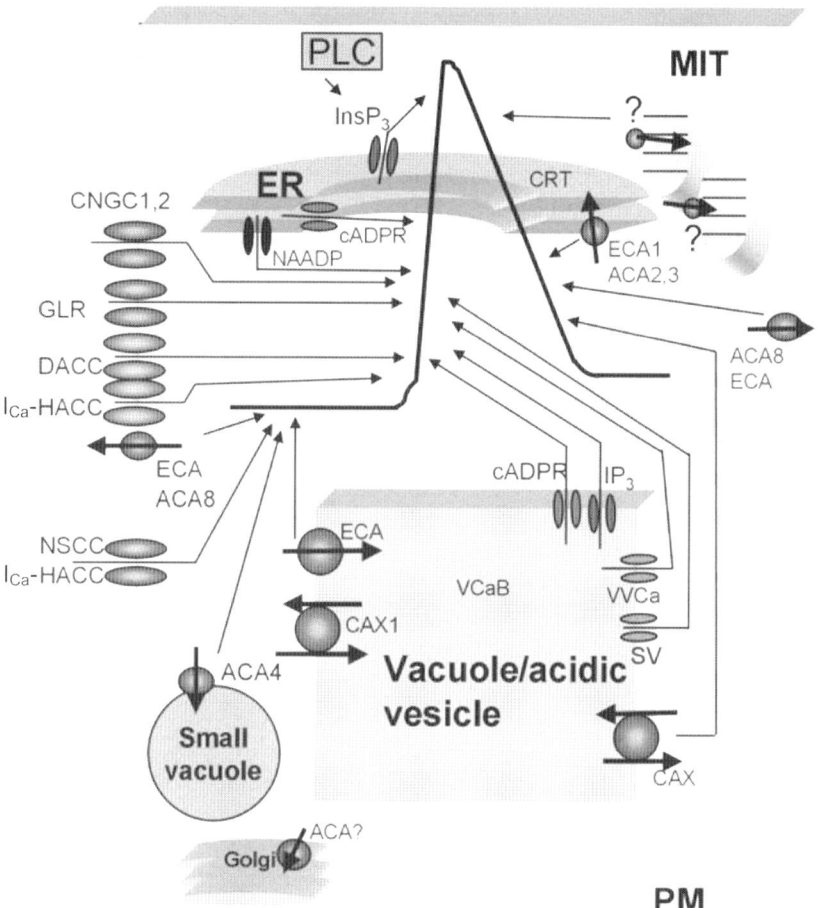

Fig. 4.1 Schematic representation of the major Ca^{2+} transport pathways in plant cells. The chloroplast and nucleus are not shown. ACA and ECA = Ca^{2+}-ATPases; CAX, H^+/Ca^{2+} exchangers; NSCC = nonselective cation channels; DACC = depolarization-activated Ca^{2+} channels; I_{Ca}-HACC = hyperpolarization-activated Ca^{2+} channels; GLR = glutamate receptor channels; CNGC = cyclic nucleotide-gated channels; PM = plasma membrane; mit = mitochondria; ER = endoplasmic reticulum; CRT = calreticulin; VCaB = vacuolar Ca^{2+}-binding protein. (From Hetherington and Brownlee 2004, reprinted with permission from the *Annual Review of Plant Biology*, volume 55. © 2004 by Annual Reviews, www.annualreviews.org.)

2000). These are currently the most widely characterized types of plant Ca^{2+} channels, and it appears likely that several distinct gene families may contribute to their functions, as described later.

2. Ca^{2+}-permeable channels that are activated by membrane depolarization (Huang et al. 1994; Marshall et al. 1994; Thuleau et al. 1994a, 1994b; Thion et al. 1998). However, to date this class of Ca^{2+} channel has not yet been characterized in *Arabidopsis*, and it is possible that these channels are either rarely used or absent in the *Arabidopsis* genome, similar to the Ca^{2+} and CaM-activated kinases (CCaMKs) found in legumes but not *Arabidopsis* (Wais et al. 2000; see also Oldroyd and Downie 2004; Harper and Harmon 2005; Yang et al. 2007).

3. Stretch-activated Ca^{2+} channels have been recorded (Cosgrove and Hedrich, 1991). These stretch channels have been proposed as candidates for mechanical responses (Monshausen et al. 2007). However, few data exist and more studies are needed to characterize the detailed regulation properties of these stretch-activated Ca^{2+} channels. The recently characterized mechanically induced elevation in cytosolic Ca^{2+} in roots should provide a system in which the underlying mechanisms can be further studied (Monshausen et al. 2007).

Thus, of the above three major general classes of Ca^{2+} channels, the hyperpolarization-activated Ca^{2+} channels have been studied in most detail, including a number of analyses with respect to stress signaling, which are further reviewed here. Hyperpolarization-activated Ca^{2+} permeable channels are activated by fungal elicitors in tomato cells (Gelli and Blumwald 1997) and by ABA in guard cells (Hamilton et al. 2000; Pei et al. 2000). ABA enhances the activity of Ca^{2+} permeable cation (I_{Ca}) channels, by shifting their activation voltage to more positive membrane potentials (Hamilton et al. 2000). In *Arabidopsis* guard cells, brown algae and root epidermal cells, these I_{Ca} channels are activated by reactive oxygen species (Pei et al. 2000; Murata et al. 2001; Coelho et al. 2002; Foreman et al. 2003). Further, reactive oxygen species (ROS) cause $[Ca^{2+}]_{cyt}$ increases in guard cells (McAinsh et al. 1996; Pei et al. 2000). Studies of ABA signaling have shown that ABA causes an increase in ROS levels (Pei et al. 2000; Zhang et al. 2001; Mustilli et al. 2002; Bright et al. 2006). ABA activation of these I_{Ca} channels in guard cells requires the presence of intracellular NAD(P)H (Murata et al. 2001). Genetic evidence for a role of ROS in ABA activation of I_{Ca} channels was obtained through a combination of genomic guard cell expression analyses using microarrays, and generation of a double mutant in two NADPH oxidase genes, that are highly expressed in guard cells, showing impairment in ABA activation of I_{Ca} channels in double mutant guard cells (Kwak et al. 2003). Phosphorylation events and Ca^{2+}-dependent protein kinases also function in ABA modulation of these I_{Ca} channels (Kohler and Blatt 2002; Mori et al. 2006).

Research in root epidermal cells and guard cells suggests that more than one class of hyperpolarization-activated Ca^{2+} channels may well exist in plants (McAinsh et al. 1995; Demidchik et al. 2002; Klusener et al. 2002). However, molecular characterization of different Ca^{2+} channels is necessary to investigate this hypothesis. Because I_{Ca} channels are mainly activated at hyperpolarized voltages and the genes encoding these ion channels remain unknown, their exact physiological functions remain to be determined. For example, in guard cells, these ion channels could function as a "fail-safe" mechanism, counteracting stomatal reopening once initial ABA-induced stomatal closing events have been initiated (Israelsson et al. 2006), and they may also function in initial ABA-induced Ca^{2+} increases when guard cells are more hyperpolarized. However, the in vivo time course of ABA-induced I_{Ca} channel activation has not been analyzed in detail and, at least in patch clamp studies, occurs within approximately 3 minutes of ABA application (Pei et al. 2000; Murata et al. 2001), indicating that the former function may be one of the physiologically relevant roles. Mutation of the AtMRP5 ATP-binding cassette (ABC) transmembrane regulator protein causes partial ABA insensitive stomatal closing (Gaedeke et al. 2001) and impairs ABA activation of guard cell I_{Ca} channels (Suh et al. 2007). However, it is unlikely that ABC transporters encode I_{Ca} channels, because *atmrp5* mutant guard cells also show impairment in Ca^{2+} activation of S-type anion channels in guard cells, leading to the suggestion that AtMRP5 may function as a membrane-bound regulator of more than one ion channel type in guard cells (Suh et al. 2007).

Although many genes have been identified that are thought to encode plasma membrane Ca^{2+} channels in plants, remarkably direct evidence is limited to some evidence for homologs to glutamate receptors in plants (Dennison and Spalding 2000; Qi et al. 2006). Currently, two classes of ion channel homologous genes have been considered as candidates for plasma membrane Ca^{2+} channels in plants: the glutamate-receptor homologous (GluR) genes and the cyclic nucleotide gated channel homologous genes (*CNGCs*). Both of these families include 20 full-length genes encoded in the *Arabidopsis* genome. Glutamate has been shown to induce relatively fast (within a few seconds) increases in $[Ca^{2+}]_{cyt}$ in aequorin-transformed whole *Arabidopsis* seedlings (Dennison and Spalding 2000). Ca^{2+} influx is implicated in this response based on the inhibitory effects of La^{3+}.

A patch clamp study in guard cells has reported activation of I_{Ca}-like channels by cyclic adenosine monophosphate (cAMP) but not cyclic guanosine monophosphate (Lemtiri-Chlieh and Berkowitz 2004). Recessive mutations in the *CNGC2* and *CNGC4* genes show altered plant pathogen-induced cell death, suggesting a role for these ion channels in modulating part of the pathogen-mediated response (Clough et al. 2000; Balague et al. 2003). Furthermore, a dominant mutation that affects the tandem genes *CNGC11* and *CNGC12* constitutively activates various defense responses, exhibits stunted growth with curly leaves, and displays enhanced resistance to the virulent pathogen *Hyaloperonospora parasitica* Emco5 (Yoshioka et al. 2006). However, how stress signaling pathways activate putative Ca^{2+}-permeable channels encoded by these genes remains unknown.

Endomembrane Ca^{2+}-Permeable Channels

Multiple pathways for the release of Ca^{2+} from endomembrane stores into the cytosol have also been identified in plant cells (see Sanders et al. 2002; Hetherington and Brownlee 2004; Pottosin and Schonknecht 2007) contributing further to the ability of plants to generate complex, and potentially stimulus-specific, Ca^{2+} signals. However, unlike plasma membrane channels, it is not possible to patch-clamp and electrophysiologically characterize endomembrane channels in intact cells, making it difficult to link the activity of specific endomembrane Ca^{2+} channels to changes in $[Ca^{2+}]_{cyt}$. Furthermore, the first characterization of a plant endomembrane Ca^{2+} channel at the molecular level has only recently been completed (Peiter et al. 2005).

At least four Ca^{2+}-permeable channels are present on the vacuolar membrane, two of which are ligand-gated channels and two voltage-dependent channels (Pottosin and Schonknecht 2007). Inositol-1,4,5-trisphosphate ($InsP_3$) and cyclic adenosine 5′–diphosphoribose (cADPR) have both been shown to cause the release of Ca^{2+} from the vacuole (Schumaker and Sze 1986; Wu et al. 1997; Leckie et al. 1998) and to cause increases in $[Ca^{2+}]_{cyt}$ in guard cells (Blatt et al. 1990; Gilroy 1990; Leckie et al. 1998). The role of these two ligand-gated Ca^{2+} channels in ABA responses has been extensively studied. ABA stimulates an increase in internal $InsP_3$ and cADPR levels (Lee et al. 1996; Wu et al. 1997), and both InsP$_3$- and cADPR-dependent pathways have been implicated in vacuolar solute loss during ABA-induced stomatal closure (MacRobbie 2000). Furthermore, antagonists of InsP$_3$- and cADPR-mediated Ca^{2+} signaling (the phospholipase C inhibitor U73122 and ryanodine, respectively) have been shown to partially inhibit ABA- or voltage-induced changes in guard cell [Ca^{2+}]$_{cyt}$ (Grabov and Blatt 1999; Staxen et al. 1999). In addition, InsP$_3$ and/or cADPR have also been implicated in the hyperosmotic shock and salinity responses (Drobak and Watkins 2000; DeWald et al. 2001), gravitropism (Perera et al. 1999), cold (Knight et al. 1996), and defense responses (Lecourieux et al. 2006). Therefore, ligand-gated endomembrane Ca^{2+} channels constitute a potential point of crosstalk between Ca^{2+} signaling pathways in plants. Although InsP$_3$ and cADPR can cause Ca^{2+} release from intracellular stores in plants in a manner similar to their role as Ca^{2+} mobilzing agents from the endoplasmic reticulum in animal cells, the exact nature of these ligand-gated Ca^{2+} channels and their organelle location is less clear. Early studies on microsomes from *Chenopodium album* (Lommel and Felle 1997) and red beet (Allen et al. 1995) suggest a vacuolar origin for InsP$_3$- and cADPR-induced Ca^{2+} release, whereas subsequent studies indicate that the Ca^{2+}-release pathway activated by InsP$_3$ also appears to be present on the plasma membrane (Muir and Sanders 1997) and that cADPR-dependent Ca^{2+} release is either mainly (Navazio et al. 2001) or exclusively (Martinec et al. 2000) from the endoplasmic reticulum. Furthermore, the *Arabidopsis* and rice genomes do not include homologues of animal InsP$_3$ and ryanodine receptor channels. Therefore, Pottosin and Schonknecht (2007) have recently suggested the need for further experimentation to clarify this issue.

Of the two voltage-dependent Ca^{2+} channels, the vacuolar voltage-gated Ca^{2+} (VVCa) channel is gated by membrane hyperpolarization and the slow-activating vacuolar (SV) channel is gated by membrane depolarization (see Sanders et al. 2002; Hetherington and Brownlee 2004; Pottosin and Schonknecht 2007). Both VVCa and SV channels have a high affinity for Ca^{2+} and a low affinity for K$^+$ (see Pottosin and Schonknecht 2007). The VVCa channel is activated by Ca^{2+} (Johannes et al. 1992) and inhibited by H$^+$ from the vacuolar side (Ward and Schroeder 1994). The SV channel is the most abundant channel in the vacuolar membrane (Hedrich and Neher 1987; Pottosin et al. 1997) and appears to be ubiquitous among terrestrial plants (Hedrich et al. 1988). In contrast to the VVCa channel, the SV channel is activated by [Ca^{2+}]$_{cyt}$ within the physiological range (Hedrich and Neher 1987) and inhibited by cytosolic H$^+$ (Schulz-Lessdorf and Hedrich 1995). The finding that SV channels are permeable to Ca^{2+} ions (Ward and Schroeder 1994), implicated the potential for Ca^{2+}-induced Ca^{2+}-release (CICR) mediated by the SV channel (Ward and Schroeder 1994; Bewell et al. 1999; Pei et al. 1999; Carpaneto et al. 2001). Several additional regulators of SV channel activity have also been identified, including CaM which increases channel sensitivity to [Ca^{2+}]$_{cyt}$ (Bethke and Jones 1994), the phosphorylation state of the channel (Allen and Sanders 1995; Bethke and Jones 1997), and 14-3-3-proteins which reduce SV currents without affecting their voltage dependence (van den Wijngaard et al. 2001).

The molecular identity of the SV channel has recently been established (Peiter et al. 2005). The two-pore channel (TPC) family of voltage-gated cation channels, which consist of two homologous domains with six transmembrane helices and one pore domain, was originally identified in rat kidney (Ishibashi et al. 2000). TPC1 is the only member of the TPC family present in *Arabidopsis* (Furuichi et al. 2001). Peiter et al. (2005) have shown that an *Arabidopsis* knockout mutant lacking TPC1 (*tpc1-2*) does not show any SV channel activity, whereas *TPC1* overexpressing lines have increased SV channel activity demonstrating that the *TPC1* gene of *Arabidopsis* encodes the SV channel. Importantly, *tpc1-2* also fails to show ABA inhibition of germination or external Ca^{2+}-induced stomatal closure, whereas these responses are unaffected in *TPC1* overexpressing lines. These studies suggest a key role for the TPC1/SV channel in plants. However, although the TPC1/SV channel has also been implicated in both abiotic and biotic plant stress responses (see Pottosin and Schonknecht 2007) in a recent study, Ranf et al. (2008) were unable to detect any effect of altered *TPC1* expression on abiotic (cold, hyperosmotic shock, salt, ROS) and biotic (elicitor) stress-induced changes in $[Ca^{2+}]_{cyt}$ in *Arabidopsis* seedlings expressing cytosolic aequorin. Therefore, although the authors could not exclude a role for localized Ca^{2+} release from the vacuole, which would not have been be detected in their study, the precise role played by the TPC1/SV channel in plant Ca^{2+} signaling still requires further clarification.

There are several additional sources of Ca^{2+} release from endomembrane stores. The intracellular signaling molecule, *myo*-inositol hexakisphosphate ($InsP_6$) (phytate) is able to activate two vacuolar Ca^{2+} currents leading to an increase in $[Ca^{2+}]_{cyt}$ (Lemtiri-Chlieh et al. 2003). However, whether $InsP_6$ interacts directly or through a regulatory agent remains to be determined. The increase in $[Ca^{2+}]_{cyt}$ in response to $InsP_6$ is at least partially derived from the vacuole, although other intracellular stores may also be involved (Lemtiri-Chlieh et al. 2003). A Ca^{2+}-permeable channel activated by and sensitive to nicotinic acid adenine dinucleotide phosphate (NAADP) has also been identified and found to be situated on the endoplasmic reticulum in plants (Navazio et al. 2000). Unlike $InsP_3$ and cADPR, the NAADP-sensitive channel does not cause a significant release of Ca^{2+} from the vacuole. Activation of the NAADP-sensitive Ca^{2+}-permeable channel is independent of $[Ca^{2+}]_{cyt}$, demonstrating that it is not involved in CICR (Navazio et al. 2000). To the contrary, treatment with the putative inhibitor of intracellular Ca^{2+}-release, ruthenium red, causes a 50% reduction in NAADP-induced Ca^{2+}-release, whereas the Ca^{2+} chelator EGTA has no affect, suggesting that NAADP initiates intracellular release and not Ca^{2+} influx (Navazio et al. 2000). There is also evidence to suggest that Ca^{2+} release from the mitochondria (Logan and Knight 2003) and nucleus (Pauly et al. 2000) play a role in shaping plant Ca^{2+} signals and a Ca^{2+}-permeable channel has been recorded in excised patches from the nuclear membrane of red beet cells (Grygorczyk and Grygorczyk 1998). Therefore, there is the potential for many different Ca^{2+}-mobiliszing pathways to contribute to an increase in $[Ca^{2+}]_{cyt}$ in response to various stimuli.

Given the recent identification of the SV channel as the product of the *TPC1* gene, together with the strong evidence for an important role of $InsP_3$- and cADPR-gated Ca^{2+} channels in plant Ca^{2+}, it is surprising that no corresponding genes have yet been identified (Nagata et al. 2004). Two explanations have recently been advanced for this discrepancy (Pottosin and Schonknecht 2007): 1) there may be limited homology between plant and animal release channels due to an early evolutionary divergence (Maathius 2004), or 2) it has also been speculated that functionally similar Ca^{2+} signaling pathways may have evolved from different molecular building blocks (Bothwell and Ng 2005). However, it is essential that the genes encoding the channels responsible for both InsP$_3$- and cADPR-induced Ca^{2+} release are identified and their products characterized at the physiological level to advance further our understanding of how Ca^{2+} is positioned within the signaling network by which plants respond to environmental stimuli and the potential for crosstalk within the network.

Ca^{2+} Efflux Pathways

Efflux transporters include high-affinity Ca^{2+}-ATPases and low-affinity, high-capacity Ca^{2+}/H$^+$ antiporters, which are secondary energized transporters dependent on the proton gradient produced by proton pumps (Geisler et al. 2000; Pittman and Hirschi 2003; Sze et al. 2000). The Ca^{2+}-ATPases are one of five subfamilies of the P-type ATPases that are characterized structurally in plants by having a single subunit, 8-12 transmembrane segments, N and C termini exposed to the cytoplasm, and a large central cytoplasmic domain including the phosphorylation and ATP binding sites (Axelsen and Palmgren 2001). There are two categories of Ca^{2+}-ATPases in plants based on their homology with animal counterparts, type IIA and type IIB. Four type IIA Ca^{2+}-ATPases are present in the *Arabidopsis* genome (Axelsen and Palmgren 2001). The main function of these ATPases is to transport of Ca^{2+} and Mn^{2+} out of the cytosol (Liang et al. 1997; Durr et al. 1998). Type IIA Ca^{2+}-ATPases are called ECAx for ER-type Ca^{2+}-ATPase as they show similarity to the animal Ca^{2+}-ATPases of the sarco- and endoplasmic reticulum (SERCA pumps) and the fungal and animal secretory pathway ATPases (ATP2C1 in animals; PMR1 in yeast) (Geisler et al. 2000; Axelsen and Palmgren 2001). However, although all four of the genes have been cloned and *ECA1-3* has been characterized (Liang et al. 1997; Pittman et al. 1999) there is only evidence for the subcellular localization of ECA1 in plants, which is present on the endoplasmic reticulum (Liang et al. 1997; Axelsen and Palmgren 1997). No direct regulation of type IIA Ca^{2+}-ATPases has been demonstrated in plants suggesting a constitutive role in maintaining resting [Ca^{2+}]$_{cyt}$ levels (Hetherington and Brownlee 2004).

Ten type IIB Ca^{2+}-ATPases have been identified in *A. thaliana* (Axelsen and Palmgren 2001). These ATPases, which are most similar to the mammalian plasma membrane Ca^{2+}-ATPases (PMCA pumps) and the PCA1 ATPase from yeast, all contain an autoinhibitory sequence within the molecule and are therefore named ACAx for autoinihibited Ca^{2+}-ATPase (Geisler et al. 2000). Four of the 10 proteins have been cloned and characterized showing that each protein is located to a specific membrane including the plasma membrane (ACA8, Bonza et al. 2000), the endoplasmic reticulum (ACA2, Liang et al. 1997), the membrane of small vacuoles (ACA4; Geisler et al. 2000), and the chloroplast envelope (ACA1; Huang et al. 1993). Studies of the control of type IIB Ca^{2+}-ATPases show that their activity is increased by binding CaM (Harper et al. 1998) and decreased by the binding of and dephosphorylation by Ca^{2+}-dependent protein kinase 1 (CPK1; Hwang et al. 2000b). There is evidence for the presence of CaM-binding sequences in the N-terminal domains of ACA2 (Harper et al. 1998), ACA4 (Geisler et al. 2000), and ACA8 (Bonza et al. 2000), and in these ATPase, the N terminus is likely to form an autoinhibitory regulatory domain (Geisler et al. 2000; Hwang et al. 2000a, 2000b). The regulation of the Ca^{2+}-ATPase activity by CaM provides strong evidence for an active role for Ca^{2+} transporters in shaping Ca^{2+} signals (Hetherington and Brownlee 2004). It is possible that an increase in $[Ca^{2+}]_{cyt}$ would activate the Ca^{2+}-ATPase through CaM, which would result in the removal of Ca^{2+} from the cytosol and, therefore, terminate or at least reduce the Ca^{2+} signal.

In *Arabidopsis* Ca^{2+}/H^+ antiport activity is encoded by a small multigene family of cation exchanger (CAX) genes. Initial genomic analysis identified 11 CAX genes (Pittman and Hirschi 2003), although five uncharacterized members (*CAX7* to *CAX11*) are more closely related to mammalian K^+-dependent Na^+/Ca^{2+} exchangers and so fall within a different subgroup of exchangers (Shigaki et al. 2006). CAX1 to CAX6 are tonoplast localized and all appear to have some Ca^{2+}/H^+ antiport activity (Pittman and Hirschi 2003), although for some of these (particularly CAX2, CAX5, and CAX6) Ca^{2+} may not be the primary cation substrate (Pittman et al. 2004). CAX1 and CAX3 the primary components in high-capacity vacuolar Ca^{2+} sequestration in most *Arabidopsis* tissues as demonstrated by the significant impairment to Ca^{2+} homeostasis and plant development following deletion of both of these transporters (Cheng et al. 2005). In addition, under some developmental conditions, CAX1 and CAX3 appear to form a transporter complex potentially with altered kinetics of transport activity (Cheng et al. 2005). CAX transporter activity can be regulated through an N-terminal autoinhibitory domain, possibly through interaction with a regulatory protein (Pittman and Hirschi 2001, 2003). However, unlike the ACA Ca^{2+}-ATPases, the N-terminal of the CAX Ca^{2+}/H^+ antiporters does not interact with CaM and the nature of the regulatory protein is unknown (Hetherington and Brownlee 2004).

ECA and *CAX* knockout and overexpression mutants provide evidence for a role for both Ca^{2+}-ATPases and Ca^{2+}/H^+ antiporters in abiotic stress response pathways in plants. *ECA1* knockout mutants exhibit reduced growth on Ca^{2+}-depletion media (Wu et al. 2002). In addition, overexpression of *CAX1* increases the sensitivity of tobacco to cold shock, whereas the *cax1* knockout has increased tolerance to freezing following cold acclimation (Hirschi 1999; Catala et al. 2003). Similarly, when *CAX1* is overexpressed in *Arabidopsis* or tobacco, sensitivity to salinity is increased, although the *cax1* knockout has no altered phenotype following salt treatment (Cheng et al. 2004; Mei et al. 2007). Although there is no direct evidence to implicate these transporters shaping Ca^{2+} signals in plants, several studies provide indirect support for this suggestion. In nonplant systems, direct inhibition of Ca^{2+} efflux transporters can have an impact on stimulus-induced changes $[Ca^{2+}]_{cyt}$. For example, knockout of the mouse endoplasmic reticulum low-affinity Ca^{2+}-ATPase *SERCA3* alters the pattern of glucose-induced oscillations (Beauvois et al. 2006). Likewise, deletion of the yeast vacuolar Ca^{2+}/H^+ antiporter *VCX1* perturbs the changes in $[Ca^{2+}]_{cyt}$ observed in response to hypertonic shock (Denis and Cyert 2002). Furthermore, the *Arabidopsis det3* mutant, which has a 60% reduction in the activity of the vacuolar H^+-ATPase, the primary energizer of Ca^{2+}/H^+ antiport activity, exhibits altered external Ca^{2+}- and H_2O_2-induced changes in guard cell $[Ca^{2+}]_{cyt}$ compared with wildtype (Allen et al. 2000). Theses studies strongly suggest a role for Ca^{2+} transporters in shaping Ca^{2+} signals in plants and therefore in controlling crosstalk and specificity within Ca^{2+} signaling networks.

Specificity in Ca^{2+} Signaling

Given the many stress and other physiological and developmental stimuli in plants that use $[Ca^{2+}]_{cyt}$ for signal transduction (see Sanders et al. 2002; Hetherington and Brownlee 2004; Oldroyd and Downie 2006) the question how specificity in Ca^{2+} signaling occurs is a central focus both in plant and animal cell signal transduction research. Several mechanisms have emerged that all can mediate and contribute to specificity in Ca^{2+} signaling, including 1) cell-type specific expression of the appropriate signaling machinery; 2) localized elevations in $[Ca^{2+}]_{cyt}$ in specific regions of the cell (e.g., plasma membrane microdomains or at the nucleus, see Meinrenken et al. 2003; Schneggenburger and Neher 2005; Oldroyd and Downie 2006); 3) the pattern of $[Ca^{2+}]_{cyt}$ elevations controls specific responses; and 4) Ca^{2+} sensitivity priming. Because of space limitations, we focus on the latter two: oscillations and transients in $[Ca^{2+}]_{cyt}$ and Ca^{2+} sensitivity priming.

Oscillations and Transients in $[Ca^{2+}]_{cyt}$

The spatial and temporal dynamics of stimulus-induced increases in plant $[Ca^{2+}]_{cyt}$ vary markedly. As discussed earlier, localized increases in $[Ca^{2+}]_{cyt}$ together with oscillations and transients in $[Ca^{2+}]_{cyt}$ have been observed in several plant cell types (see Evans et al. 2001) including pollen tubes (Holdaway-Clark and Hepler 2003), root hairs (Shaw and Long 2003), and stomatal guard cells (McAinsh et al. 1992; Fan et al. 2004; McAinsh 2007). Studies in animal cells suggest that signaling information may be encoded in the pattern of changes in $[Ca^{2+}]_{cyt}$, contributing to the generation of stimulus-specific Ca^{2+} signals; in plants these have been referred to by some authors as a stimulus-specific a "Ca^{2+} signature" (Knight et al. 1998; McAinsh and Hetherington 1998). Spatial heterogeneities and localized increases in $[Ca^{2+}]_{cyt}$ are known to play an important role in the encryption of stimulus-specific signaling information (Berridge et al. 2000, 2003; Berridge 2007). For example, in mouse pituitary cells, Ca^{2+}-activated gene expression is differentially regulated through elevations in nuclear Ca^{2+} and $[Ca^{2+}]_{cyt}$ through the cAMP and the serum response elements, respectively (Hardingham et al. 1997), whereas in the brown alga *Fucus* variations in the spatiotemporal dynamics of hypo-osmotic shock-induced Ca^{2+} waves in embryos results in the differential regulation of cell volume changes and the rate of cell division (Goddard et al. 2000). Temporal heterogeneities in $[Ca^{2+}]_{cyt}$, including transients, spikes and oscillations, have also been shown to be important in encoding specificity in Ca^{2+} signals (Berridge et al. 2000, 2003; Berridge 2007). In mammalian cells, the pattern of oscillations in $[Ca^{2+}]_{cyt}$ is dependent on the cell type and varies depending on the strength and nature of the stimulus applied. For example, the activation of transcriptional regulators is dependent on the amplitude and duration of Ca^{2+} signals (Dolmetsch et al. 1998), whereas in pancreatic acinar cells, Ca^{2+} spikes in the micromolar range and submicromolar range are associated with the induction of exocytosis or the activation of luminal and basal ion channels, respectively (Ito et al. 1997). The mechanisms by which these complex patterns of changes in $[Ca^{2+}]_{cyt}$ are generated and maintained involve both positive and negative feedback and often invoking the release of Ca^{2+} from intracellular stores through the action of additional second messengers such as $InsP_3$ and fluxes of Ca^{2+} across the plasma membrane or between intracellular stores (Berridge et al. 2000, 2003; Berridge 2007).

Several studies in plant systems provide evidence that signaling information can also be encoded in the spatial-temporal characteristics of stimulus-induced changes in plant $[Ca^{2+}]_{cyt}$. A change in guard cell $[Ca^{2+}]_{cyt}$ is a key component of the signaling pathway(s) by which stomata respond to stimuli such as light, elevated $[CO_2]$, temperature, and plant hormones (Schroeder et al. 2001; Sanders et al. 2002; Hetherington and Brownlee 2004). Therefore, to formulate the optimal stomatal aperture for a given set of environmental conditions, guard cells must possess all of the signaling machinery necessary to enable 1) the integration of signaling information from a number of Ca^{2+} signaling pathways, which can have opposite effects on stomatal aperture (e.g., auxin, which promotes opening [Irving et al. 1992] and ABA, which promotes closure [McAinsh et al. 1990]) acting in concert and 2) the generation of a graded rather than an "all or nothing" response to different magnitudes of stimuli. Consequently, the potential for crosstalk between individual signaling pathways makes the issue of signal specificity fundamental to the correct functioning of guard cells and the regulation of gas exchange in plants.

As discussed previously, stimulus-induced changes in guard cell $[Ca^{2+}]_{cyt}$ show marked spatial and temporal heterogeneities. Spatial heterogeneities in $[Ca^{2+}]_{cyt}$ take the form of "hot spots" and Ca^{2+} quiescent regions within guard cells (Gilroy et al. 1991; Irving et al. 1992; McAinsh et al. 1992, 1995; Allen et al. 1999). It is possible that these could result from 1) differential accessibility of the stimulus to only a subset of signaling machinery or 2) the nonuniform distribution of the intracellular signaling machinery (McAinsh and Hetherington 1998; Ng and McAinsh 2003). In animals, global Ca^{2+} signals such as Ca^{2+} waves result from a series of localized "elemental events" termed "quarks," "blips," "puffs," and "sparks" (Berridge et al. 2000, 2003; Berridge 2007). A similar phenomenon has also been described in *Fucus* in response to osmotic stress (Goddard et al. 2000). Temporal heterogeneities in $[Ca^{2+}]_{cyt}$ take the form of transients and oscillations in guard cell $[Ca^{2+}]_{cyt}$ (see McAinsh 2007). The pattern of transients and oscillations (i.e., the period, frequency, and amplitude) is dependent on both the nature and strength of the stimulus (McAinsh et al. 1995; Staxen et al. 1999; Allen et al. 2000, 2001; Young et al. 2006) and therefore provides a potential mechanism for differentiating between Ca^{2+} signals through the generation of stimulus-specific changes in $[Ca^{2+}]_{cyt}$ (McAinsh and Hetherington 1998; Evans et al. 2001). This also allows for crosstalk between Ca^{2+} signaling pathways or the integration of signaling information through the generation of a novel pattern of changes in $[Ca^{2+}]_{cyt}$ that reflects the range of environmental stimuli to which the guard cells are exposed (Hetherington et al. 1998).

Clearly, there is the potential for signaling information to be encoded in the spatiotemporal dynamics of changes in guard cell $[Ca^{2+}]_{cyt}$. In addition, a causal relationship between oscillations and transients in guard cell $[Ca^{2+}]_{cyt}$ and the control of stomatal aperture has also been established. Allen et al. (2000) have shown that guard cells of wildtype *Arabidopsis* exhibit oscillations in guard cell $[Ca^{2+}]_{cyt}$ and long-term steady-state stomatal closure in response to ABA, cold, external Ca^{2+}, and H_2O_2. In contrast, in the de-etiolated 3 (*det3*) mutant, which exhibits a reduction in the V-type H^+-ATPase activity (Schumacher et al. 1999) associated with Ca^{2+} sequestration into endomembrane stores (Rooney et al. 1994; Xie et al. 1996; Hirschi 1999; Sze et al. 1999; Camello-Almaraz et al. 2000) and which therefore may be impaired in Ca^{2+} signaling, only ABA- and cold-produced oscillations in guard cell $[Ca^{2+}]_{cyt}$ and long-term steady state stomatal aperture similar to the wildtype. In contrast, external Ca^{2+} and H_2O_2 both resulted in prolonged increases in $[Ca^{2+}]_{cyt}$, which failed to induce stomatal closure even though the total integrated increase in $[Ca^{2+}]_{cyt}$ was higher in *det3* than in wildtype guard cells (Figure 4.2). Experimentally imposed oscillations in guard cell $[Ca^{2+}]_{cyt}$ generated using a "Ca^{2+} clamp" protocol to hyperpolarize repeatedly the plasma membrane promoting Ca^{2+} influx (Gilroy et al. 1991; Grabov and Blatt 1998; Allen et al. 2000; Hamilton et al. 2000) resulted in the rescue of external Ca^{2+}-induced long-term stomatal closure in *det3*. Importantly, imposing prolonged increases in $[Ca^{2+}]_{cyt}$ in wildtype guard cells, similar to the external Ca^{2+}-induced

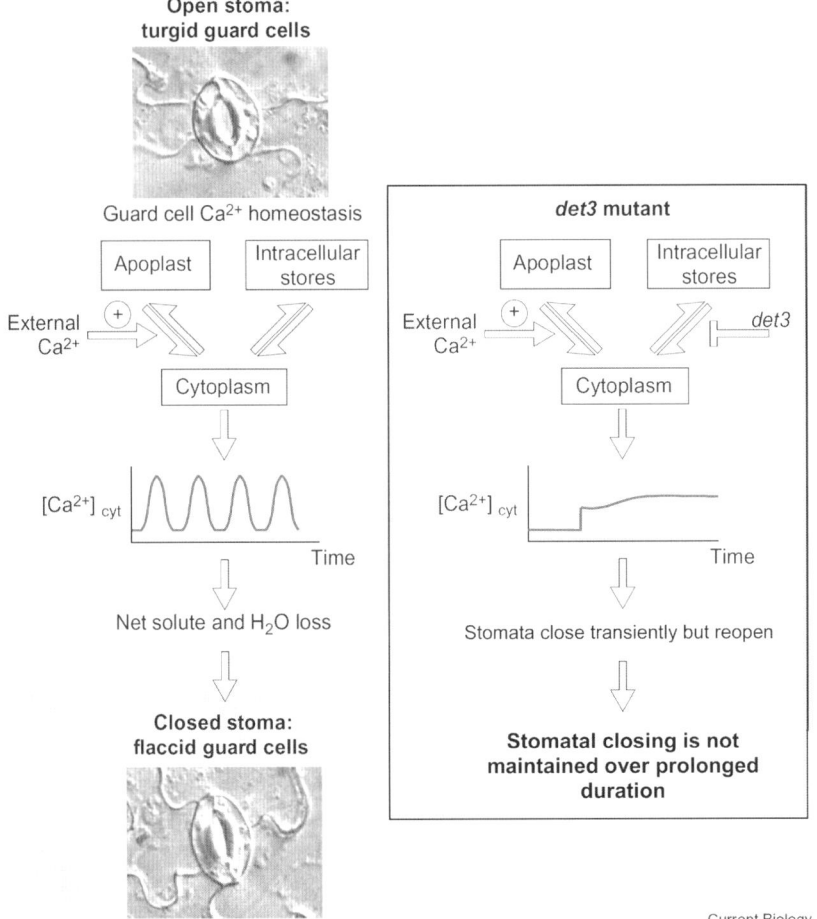

Fig. 4.2 A schematic representation of the effect of *det3* mutant on external Ca^{2+}-induced oscillations in $[Ca^{2+}]_{cyt}$ and stomatal closure in *Arabidopsis*. (Reprinted from *Current Opinion in Plant Biology*, 4, Evans et al., Calcium oscillations in higher plants, 415–420, 2001, with permission from Elsevier.)

increases in guard cell $[Ca^{2+}]_{cyt}$ observed in *det3*, failed to elicit long-term steady-state stomatal closure.

These data position oscillations in $[Ca^{2+}]_{cyt}$ as a key component in the guard cell signaling pathway maintaining stomatal closure. In addition, Allen and coworkers (2001) also used the Ca^{2+} clamp protocol (Allen et al. 2000) to establish the parameters of the oscillations in guard cell $[Ca^{2+}]_{cyt}$ that encode the signaling information responsible for triggering steady-state stomatal closure. They showed that only oscillations within a defined window of frequency, transient number, durations, and amplitude resulted in long-term steady-state, also named "Ca^{2+} programmed" stomatal closure. However, oscillations outside of this window of parameters resulted in immediate short-term or "Ca^{2+}-reactive" stomatal closure but did not result in a steady-state change in aperture. Therefore a range of experimentally imposed Ca^{2+} elevations thus functioned in maintaining closed stomatal apertures or inhibiting the reopening of stomatal pores after they had been closed by the Ca^{2+} reactive response (Allen et al. 2001). The physiological significance of this observation was further investigated by examining the kinetics of stimulus-induced oscillations in guard cell $[Ca^{2+}]_{cyt}$ in wildtype *Arabidopsis* and the ABA-insensitive mutant *gca2*, the stomata of which exhibit a differential response to both external Ca^{2+} and ABA (Allen et al. 2001). In wildtype plants, the kinetics of external Ca^{2+}- and ABA-induced $[Ca^{2+}]_{cyt}$ oscillations fell within the window of parameters resulting in long-term steady-state Ca^{2+}-programmed stomatal closure, whereas in *gca2* the period and duration were significantly shorter and steady-state stomatal closure was abolished. Imposing oscillations in $[Ca^{2+}]_{cyt}$ in guard cells of *gca2* with kinetics that fell within the window of parameters that induces long-term steady-state stomatal closure in wildtype plants recovered approximately 60% of the wildtype long-term steady-state stomatal closure in *gca2* showing that the guard cells of this mutant remain partially competent to respond to $[Ca^{2+}]_{cyt}$ oscillations. Further analyses of the CO_2-regulated $[Ca^{2+}]_{cyt}$ dynamics in *gca2* guard cells, led to the suggestion that the shorter period of Ca^{2+} transients in *gca2* guard cells after stimulation by CO_2 or ABA may be a result of continued hyperpolarization of *gca2* guard cells, whereas wildtype guard cells show ABA- and CO_2-induced depolarization (Young et al. 2006). This is consistent with findings that hyperpolarization of guard cells increases the frequency of Ca^{2+} transients in guard cells (Grabov and Blatt 1998; Staxen et al. 1999; Klusener et al. 2002). These data provide experimental evidence for a role for oscillations in $[Ca^{2+}]_{cyt}$ in the guard cell signaling pathway, leading to maintaining stomatal closure, after Ca^{2+} reactive mechanisms have caused the initial stomatal closure response.

A number of studies have reported stimulus-induced stomatal closure in the absence of oscillations in guard cell $[Ca^{2+}]_{cyt}$ (Schroeder and Hagiwara 1990; Gilroy et al. 1991; McAinsh et al. 1990, 1992, 1996; Allan et al. 1994; Webb et al. 1996; Allen et al. 1999; Romano et al. 2000; Levchenko et al. 2005). Furthermore, spontaneous transients in guard cell $[Ca^{2+}]_{cyt}$ were revealed which do not necessarily result in stomatal closure (Allen et al. 1999; Klusener et al. 2002; Young et al. 2006). Results from guard cells in which no Ca^{2+} elevations have been observed (by definition "negative results"), need to be followed up with quantitative studies in which $[Ca^{2+}]_{cyt}$ elevations are experimentally definitely inhibited in cells to quantify directly a relative contribution of the proposed Ca^{2+}-independent pathway for stomatal closing.

The Ca^{2+} Sensitivity Priming Hypothesis

Recently, an additional mechanism that can mediate specificity in Ca^{2+} signaling in eukaryotic cells has emerged from research in guard cells, in which physiological stimuli are proposed to enhance the Ca^{2+} sensitivity of specific responses, thus enabling a defined response to occur in cells that express many different Ca^{2+} sensors (Young et al. 2006). This Ca^{2+} sensitivity priming hypothesis was proposed on the basis of the confluence of several findings, including the following:

1. Guard cells can exhibit continual "spontaneous" Ca^{2+} transients, particularly when the membrane voltage is hyperpolarized (Grabov and Blatt 1998; Allen et al. 1999; Staxen et al. 1999; Klusener et al. 2002; Young et al. 2006; Yang et al. 2008). However, these Ca^{2+} transients do not always cause stomatal closing (Young et al. 2006).
2. Both ABA- and CO_2-induced stomatal closing depend on $[Ca^{2+}]_{cyt}$ (Webb et al. 1996; Young et al. 2006).
3. The Ca^{2+} sensitivity of $[Ca^{2+}]_{cyt}$ activation of S-type anion channels can be turned "on" or "off" depending on the preceding conditions (see Figure 3 in Allen et al. 2002).
4. Spontaneous Ca^{2+} transients occur in intact plants (Yang et al. 2008).

The Ca^{2+} sensitivity priming hypothesis could provide a mechanism that contributes to Ca^{2+} signaling specificity in eukaryotes, in light of the finding that, for instance, more than 250 putative Ca^{2+} binding proteins are expressed in the *Arabidopsis* genome, and indeed numerous of these are expressed even in single cell types (McCormack et al. 2005; Leonhardt et al. 2004; Yang et al. 2008). Nevertheless, further research is needed to test the Ca^{2+} sensitivity priming hypothesis and to determine the mechanisms that can sensitize and desensitize Ca^{2+} responses, thus allowing priming and depriming. Recent advances at identifying Ca^{2+}-dependent protein kinases (CDPKs) that function in ABA- and Ca^{2+}-induced stomatal closing and activation of guard cell anion channels (Mori et al. 2006; Zhu et al. 2007) and the identification of a target of Ca^{2+} signaling, the membrane protein SLAC1 that is required for the Ca^{2+}-activated anion channel activity in guard cells (Negi et al. 2008; Vahisalu et al. 2008) provide tools to analyze further the biochemical mechanisms mediating Ca^{2+} sensitivity priming. Furthermore, the identification of SLAC1 should also allow analyses of the question of how many additional types of Ca^{2+}-regulated efflux transporters and channels may exist in the plasma membrane of guard cells (De Angeli et al. 2006; Schroeder 2006; Vahisalu et al. 2008).

Note that the priming of the stomatal closing Ca^{2+} response discussed here defines the early rapid Ca^{2+}-reactive response to Ca^{2+} elevations leading to Ca^{2+} activation of anion channels and early induction of stomatal closing (Mori et al. 2006; Young et al. 2006; Vahisalu et al. 2008). In contrast, to the long-term Ca^{2+}-programmed stomatal response (inhibition of stomatal reopening) depends on the preceding Ca^{2+} elevation pattern in guard cells (Allen et al. 2001; Li et al. 2004).

Decoding Ca^{2+} Signals

It is clear that plants have the ability to generate complex Ca^{2+} signals with the potential to encode information about both the nature and strength of the stimulus. However, it is essential that they also possess the necessary machinery to decode this information. Several models have been proposed in animal cells to explain how signaling information encrypted in transients, spikes, and oscillations in [Ca^{2+}]$_{cyt}$ might be decoded based largely on Ca^{2+}-dependent activation of protein kinases (Berridge et al. 2000, 2003; Berridge 2007). *Arabidopsis* has 1,019 protein kinases of which 67 have been implicated in Ca^{2+} signaling (Hrabak et al. 2003; Harper et al. 2004). These fall into four families: the Ca^{2+}/calmodulin-domain protein kinases (CDPKs), the Snf1-related kinases (SnRK) of which the calcineurin-B-like interacting protein kinases (CIPKs) are a subgroup, the CDPK-related kinases (CRKs), and the Ca^{2+} and CaM-activated kinases (CCaMKs; Harper et al. 2004). The activity of plant protein kinases is regulated directly by Ca^{2+}, enabling the kinase to act as a Ca^{2+} sensor, and/or through interactions with additional Ca^{2+} sensor proteins, that is, Ca^{2+}-binding proteins that change their conformation in a Ca^{2+}-dependent manner, including calmodulin (CaM) and CaM-related proteins, and calcineurin-B-like (CBL) proteins (see Luan et al. 2002; Harper et al. 2004; Bouche et al. 2005; Yang et al. 2007). Although the regulation, function, and interconnections of many of these Ca^{2+} sensor proteins remains to be determined they are likely to recognize specific Ca^{2+} signals and relay these into specific downstream responses (see Luan et al. 2002; Harper et al. 2004; Bouche et al. 2005; Yang et al. 2007). Because of space limitations, we focus on the role of CDPKs, CBL–CIPK interactions, and the CCaMKs together with a novel family of CaM-binding transcription activators (CAMTAs), which have recently been implicated in crosstalk between Ca^{2+} signaling pathways (see Finkler et al. 2007; Yang et al. 2007).

CBL-CIPKs

The calcineurin B-like Ca^{2+}-binding proteins (CBLs) were initially identified on the basis of their similarity to both the regulatory subunit of calcineurin B and neuronal Ca^{2+} sensors (NCS) of animals and their function in salinity stress (Liu and Zhu 1998; Kudla et al. 1999). CBL proteins are characterized by the presence of Ca^{2+}-binding EF-hand motifs and by their interaction with a defined set of serine-threonine protein kinases referred to as CBL-interacting protein kinases (CIPKs; Shi et al. 1999; Kim et al. 2000). These kinases are structurally most closely related to the SNF1 (sucrose non-fermenting) kinase from yeast and cAMP-dependent kinases from animals (Shi et al. 1999), and therefore the CIPKs have been assigned to the SnRK3 subgroup of plant SNF-like kinases (Hrabak et al. 2003). There are 10 CBLs and 25 CIPKs in the *Arabidopsis* genome (Kolukisaoglu et al. 2004). This large number of CBLs and CIPKs allows for many different CBL–CIPK interactions highlighting the potential for crosstalk within the CBL–CIPK signaling network and hence the need for specificity in the signaling network.

There are several levels at which CBL–CIPK interactions may contribute to response specificity within Ca^{2+} signaling pathways. 1) Specificity appears to be encoded, at least in part, through the interaction of individual CBLs with defined subsets of CIPKs. Only a few CBLs, for example CBL2, have a very high number of potential partners acting as a potential node for crosstalk within CBL–CIPK signaling networks (see Luan et al. 2002; Batistic and Kudla 2004; Harper et al. 2004; Kolukisaoglu et al. 2004). 2) For CBL–CIPK interactions to occur, the individual partners must both be expressed at the same developmental stage and in the same tissue. Therefore, differential gene expression and differences in the subcellular localization may influence specificity and crosstalk in CBL–CIPK signaling. For example, CBL1 and CBL9, and the interacting kinase CIPK1, are all localized to the root tip and vascular stele; physical interaction of CBL1 and CBL9 with CIPK1 result in the kinase being targeted to the plasma membrane (D'Angelo et al. 2006). Interestingly, loss of CIPK1 gene function in either a *cbl1* or *cbl9* mutant background affects sensitivity to osmotic stress/salinity and enhances ABA responses differentially, suggesting that CIPK1 regulates distinct stress-response pathways depending on its CBL partner and, therefore, represents a point of crosstalk between individual signaling pathways (D'Angelo et al. 2006). CIPK3 is upregulated during early seedling development, and loss of CIPK3 function affects seed germination and seedling development (Kim et al. 2003). Furthermore, there are many examples of stimulus-specific regulation of both CBLs and CIPKs expression patterns (see Luan et al. 2002; Batistic and Kudla 2004; Harper et al. 2004; Kolukisaoglu et al. 2004). The specific localization of CBL Ca^{2+} sensors and their interacting CIPKs may be essential components of the machinery by which cells decode the signaling information encrypted in localized Ca^{2+} signals. 3) CBL proteins can be subdivided on the basis of their EF-hand composition (Kolukisaoglu et al. 2004), which, in turn, affects the affinity of individual CBLs for Ca^{2+}. In *Arabidopsis*, CBL1, which has two canonical EF-hands, exhibits a high affinity for Ca^{2+} in vitro, whereas the affinity of CBL4, which lacks a canonical EF-hand, is markedly lower (Kudla et al. 1999). The Ca^{2+} dependence of individual CBL–CIPK interactions may also vary. Therefore, the affinity of CBLs for Ca^{2+} may enable them to decode very specific Ca^{2+} signals in a manner similar to NCS proteins in animals (Burgoyne and Weiss 2001).

Many CBL–CIPK proteins are known to regulate Ca^{2+}-dependent abiotic stress responses (Cheong et al. 2003; Kim et al. 2003). Perhaps the best characterized example is the CBL–CIPK complex comprising SOS2 (also named CIPK24) and its Ca^{2+}-dependent activator SOS3 (CBL4). Activated SOS2 regulates a plasma membrane Na^+/H^+ antiporter SOS1, which provides salt tolerance by transporting Na^+ out of the cell (Zhu 2003). In addition, CIPK24 (SOS2) also interacts with CBL10 (Kim et al. 2007). The CBL10–CIPK24 complex is associated with the vacuolar compartment, and *cbl10* mutants exhibit marked growth defects and hypersensitive cell death in leaf tissues in response to salt stress. This indicates a functional role for CBL10 in salt tolerance through the regulation of the storage and detoxification of salt in the vacuole (Kim et al. 2007). This identifies CIPK24 (SOS2) as a multifunctional protein kinase that regulates various aspects of salt tolerance by interacting with distinct Ca^{2+} sensor CBL proteins and provides an example of talk within Ca^{2+}-mediated ion transport. There are many examples of ion transporter regulation by CBL–CIPK interactions in response to abiotic stress (Cheng et al. 2004; Li et al. 2006; Xu et al. 2006; Fuglsang et al. 2007), suggesting a central role for CBL–CIPK complexes in regulating ion transport. However, although the regulation, function, and interconnections of most of these sensor proteins remains to be determined they are likely to recognize specific Ca^{2+} signals and relay these into downstream responses (Luan et al. 2002; Batistic and Kudla 2004; Harper et al. 2004; Kolukisaoglu et al. 2004).

CDPKs

The Ca^{2+}/CaM-domain protein kinases (CDPKs) are Ca^{2+}-activated protein kinases that have been characterized in many plant species (Harper et al. 1991; Harmon et al. 2000; Cheng et al. 2002). CDPKs include a Ca^{2+}-binding CaM domain together with a kinase domain within the same protein. Biochemical studies of kinase activities have provided evidence that CDPKs can function in many plant signal transduction cascades downstream of intracellular $[Ca^{2+}]$ elevations (Bachmann et al. 1995; Cheng et al. 2002). There are 34 *CDPK* gene family members encoded in the *Arabidopsis* genome alone (Cheng et al. 2002; Hrabak et al. 2003), rendering genetic analyses of *CDPK* functions in stress signaling more challenging because of possible overlapping functions of *CDPK* genes.

Genetic loss-of-function studies in *CDPK* genes have recently emerged, allowing linkage of individual *CDPK* genes to physiological responses. Virus-induced gene silencing of a *CDPK* allowed demonstration of a role of CDPKs in tobacco defense responses (Romeis et al. 2000, 2001). These studies further provided biochemical evidence that fungal pathogens modulate CDPKs, such that their activity is enhanced. Antisense repression of CDPK in *M. truncatula* impairs root hair development and the rate of formation of nodules that function in bacterial symbiosis (Ivashuta et al. 2005).

CDPKs include an autoinhibitory domain (Harper et al. 1994; Yoo and Harmon 1996). Thus, truncated dominant constitutively active mutant forms of CDPKs can be generated through deletion of the calmodulin domain and the autoinhibitory linker domain. Overexpression of the dominant active truncated isoforms of *CPK10/AtCDPK1* (At1g74740) or *CPK30/AtCDPK1a* (At1g18890) cause constitutive stress and ABA signaling in *Arabidopsis* (Sheen 1996). Dominant mutant CDPK expression impairs pollen tube growth in *Petunia* (Yoon et al. 2006).

Biochemical studies suggest roles for CDPKs in stomatal guard cell ion channel regulation. Recombinant *Arabidopsis* CDPK activates chloride and malate influx channels in *Vicia faba* vacuoles (Pei et al. 1996). A Ca^{2+}-induced protein kinase activity phosphorylates the K^+ uptake channel KAT1 in vitro (Li and Assmann 1998; Mori et al. 2000).

Although many biochemical studies have shown functions of CDPKs, gene knock-out phenotypes of CDPKs had not yet been reported in *Arabidopsis* and would enable characterization of cellular functions of defined individual *CDPK* genes. Recently, first gene disruption phenotypes for *CDPK* genes have been identified in plants. By generation of *cdpk* double mutants, guard cell signal transduction and abscisic acid signaling phenotype were resolved (Mori et al. 2006; Zhu et al. 2007). Disruption of the *CPK3* and *CPK6* genes caused reduced $[Ca^{2+}]_{cyt}$ activation, as well as ABA activation, of S-type anion channels in the plasma membrane of guard cells (Mori et al. 2006). Unexpectedly, single and double mutants in these *CDPK* genes also impaired ABA activation of the plasma membrane Ca^{2+}-permeable I_{Ca} channels in the plasma membrane of guard cells (Mori et al. 2006). Consistent with the findings, ABA- and Ca^{2+}-induced stomatal closing were partially impaired in *cpk3cpk6* and in *cpk4cpk11* double mutants. Thus, these recent studies of guard cell-expressed CDPKs have identified Ca^{2+} sensors that mediate anion channel regulation and stomatal closing.

CCaMKs

Ca^{2+} and CaM-activated kinases (CCaMKs) contain a kinase domain, a CaM-binding domain overlapping an autoinhibitory domain, and a C-terminal Ca^{2+}-binding regulatory domain with three EF-hands, which is more similar to visinin than CaM (Patil et al. 1995). It is the latter that distinguish CCaMKs from CDPKs. CCaMKs are present in both monocots (Patil et al. 1995), Dicots (Poovaiah et al. 1999) and mosses (Okada et al. 2003). However, there is no evidence of CCaMKs in *Arabidopsis* or other *Brassica* species (Yang et al. 2007). It has been proposed that CCaMKs "sense" Ca^{2+} through the binding of Ca^{2+} to the visinin-like domain, resulting in autophosphorylation, Ca^{2+}/CaM binding to the CaM binding domain and therefore the activation of the kinase (Sathyanarayanan et al. 2000). Importantly, in animal cells, CCaMKII has been shown to be activated in a Ca^{2+}-spike frequency-dependent manner (Hudmon and Schulman 2002). This property is dependent, at least in part, on the ability of the kinase to autophosphorylate (Hudmon and Schulman 2002).

CCaMKs have been implicated in decoding Ca^{2+} signals at the initiation of bacterial and fungal symbiosis in the roots of a legume, *M. truncatula*, and other plants (see Oldroyd and Downie 2004; Harper and Harmon 2005; Yang et al. 2007). For example, genetic analysis has led to the identification of genes of at least three genes in *M. truncatula*, *DMI1*, *DMI2*, and *DMI3*, which are involved in Nod factor Ca^{2+} signaling (Oldroyd and Downie 2004). Mutants of *DMI1*, *DMI2*, and *DMI3* all fail to exhibit induction of root hair branching, nodulation gene expression, and cortical cell division in response to Nod factors (Catoira et al. 2000). However, *dmi1*, *dmi2*, and *dmi3* mutants differ in their ability to generate Ca^{2+} spikes in response to Nod factors; Nod-factor-induced Ca^{2+} spiking is abolished in *dmi1* and *dmi2* but not *dmi3* mutants (Ehrhardt et al. 1996; Wais et al. 2000). This suggests that *DMI1* and *DMI2* are required the generation Ca^{2+} spikes in response to Nod factors, whereas *DMI3* acts downstream of Ca^{2+} spiking. Interestingly, *DMI3* has been shown to be a member of the CCaMK family of Ca^{2+} sensors (Levy et al. 2004; Mitra et al. 2004). Recent studies have shown that removal of the autoinhibition domain of CCaMK/DMI3 leads to the activation of the nodulation signaling pathway and the spontaneous induction of nodulation gene expression and nodule formation (Gleason et al. 2006) and that point mutations in the autophosphorylation site, corresponding to the Thr^{267} in lily (Sathyanarayanan et al. 2001), result in spontaneous nodule initiation (Tirichine et al. 2006). Therefore, CCaMK appears to play a critical role in transducing the information encoded in Nod-factor-induced Ca^{2+} spiking to the downstream effector proteins that are required for initiating the nodulation response.

CAMTAs

Ca^{2+} plays a key role in regulating gene transcription (Ikura et al. 2002). Ca^{2+} and CaM have both been shown to bind to and regulate certain transcription factors (Corneliussen et al. 1994; Carrion et al. 1999). A novel family of CaM-binding transcription activators (CAMTAs) has been described (Bouche et al. 2002). These contain four conserved domains: a unique DNA-binding motif in the N-terminal domain, a TIG (transcription factor immunoglobulin-like DNA-binding) domain implicated in nonspecific DNA contacts in transcription factors, an ankyrin-like repeat, and a CaM-binding site in the C-terminal domain (see Finkler et al. 2007; Yang et al. 2007). There are six CAMTAs in *Arabidopsis* (Bouche et al. 2002). *Arabidopsis CAMTA* genes respond differentially (within 15 minutes) to environmental stimuli (Yang and Poovaiah 2002), which are known to stimulate changes in $[Ca^{2+}]_{cyt}$ (Rudd and Franklin-Tong 2001; Sanders et al. 2002; Hetherington and Brownlee 2004). ABA, H_2O_2, and salicylic acid induce the expression of *CAMTA*,2,4,5 and 6 but have no effect on *CAMTA1* and 3. In contrast, heat and cold shock, and ultraviolet B induce the expression of all *CAMTA* genes except *CAMTA2*, whereas salt stress has a similar effect, inducing the expression of all *CAMTA* genes except *CAMTA5*. Furthermore, wounding induced the expression of *CAMTA1* through to 6; *CAMTA1* and 3 are induced the fastest and to the highest level their expression increasing by more than 80-fold within 2 hours compared with just a 12-fold increase for *CAMTA2* and 6 (Yang and Poovaiah 2002). Importantly, CAMTAs have been shown to bind to core *cis*-elements in plant gene promoters, including two known ABA-responsive *cis*-elements (ABREs; Kaplan et al. 2006). This provides additional support that for a role of CAMTAs in ABA signaling. The rapid response of *CAMTA* genes to such a range of stimuli suggests that they may contribute to crosstalk between Ca^{2+} signaling pathways involved in plant stress response.

Conclusions

It is clear that the potential for crosstalk between Ca^{2+} signaling pathways is enormous. The existence of multiple Ca^{2+} signals within plants and also different cell types, which are generated through complex interactions between Ca^{2+} channels and transporters and which are subsequently perceived and decoded by a range of Ca^{2+} sensor proteins, Ca^{2+}-regulated kinase, and transcription factors highlights the many points at which crosstalk may occur in Ca^{2+} signaling networks. However, although many Ca^{2+} signaling genes have already been isolated, the number of components of the Ca^{2+} signaling network that have been functionally assigned is still limited. The progress achieved in recent years at identifying physiological, genetic, biochemical/electrophysiological, molecular biological, and cell biological mechanisms that mediate Ca^{2+} signaling holds much promise and suggests that major advances at understanding Ca^{2+} signaling in plants can be made through further analyses of cellular and stimulus specific model systems.

References

Allan, A.C., Fricker, M.D., Ward, J.L., Beale, M.H., and Trewavas, A.J. 1994. 2 transduction pathways mediate rapid effects of abscisic acid in *Commelia* guard cell. *Plant Cell* 6:1319–1328.

Allen, G.J., Chu, S.P., Schumacher, K., Shimazaki, C.T., Vafeados, D., Kemper, A., Hawke, S.D., Tallman, G., Tsien, R.Y., Harper, J.F., Chory, J., and Schroeder, J.I. 2000. Alteration of stimulus-specific guard cell calcium oscillations and stomatal closing in *Arabidopsis det3* mutants. *Science* 289:2338–2342.

Allen, G.J., Chu, S.P., Harrington, C.L., Schumacher, K., Hoffmann, T., Tang, Y.Y., Grill, E., and Schroeder, J.I. 2001. A defined range of guard cell calcium oscillation parameters encodes stomatal movements. *Nature* 411:1053–1057.

Allen, G.J., Kwak, J.M., Chu, S.P., Llopis, J., Tsien, R.Y., Harper, J.Y., and Schroeder, J.I. 1999. Cameleon calcium indicator reports cytoplasmic calcium dynamics in *Arabidopsis* guard cells. *Plant J* 19:735–747.

Allen, G.J., Muir, S.R., and Sanders, D. 1995. Release of Ca^{2+} from individual plant vacuoles by both InsP3 and cyclic ADP-ribose. *Science* 268:735–737.

Allen, G.J., Murata, Y., Chu, S.P., Nafisi, M., and Schroeder, J.I. 2002. Hypersensitivity of abscisic acid-induced cytosolic calcium increases in *Arabidopsis* farnesyltransferase mutant *era1–2*. *Plant Cell* 14:1649–1662.

Allen, G.J., and Sanders, D. 1995. Calcineurin, A type 2B protein phosphatase, modulates the Ca^{2+}-permeable slow vacuolar ion-channel of stomatal guard-cells. *Plant Cell* 7:1473–1483.

Axelsen, K.B., and Palmgren, M.G. 2001. Inventory of the superfamily of P-type ion pumps in *Arabidopsis*. *Plant Physiol* 126: 696–706.

Bachmann, M., McMicheal, R.W.J., Huber, J.L., Kaiser, W.M., and Huber, S.C. 1995. Partial purification and characterization of a calcium-dependent protein kinase and an inhibitor protein required for inactivation of spinach leaf nitrate reductase. *Plant Physiol* 108:1083–1091.

Balague, C., Lin, B., Alcon, C., Flottes, G., Malstrom, S., Kohler, C., Neuhaus, G., Pelletier, G., Gaymard, F., and Roby, D. 2003. HLM1, an essential signaling component in the hypersensitive response, is a member of the cyclic nucleotide-gated channel ion channel family. *Plant Cell* 15:365–379.

Batistic, O., and Kudla, J. 2004. Integration and channeling of calcium signaling through the CBL calcium sensor/CIPK protein kinase network. *Planta* 210:916–924.

Beauvois, M.C., Merezak, C., Jonas, J.C., Ravier, M.A., Henquin, J.C., and Gilon, P. 2006. Glucose-induced mixed $[Ca^{2+}]_c$ oscillations in mouse beta-cells are controlled by the membrane potential and the SERCA3 Ca^{2+}-ATPase of the endoplasmic reticulum. *Am J Physiol Cell Physiol* 290:C1503–C1511.

Benedict, C., Geisler, M., Trygg, J., Huner, N., and Hurry, V. 2006. Consensus by democracy. Using meta-analyses of microarray and genomic data to model the cold acclimation signaling pathway in *Arabidopsis*. *Plant Physiol* 141:1219–1232.

Berridge, M.J. 2007. *Cell Signalling Biology*. Colchester, UK: Portland Press.

Berridge, M.J. Bootman, M.D., and Roderick, H.L. 2003. Calcium signalling: dynamics, homeostasis and remodelling. *Nat Rev Mol Cell Biol* 4:517–529.

Berridge, M.J., Lipp, P., and Bootman, M.D. 2000. The versatility and universality of calcium signalling. *Nat Rev Mol Cell Biol* 1:11–21.

Bethke, P.C., and Jones, R.L. 1994. Ca^{2+}-calmodulin modulates ion-channel activity in storage protein vacuoles of barley aleurone cells. *Plant Cell* 6:277–285.

Bethke, P.C., and Jones, R.L. 1997. Reversible protein phosphorylation regulates the activity of the slow-vacuolar ion channel. *Plant J* 11:1227–1235.

Bewell, M.A., Maathuis, F.J.M., Allen, G.J., Sanders, D. 1999. Calcium induced calcium release mediated by a voltage-activated cation channel in vacuolar vesicles from red beet. *FEBS Lett* 458:41–44.

Blatt, M.R., Thiel, G., Trentham, D.R. 1990. Reversible inactivation of K^+ channels of *Vicia* stomatal guard cells following the photolysis of caged inositol 1,4,5-trisphosphate. *Nature* 346:766–769.

Blume, B., Nurnberger, T., Nass, N., Scheel, D. 2000. Receptor-mediated increase in cytoplasmic free calcium required for activation of pathogen defense in parsley. *Plant Cell* 12:1425–1440.

Bonza, M.C., Morandini, P., Luoni, L., Geisler, M., Palmgren, M.G., and De Michelis, M.I. 2000. At-ACA8 encodes a plasma membrane-localized calcium-ATPase of *Arabidopsis* with a calmodulin-binding domain at the N terminus. *Plant Physiol* 123:1495–1505.

Bothwell, J.H.F., and Ng, C.K.Y. 2005. The evolution of Ca^{2+} signalling in photosynthetic eukaryotes. *New Phytol* 166:21–38.

Bouche, N., Scharlat, A., Snedden, W., Bouchez, D., and Fromm, H. 2002. A novel family of calmodulin-binding transcription activators in multicellular organisms. *J Biol Chem* 277:21851–21861.

Bouche, N., Yellin, A., Snedden, W.A., and Fromm H. 2005. Plant-specific calmodulin-binding proteins. *Annu Rev Plant Biol* 56:435–466.

Bright, J., Desikan, R., Hancock, J.T., Weir, I.S., and Neill, S.J. 2006. ABA-induced NO generation and stomatal closure in *Arabidopsis* are dependent on H_2O_2 synthesis. *Plant J* 45:113–122.

Burgoyne, R.D., and Weiss, J.L. 2001. The neuronal calcium sensor family of Ca^{2+}-binding proteins. *Biochem J* 353:1–12.

Camello-Almaraz, C., Pariente, J.A., Salido, G., Camello, P.J. 2000. Differential involvement of vacuolar H$^+$-ATPase in the refilling of thapsigargin- and agonist-mobilized Ca^{2+} stores. *Biochem Biophys Res Comm* 271:311–317.

Cardenas, L., Vidali, L., Dominguez, J., Perez, H., Sanchez, F., Hepler, P.K., and Quinto, C. 1998. Rearrangement of actin microfilaments in plant root hairs responding to Rhizobium *etli* nodulation signals. *Plant Physiol* 116:871–877.

Carrion, A.M., Link, W.A., Ledo, F., Mellstrom, B., and Naranjo, J.R. 1999. DREAM is a Ca^{2+}-regulated transcriptional repressor. *Nature* 398:80–84.

Carpaneto, A., Cantu, A.M., and Gambale, F. 2001. Effects of cytoplasmic Mg^{2+} on slowly activating channels in isolated vacuoles of *Beta vulgaris*. *Planta* 213:457–468.

Catala, R., Santos, E., Alonso, J.M., Ecker, J.R., Martinez-Zapater, J.M., and Salinas, J. 2003. Mutations in the Ca^{2+}/H$^+$ transporter CAX1 increase CBF/DREB1 expression and the cold-acclimation response in *Arabidopsis*. *Plant Cell* 15:2940–2951.

Catoira, R., Galera, C., de Billy, F., Penmetsa, R.V., Journet, E.P., Maillet, F., Rosenberg, C., Cook, D., Gough, C., and Denarie J. 2000. Four genes of *Medicago truncatula* controlling components of a nod factor transduction pathway. *Plant Cell* 12:1647–1665.

Cheng, N.H., Pittman, J.K., Shigaki, T., Lachmansingh, J., LeClere, S., Lahner, B., Salt, D.E., and Hirschi, K.D. 2005. Functional association of *Arabidopsis* CAX1 and CAX3 is required for normal growth and ion homeostasis. *Plant Physiol* 138:2048–2060.

Cheng, N.H., Pittman, J.K., Zhu, J.K., and Hirschi, K.D. 2004. The protein kinase SOS2 activates the *Arabidopsis* H$^+$/Ca^{2+} antiporter CAX1 to integrate calcium transport and salt tolerance. *J Biol Chem* 279:2922–2926.

Cheng, S-H., Willmann, M.R., Chen H-C., and Sheen, J. 2002. Calcium signaling through protein kinases. The *Arabidopsis* calcium-dependent protein kinase gene family. *Plant Physiol* 129:469–485.

Cheong, Y.H., Kim, K.N., Pandey, G.K., Gupta, R., Grant, J.J., and Luan, S. 2003. CBL1, a calcium sensor that differentially regulates salt, drought, and cold responses in *Arabidopsis*. *Plant Cell* 15:1833–1845.

Clayton, H., Knight, M.R., Knight, H., McAinsh, M.R., and Hetherington, A.M. 1999. Dissection of the ozone-induced calcium signature. *Plant J* 17:575–579.

Clough, S.J., Fengler, K.A., Yu, I.C., Lippok, B., Smith, R.K., and Bent, A.F. 2000. The *Arabidopsis* dnd1 "defense, no death" gene encodes a mutated cyclic nucleotide-gated ion channel. *Proc Natl Acad Sci USA* 97:9323–9328.

Coelho, S.M., Taylor, A.R., Ryan, K.P., Sousa-Pinto, I., Brown, M.T., and Brownlee, C. 2002. Spatiotemporal patterning of reactive oxygen production and Ca^{2+} wave propagation in *Fucus rhizoid* cells. *Plant Cell* 14:2369–2381.

Cosgrove, D.J., and Hedrich, R. 1991. Stretch-activated chloride, potassium, and calcium channels coexisting in plasma-membranes of guard cells of *Vicia faba* L. *Planta* 186:143–153.

Corneliussen, B., Holm, M., Waltersson, Y., Onions, J., Hallberg, B., Thornell, A., and Grundstrom, T. 1994. Calcium/calmodulin inhibition of basic-helix–loop–helix transcription factor domains. *Nature* 368:760–764.

D'Angelo, C., Weinl, S., Batistic, O., Pandey, G.K., Cheong, Y.H., Schultke, S., Albrecht, V., Ehlert, B., Schulz, B., Harter, K., Luan, S., Bock, R., and Kudla, J. 2006. Alternative complex formation of the Ca^{2+}-regulated protein kinase CIPK1 controls abscisic acid-dependent and independent stress responses in *Arabidopsis*. *Plant J* 48:857–872.

De Angeli, A., Monachello, D., Ephritikhine, G., Frachisse, J.M., Thomine, S., Gambale, F., and Barbier-Brygoo, H. 2006. The nitrate/proton antiporter AtCLCa mediates nitrate accumulation in plant vacuoles. *Nature* 442:939–942.

Demidchik, V., Bowen, H.C., Maathuis, F.J.M., Shabala, S.N., Tester, M.A., White, P.J., and Davies, J.M. 2002. *Arabidopsis thaliana* root non-selective cation channels mediate calcium uptake and are involved in growth. *Plant J* 32:799–808.

Demidchik, V., Davenport, R.J., and Tester, M. 2002. Nonselective cation channels in plants. *Annu Rev Plant Biol* 53:67–107.

Dennison, K.L., and Spalding E.P. 2000. Glutamate-gated calcium fluxes in *Arabidopsis*. *Plant Physiol* 124:1511–1514.

Denis, V., and Cyert, M.S. 2002. Internal Ca^{2+} release in yeast is triggered by hypertonic shock and mediated by a TRP channel homologue. *J Biol Chem* 156:29–34.

DeWald, D.B., Torabinejad, J., Jones, C.A., Shope, J.C., Cangelosi, A.R., Thompson, J.E., Prestwich, G.D., and Hama, H. 2001. Rapid accumulation of phosphatidylinositol 4,5-bisphosphate and inositol 1,4,5-trisphosphate correlates with calcium mobilization in salt-stressed *Arabidopsis*. *Plant Physiol* 126:759–769.

Dodd, A.N., Gardner, M.J., Hotta, C.T., Hubbard, K.E., Dalchau, N., Love, J., Assie, J.M., Robertson, F.C., Jakobsen, M.K., Goncalves, J., Sanders, D., and Webb, A.A.R. 2007. The *Arabidopsis* circadian clock incorporates a cADPR-based feedback loop. *Science* 318:1789–1792.

Dodd, A.N., Jakobsen, M.K., Baker, A.J., Telzerow, A., Hou, S.W., Laplaze, L., Barrot, L., Poethig, R.S., Haseloff, J., and Webb, A.A.R. 2006. Time of day modulates low-temperature Ca^{2+} signals in *Arabidopsis*. *Plant J* 48:962–973.

Dolmetsch, R.E., Xu, K., and Lewis, R.S. 1998. Calcium oscillations increase the efficiency and specificity of gene expression. *Nature* 392:933–936.

Drobak, B.K., and Watkins, P.A.C. 2000. Inositol(1,4,5.trisphosphate production in plant cells: an early response to salinity and hyperosmotic stress. *FEBS Lett* 481:240–244.

Dumas, C., and Gaude, T. 2006. Fertilization in plants: Is calcium a key player? *Semin Cell Dev Biol* 17:244–253.

Durr, G., Strayle, J., Plemper, R., Elbs, S., Klee, S.K., Catty, P., Wolf, D.H., and Rudolph, H.K. 1998. The medial-Golgi ion pump Pmr1 supplies the yeast secretory pathway with Ca^{2+} and Mn^{2+} required for glycosylation, sorting, and endoplasmic reticulum associated protein degradation. *Mol Biol Cell* 9:1149–1162.

Dutta, R., and Robinson K.R. 2004. Identification and characterization of stretch-activated ion channels in pollen protoplasts. *Plant Physiol* 135: 1398–1406.

Ehrhardt, D.W., Wais, R., and Long, S.R. 1996. Calcium spiking in plant root hairs responding to Rhizobium nodulation signals. *Cell* 85:673–681.

Evans, N.H., McAinsh, M.R., and Hetherington, A.M. 2001. Calcium oscillations in higher plants. *Curr Opin Plant Biol* 4:415–420.

Evans, N.H., McAinsh, M.R., Hetherington, A.M., and Knight, M.R. 2005. ROS perception in *Arabidopsis thaliana*: the ozone-induced calcium response. *Plant J* 41:615–626.

Fan L-.M., Zhao, Z., and Assmann, S.M. 2004. Guard cells: a dynamic signalling model. *Curr Opin Plant Biol* 7:537–546.

Fellbrich, G., Blume, B., Brunner, F., Hirt, H., Kroj, T., Ligterink, W., Romanski, A., and Nurnberger, T. 2000. *Phytopthora parasitica* elicitor induced reactions in cells of *Petroselinum crispum*. *Plant Cell Physiol* 41:692–701.

Felle, H.H., Kondorosi, E., Kondorosi, A., and Schultze, M. 1998. The role of ion fluxes in Nod factor signalling in *Medicago sativa*. *Plant J* 13:455–463.

Finkler, A., Ashery-Padan, R., and Fromm, H. 2007. CAMTAs: Calmodulin-binding transcription activators from plants to human. *FEBS Lett* 581:3893–3898.

Foreman, J., Demidchik, V., Bothwell, J.H.F., Mylona, P., Miedema, H., Torres, M.A., Linstead, P., Costa, S., Brownlee, C., Jones, J.D.G., Davies, J.M., Dolan L. 2003. Reactive oxygen species produced by NADPH oxidase regulate plant cell growth. *Nature* 422:442–446.

Franklin-Tong, V.E., Hackett, G., Hepler, P.K. 1997. Ratio-imaging of Ca^{2+} in the self-incompatibility response in pollen tubes of *Papaver rhoeas*. *Plant J* 12:1375–1386.

Franklin-Tong, V.E., Holdaway-Clarke, T.L., Straatman, K.R., Kunke, J.G., Hepler, P.K. 2002. Involvement of extracellular calcium influx in the self-incompatibility response of *Papaver rhoeas*. *Plant J* 29:333–345.

Franklin-Tong, V.E., Ride, J.P., Franklin, F.C.H. 1995. Recombinant stigmatic self-incompatibility. S− protein elicits a Ca^{2+} transient in pollen of *Papaver rhoeas*. *Plant J* 8:299–307.

Franklin-Tong, V.E., Ride, J.P., Read, N.D., Trewavas, A.J., and Franklin, F.C.H. 1993. The self-incompatibility reaction in *Papaver rhoeas* is mediated by cytosolic free calcium. *Plant J* 4:163–177.

Frietsch, S., Wang, Sladek, C., Poulsen, L.R., Romanowsky, S.M., Schroeder, J.I., and Harper, J.F. 2007. A cyclic nucleotide-gated channel is essential for polarized tip growth of pollen. *Proc Natl Acad Sci USA* 104:14531–14536.

Furuichi, T., Cunningham, K.W., and Muto, S. 2001. A putative two pore channel AtTPC1 mediates Ca^{2+} flux in *Arabidopsis* leaf cells. *Plant Cell Physiol* 42:900–905.

Fuglsang, A.T., Guo, Y., Cuin, T.A., Qiu, Q.S., Song, C.P., Kristiansen, K.A., Bych, K., Schulz, A., Shabala, S., Schumaker, K.S., Palmgren, M.G., and Zhu, J.K. 2007. *Arabidopsis* protein kinase PKS5 inhibits the plasma membrane H^+-ATPase by preventing interaction with 14–3–3 protein. *Plant Cell* 19:1617–1634.

Gaedeke, N., Klein, M., Kolukisaoglu, U., Forestier, C., Muller, A., Ansorge, M., Becker, D., Mamnun, Y., Kuchler, K., Schulz, B., Mueller-Roeber, B., and Martinoia, E. 2001. The *Arabidopsis thaliana* ABC transporter AtMRP5 controls root development and stomata movement. *EMBO J* 20:1875–1887.

Geisler, M., Frangne, N., Gomes, E., Martinoia, E., and Palmgren, M.G. 2000. The ACA4 gene of *Arabidopsis* encodes a vacuolar membrane calcium pump that improves salt tolerance in yeast. *Plant Physiol* 124:1814–1827.

Gelli, A., and Blumwald, E. 1997. Hyperpolarization-activated Ca^{2+}-permeable channels in the plasma membrane of tomato cells. *J Membr Biol* 155:35–45.

Gilroy, S., Fricker, M.D., Read, N.D., and Trewavas, A.J. 1991. Role of calcium in signal transduction in *Commelina* guard cells. *Plant Cell* 3:333–343.

Gilroy, S., Read, N.D., and Trewavas, A.J. 1990. Elevation in cytoplasmic calcium by caged calcium or caged inositol trisphosphate initiates stomatal closure. *Nature* 346:769–771.

Gehring, C.A., Irving, H.R., Kabbara, A.A., Parish, R.W., Boukli, N.M., and Broughton, W.J. 1997. Rapid, plateau-like increases in intracellular free calcium are associated with Nod-factor-induced root hair deformation. *Mol Plant–Microbe Interact* 10:791–802.

Geurts, R., and Bisseling, T. 2002. Rhizobium nod factor perception and signaling. *Plant Cell* 14:S239–S249.

Gleason, C., Chaudhuri, S., Yang, T.B., Munoz, A., Poovaiah, B.W., and Oldroyd, G.E.D. 2006. Nodulation independent of rhizobia induced by a calcium-activated kinase lacking autoinhibition. *Nature* 441:1149–1152.

Goddard, H., Manison, N.F.H., Tomos, D., and Brownlee, C. 2000. Elemental propagation of calcium signals in response-specific patterns determined by environmental stimulus strength. *Proc Natl Acad Sci USA* 97:1932–1937.

Grabov, A., and Blatt, M.R. 1998. Membrane voltage initiates Ca^{2+} waves and potentiates Ca^{2+} increases with abscisic acid in stomatal guard cells. *Proc Natl Acad Sci USA* 95:4778–4783.

Grabov, A., and Blatt, M.R. 1999. A steep dependence of inward-rectifying potassium channels on cytosolic free calcium concentration increase evoked by hyperpolarisation in guard cells. *Plant Physiol* 119:277–287.

Grant, M., Brown, I., Adams, S., Knight, M., Ainslie, A., and Mansfield, J. 2000. The *RPM1* plant disease resistance gene facilitates a rapid and sustained increase in cytosolic calcium that is necessary for the oxidative burst and hypersensitive cell death. *Plant J* 23:1–11.

Grygorczyk, C., and Grygorczyk, R. 1998. A Ca^{2+}- and voltage-dependent cation channel in the nuclear envelope of red beet. *Biochim Biophys Acta Biomembr* 1375:117–130.

Haley, A., Russell, A.J., Wood, N., Allan, A.C., Knight, M., Campbell, A.K., and Trewavas, A.J. 1995. Effects of mechanical signaling on plant cell cytosolic calcium. *Proc Natl Acad Sci USA* 92:4124–4128.

Hamilton, D.W.A., Hills, A., Kohler, B., and Blatt, M.R. 2000. Ca^{2+} channels at the plasma membrane of stomatal guard cells are activated by hyperpolarization and abscisic acid. *Proc Natl Acad Sci USA* 97:4967–4972.

Han, S.C., Tang, R.H., Anderson, L.K., Woerner, T.E., and Pei, Z.M. 2003. A cell surface receptor mediates extracellular Ca2+ sensing in guard cells. *Nature* 425:196–200.

Hardingham, G.E., Chawla, S., Johnson, C.M., Bading, H. 1997. Distinct functions of nuclear and cytoplasmic calcium in the control of gene expression. *Nature* 385:260–265.

Harmon, A.C., Gribskov, M., and Harper, J.F. 2000. CDPKs—a kinase for every Ca^{2+} signal? *Trends Plant Sci* 5:154–159.

Harper, J.F. 2001. Dissecting calcium oscillators in plants. *Trends Plant Sci* 6:395–397.

Harper J.F., Breton, G., and Harmon A. 2004. Decoding Ca^{2+} signals through plant protein kinases. *Annu Rev Plant Biol* 55:263–288.

Harper, J.F., and Harmon A. 2005. Plants, symbiosis and parasites: a calcium signalling connection. *Nat Rev Mol Cell Biol* 6:555–566.

Harper, J.F., Hong, B.M., Hwang, I.D., Guo, H.Q., Stoddard, R., Huang, J.F., Palmgren, M.G., and Sze, H. 1998. A novel calmodulin-regulated Ca^{2+}-ATPase ACA2 from *Arabidopsis* with an N-terminal autoinhibitory domain. *J Biol Chem* 273:1099–1106.

Harper, J.F., Hung, J.F., and Lloyd, S.J. 1994. Genetic identification of an autoinhibitor in CDPK, a protein-kinase with a calmodulin-like domain. *Biochemistry* 33:7267–7277.

Harper, J.F., Sussman, M.R., Schaller, G.E., Putnam-Evans, C., Charbonneau, H., and Harmon, A.C. 1991. A calcium-dependent protein kinase with a regulatory domain similar to calmodulin. *Science* 252:951–954.

Harris, J.M., Wais, R., and Long, S.R. 2003. Rhizobium-induced calcium spiking in *Lotus japonicus*. *Mol Plant Micr Interact* 16:335–341.

Hedrich, R., Barbierbrygoo, H., Felle, H., Flugge, U.I., Luttge, U., Maathuis, F.J.M., Marx, S., Prins, H.B.A., Raschke, K., Schnabl, H., Chroeder, J.I., Struve, I., Taiz, L., and Zeiger P. 1988. General mechanisms for solute transport across the tonoplast of plant vacuoles—a patch-clamp survey of ion channels and proton pumps. *Botan Acta* 101:7–13.

Hedrich, R., and Neher E. 1987. Cytoplasmic calcium regulates voltage-dependent ion channels in plant vacuoles. *Nature* 329:833–836.

Hetherington, A.M., and Brownlee C. 2004. The generation of Ca^{2+} signals in plants. *Annu Rev Plant Biol* 55:401–427.

Hetherington, A.M., Gray, J.E., Leckie, C.P., McAinsh, M.R., Ng, C., Pical, C., Priestley, A.J., Staxen, I., and Webb, A.A.R. 1998. The control of specificity in guard cell signal transduction. *Phil Trans Royal Soc London B* 353:1489–1494.

Hetherington, A.M., and Woodward, F.I. 2003. The role of stomata in sensing and driving environmental change. *Nature* 424:901–908.

Hirschi, K.D. 1999. Expression of *Arabidopsis* CAX1 in tobacco: altered calcium homeostasis and increased stress sensitivity. *Plant Cell* 11:2113–2122.

Holdaway-Clarke, T.L., Feijo, J.A., Hackett, G.R., Kunkel, J.G., and Hepler, P.K. 1997. Pollen tube growth and the intracellular cytosolic calcium gradient oscillate in phase while extracellular calcium influx is delayed. *Plant Cell* 9:1999–2010.

Holdaway-Clarke, T.L., and Hepler, P.K. 2003. Control of pollen tube growth: the role of ion gradients and fluxes. *New Phytol* 159:539–563.

Hrabak, E.M., Chan, C.W.M., Gribskov, M., Harper, J.F., Choi, J.H., Halford, N., Kudla, J., Luan, S., Nimmo, H.G., Sussman, M.R., Thomas, M., Walker-Simmons, K., Zhu, J.K., and Harmon, A.C. 2003. The *Arabidopsis* CDPK-SnRK superfamily of protein kinases. *Plant Physiol* 132:666–680.

Hu, X.Y., Neill, S.J., Cai, W.M., and Tang, Z.C. 2004. Induction of defence gene expression by oligogalacturonic acid requires increases in both cytosolic calcium and hydrogen peroxide in *Arabidopsis thaliana*. *Cell Res* 14:234–240.

Huang, J.W., Grunes, D.L., and Kochian, L.V. 1993. Calcium-transport in right-side-out plasma membrane vesicles of wheat roots—characterization of vacuole-gated calcium channels. *Plant Physiol* S102:22–22.

Huang, J.W.W., Grunes, D.L., and Kochian, L.V. 1994. Voltage-dependent Ca^{2+} influx into right-side-out plasma-membrane vesicles isolated from wheat roots—characterization of a putative Ca^{2+} channel. *Proc Natl Acad Sci USA* 91:3473–3477.

Hudmon, A., and Schulman, H. 2002. Neuronal Ca^{2+}/calmodulin-dependent protein kinase II: The role of structure and autoregulation in cellular function. *Annu Rev Biochem* 71:473–510.

Hwang, I., Harper, J.F., Liang, F., and Sze, H. 2000a. Calmodulin activation of an endoplasmic reticulum-located calcium pump involves an interaction with the N-terminal autoinhibitory domain. *Plant Physiol* 122:157–167.

Hwang, I., Sze, H., and Harper J.F. 2000b. A calcium-dependent protein kinase can inhibit a calmodulin-stimulated Ca^{2+} pump ACA2 located in the endoplasmic reticulum of *Arabidopsis*. *Proc Natl Acad Sci USA* 97:6224–6229.

Ikura, M., Osawa, M., and Ames, J.B. 2002. The role of calcium-binding proteins in the control of transcription: structure to function. *Bioessays* 24:625–636.

Irving, H.R., Gehring, C.A., and Parish, R.W. 1992. Changes in cytosolic pH and calcium of guard cells precede stomatal movements. *Proc Natl Acad Sci USA* 89:1790–1794.

Ishibashi, K., Suzuki, M., and Imai, M. 2000. Molecular cloning of a novel form, two-repeat, protein related to voltage-gated sodium and calcium channels. *Biochem Biophys Res Comm* 270: 370–376.

Israelsson, M., Siegel, R.S., Young, J., Hashimoto, M., Iba, K., and Schroeder, J.I. 2006. Guard cell ABA and CO_2 signaling network updates and Ca^{2+} sensor priming hypothesis. *Curr Opin Plant Biol* 9:654–663.

Ito, K., Miyashita, Y., and Kasai, H. 1997. Micromolar and submicromolar Ca^{2+} spikes regulating distinct cellular functions in pancreatic acinar cells. *EMBO J* 16:242–251.

Ivashuta, S., Liu, J., Liu, J., Lohar, D.P., Haridas, S., Bucciarelli, B., VandenBosch, K.A., Vance, C.P., Harrison, M.J., and Gantt, J.S. 2005. RNA interference identifies a calcium-dependent protein kinase involved in *Medicago truncatula* root development. *Plant Cell* 17:2911–2921.

Johannes, E., Brosnan, J.M., and Sanders, D. 1992. Parallel pathways for intracellular Ca^{2+} release from the vacuole of higher-plants. *Plant J* 2:97–102.

Kaplan, B., Davydov, O., Knight, H., Galon, Y., Knight, M.R., Fluhr, R., and Fromm, H. 2006. Rapid transcriptome changes induced by cytosolic Ca^{2+} transients reveal ABRE-related sequences as Ca^{2+}-responsive *cis*-elements in *Arabidopsis*. *Plant Cell* 18:2733–2748.

Kiegle, E., Moore, C., Haseloff, J., Tester, M.A., and Knight, M.R. 2000. Cell-type specific calcium responses to drought, salt and cold in the *Arabidopsis* root. *Plant J* 23:267–278.

Kim, B.G., Waadt, R., Cheong, Y.H., Pandey, G.K., Dominguez-Solis, J.R., Schultke, S., Lee, S.C., Kudla, J., and Luan, S. 2007. The calcium sensor CBL10 mediates salt tolerance by regulating ion homeostasis in *Arabidopsis*. *Plant J* 52:473–484.

Kim, K.N., Cheong, Y.H., Grant, J.J., Pandey, G.K., and Luan, S. 2003. CIPK3, a calcium sensor-associated protein kinase that regulates abscisic acid and cold signal transduction in *Arabidopsis*. *Plant Cell* 15:411–423.

Kim, K.N., Cheong, Y.H., Gupta, R., and Luan, S. 2000. Interaction specificity of *Arabidopsis* calcineurin B-like calcium sensors and their target kinases. *Plant Physiol* 124:1844–1853.

Klusener, B., Young, J.J., Murata, Y., Allen, G.J., Mori IC, Hugouvieux, V., and Schroeder, J.I. 2002. Convergence of calcium signalling pathways of pathogenic elicitors and abscisic acid in *Arabidopsis* guard cells. *Plant Physiol* 130:2152–2163.

Knight, H., Brandt, S., and Knight, M.R. 1998. A history of stress alters drought calcium signalling pathways in *Arabidopsis*. *Plant J* 16:681–687.

Knight, H., Trewavas, A.J., and Knight, M.R. 1996. Cold calcium signaling in *Arabidopsis* involves two cellular pools and a change in calcium signature after acclimation. *Plant Cell* 3:489–503.

Knight, H., Trewavas, A.J., and Knight, M.R. 1997. Calcium signalling in *Arabidopsis thaliana* in response to drought and salinity. *Plant J* 12:1067–1078.

Knight, M.R., Campbell, A.K., Smith, S.M., and Trewavas, A.J. 1991. Transgenic plant aequorin reports the effects of touch and cold-shock and elicitors on cytoplasmic calcium. *Nature* 352:524–526.

Knight, M.R., Smith, S.M., and Trewavas, A.J. 1992. Wind-induced plant motion immediately increases cytosolic calcium. *Proc Natl Acad Sci USA* 89:4967–4971.

Kohler, B., and Blatt, M.R. 2002. Protein phosphorylation activates the guard cell Ca^{2+} channel and is a prerequisite for gating by abscisic acid. *Plant J* 32:185–194.

Kolukisaoglu, U., Weinl, S., Blazevic, D., Batistic, O., and Kudla, J. 2004. Calcium sensors and their interacting protein kinases: genomics of the *Arabidopsis* and rice CBL-CIPK signaling networks. *J Plant Physiol* 134:43–58.

Kudla, J., Xu, Q, Harter, K., Gruissem, W., and Luan, S. 1999. Genes for calcineurin B-like proteins in *Arabidopsis* are differentially regulated by salt stress. *Proc Natl Acad Sci USA* 96:4718–4723.

Kwak, J.M., Mori, I.C., Pei, Z.M., Leonhardt, N., Torres, M.A., Dangl, J.L., Bloom, R.E., Bodde, S., Jones, J.D.G., and Schroeder, J.I. 2003. NADPH oxidase AtrbohD and AtrbohF genes function in ROS-dependent ABA signalling in *Arabidopsis*. *EMBO J* 22:2623–2633.

Leckie, C.P., McAinsh, M.R., Allen, G.J., Sanders, D., and Hetherington, A.M. 1998. Abscisic acid-induced stomatal closure mediated by cyclic ADP-ribose. *Proc Natl Acad Sci USA* 95:15837–15842.

Lecourieux, D., Lamotte, O., Bourque, S., Wendehenne, D., Mazars, C., Ranjeva, R., and Pugin A. 2005. Proteinaceous and oligosaccharidic elicitors induce different calcium signatures in the nucleus of tobacco cells. *Cell Calcium* 38:527–538.

Lecourieux, D., Mazars, C., Pauly, N., Ranjeva, R., and Pugin A. 2002. Analysis and effects of cytosolic free calcium increases in response to elicitors in *Nicotiana plumbaginifolia* cells. *Plant Cell* 14:2627–2641.

Lecourieux, D., Raneva, R., and Pugin A. 2006. Calcium in plant defence-signalling pathways. *New Phytol* 171:249–269.

Lee, Y.S., Choi, Y.B., Suh, S., Lee, J., Assmann, S.M., Joe, C.O., Kelleher, J.F., and Crain, R.C. 1996. Abscisic acid-induced phosphoinositide turnover in guard cell protoplasts of *Vicia faba*. *Plant Physiol* 110:987–996.

Leonhardt, N., Kwak, J.M., Robert, N., Waner, D., Leonhardt, G., and Schroeder, J.I. 2004. Microarray expression analyses of *Arabidopsis* guard cells and isolation of a recessive abscisic acid hypersensitive protein phosphatase 2C mutant. *Plant Cell* 16:596–615.

Lemtiri-Chlieh, F., and Berkowitz, G.A. 2004. Cyclic adenosine monophosphate regulates calcium channels in the plasma membrane of *Arabidopsis* leaf guard and mesophyll cells. *J Biol Chem* 279:35306–35312.

Lemtiri-Chlieh, F., MacRobbie, E.A.C., Webb, A.A.R., Manison, N.F., Brownlee, C., Skepper, J.N., Chen, J., Prestwich, G.D., and Brearley, C.A. 2003. Inositol hexakisphosphate mobilizes an endomembrane store of calcium in guard cells. *Proc Natl Acad Sci USA* 100:10091–10095.

Levchenko, V., Konrad, K.R., Dietrich, P., Roelfsema, M.R.G., and Hedrich, R. 2005. Cytosolic abscisic acid activates guard cell anion channels without preceding Ca^{2+} signals. *Proc Natl Acad Sci USA* 102:4203–4208.

Levy, J., Bres, C., Geurts, R., Chalhoub, B., Kulikova, O., Duc, G., Journet, E.P., Ane, J.M., Lauber, E., Bisseling, T., Denarie, J., Rosenberg, C., and Debelle, F. 2004. A putative Ca^{2+} and calmodulin-dependent protein kinase required for bacterial and fungal symbioses. *Science* 303:1361–1364.

Li, J., Lee, Y.R., and Assmann, S.M. 1998. Guard cells possess a calcium-dependent protein kinase that phosphorylates the KAT1 potassium channel. *Plant Physiol* 116:785–795.

Li, S., Assmann, S.M., and Albert, R. 2006. Predicting essential components of signal transduction networks: a dynamic model of guard cell abscisic acid signaling. *PLoS Biol* 4:e312.

Li, Y., Wang, G.X., Xin, M., Yang, H.M., Wu, X.J., and Li, T. 2004. The parameters of guard cell calcium oscillation encodes stomatal oscillation and closure in *Vicia faba*. *Plant Sci* 166:415–421.

Li, Y.Q., Zhang, H.Q., Pierson, E.S., Huang, F.Y., Linskens, H.F., Hepler, P.K., and Cresti, M. 1996. Enforced growth-rate fluctuation causes pectin ring formation in the cell wall of *Lilium longiflorum* pollen tubes. *Planta* 200:41–49.

Liang, F., Cunningham, K.W., Harper, J.F., and Sze, H. 1997. ECA1 complements yeast mutants defective in Ca^{2+} pumps and encodes an ER-type Ca^{2+}-ATPase in *Arabidopsis thaliana*. *Plant Physiol* s114:54–54.

Liu, J., and Zhu, J-.K. 1998. A calcium sensor homolog required for plant salt tolerance. *Science* 280:1943–1945.

Logan, D.C., and Knight, M.R. 2003. Mitochondrial and cytosolic calcium dynamics are differentially regulated in plants. *Plant Physiol* 133:21–24.

Lommel, C, and Felle, H. 1997. Transport of Ca^{2+} across the tonoplast of intact vacuoles from *Chenopodium album* L suspension cells: ATP-dependent import and inositol–1,4,5-trisphosphate-induced release. *Planta* 201:477–486.

Luan, S., Kudla, J., Rodriguez-Concepcion, M., Yalovsky, S., and Gruissem, W. 2002. Calmodulins and calcineurin B-like proteins: calcium sensors for specific signal response coupling in plants. *Plant Cell* 14:S389–S400.

Maathius, F. 2004. Ligand-gated ion channels. In: Blatt, M.R. (ed.), Membrane Transport in Plants, Vol. 15. Oxford: Blackwell, pp. 193–220.

MacRobbie, E.A.C. 2000. ABA activates multiple Ca^{2+} fluxes in stomatal guard cells, triggering vacuolar $K^+(Rb^+)$ release. *Proc Natl Acad Sci USA* 97:12361–12368.

Malho, R., Read, N.D., Pais, M., and Trewavas, A.J. 1994. Role of cytosolic calcium in the reorientation of pollen tube growth. *Plant J* 5:331–341.

Malho, R., Read, N.D., Trewavas, A.J., and Pais, M. 1995. Calcium channel activity during pollen tube growth and re-orientation. *Plant Cell* 7:1173–1184.

Malho, R., and Trewavas, A.J. 1996. Localized apical increases of cytosolic free calcium control pollen tube orientation. *Plant Cell* 8:1935–1949.

Marshall, J., Corzo, A., Leigh, R.A., and Sanders, D. 1994. Membrane potential-dependent calcium-transport in right-side-out plasma-membrane vesicles from zea-mays l roots. *Plant J* 5:683–694.

Martinec, J., Feltl, T., Scanlon, C.H., Lumsden, P.J., and Machackova I. 2000. Subcellular localization of a high affinity binding site for D-myo-inositol 1,4,5-trisphosphate from *Chenopodium rubrum*. *Plant Physiol* 124:475–483.

McAinsh, M.R. 2007. Calcium oscillations in guard cell adaptive responses to the environment. In: Mancuso, S., and Shabala, S. (eds.), Rhythms in Plants: Phenomenology, Mechanisms, and Adaptive Significance. Berlin/Heidelberg: Springer-Verlag, 135–155.

McAinsh, M.R., Brownlee, C., and Hetherington, A.M. 1990. Abscisic acid-induced elevation of guard cell cytosolic Ca^{2+} precedes stomatal closure. *Nature* 343:186–188.

McAinsh, M.R., Brownlee, C., and Hetherington, A.M. 1992. Visualising changes in cytosolic-free Ca^{2+} during the response of stomatal guard cells to abscisic acid. *Plant Cell* 4:1113–1122.

McAinsh, M.R., Clayton, H., Mansfield, T.A., and Hetherington, A.M. 1996. Changes in stomatal behaviour and guard cell cytosolic free calcium in response to oxidative stress. *Plant Physiol* 111:1031–1042.

McAinsh, M.R., Evans, N.H., Montgomery, L.T., and North, K.A. 2002. Calcium signalling in stomatal responses to pollutants. *New Phytol* 153:441–447.

McAinsh, M.R., and Hetherington, A.M. 1998. Encoding specificity in Ca^{2+} signalling systems. *Trends Plant Sci* 3:32–36.

McAinsh, M.R., Webb, A.A.R., Taylor, J.E., and Hetherington, A.M. 1995. Stimulus-induced oscillations in guard cell cytosolic-free calcium. *Plant Cell* 7:1207–1219.

McCormack, E., Tsai, Y.C., and Braam, J. 2005. Handling calcium signaling: *Arabidopsis* CaMs and CMLs. *Trends Plant Sci* 10:383–389.

Mei, H., Zhao, J., Pittman, J.K., Lachmansingh, J., Park, S., and Hirschi, K.D. 2007. In planta regulation of the *Arabidopsis* Ca^{2+}/H^+ antiporter CAX1. *J Exp Biol* 58:3419–3427.

Meinrenken, C.J., Borst, J.G.G., and Sakmann, B. 2003. Local routes revisited: the space and time dependence of the Ca^{2+} signal for phasic transmitter release at the rat calyx of Held. *J Physiol London* 547:665–689.

Messerli, M., and Robinson, K.R. 1997. Tip localized Ca^{2+} pulses are coincident with peak pulsatile growth rates in pollen tubes of *Lilium longiflorum*. *J Cell Sci* 110:1269–1278.

Messerli, M.A., Creton, R., Jaffe, L.F., and Robinson, K.R. 2000. Periodic increases in elongation rate precede increases in cytosolic Ca^{2+} during pollen tube growth. *Dev Biol* 222:84–98.

Messerli, M.A., and Robinson, K.R. 2003. Ionic and osmotic disruptions of the lily pollen tube oscillator: testing proposed models. *Planta* 217:147–157.

Miller, D.D., Callaham, D.A., Gross, D.J., and Hepler, P.K. 1992. Free Ca^{2+} gradient in growing pollen tubes of *Lilium*. *J Cell Sci* 101:7–12.

Mithofer, A., Ebel, J., Bhagwat, A.A., Boller, T., and Neuhaus-Url, G. 1999. Transgenic aequorin monitors cytosolic calcium transients in soybean cells challenged with β-glucan or chitin elicitors. *Planta* 207:566–574.

Mitra, R.M., Shaw, S.L., and Long, S.R. 2004. Six nonnodulating plant mutants defective for Nod factor-induced transcriptional changes associated with the legume-rhizobia symbiosis. *Proc Natl Acad Sci USA* 101:10217–10222.

Miwa, H., Sun, J., Oldroyd, G.E.D., and Downie, J.A. 2006. Analysis of calcium spiking using a cameleon calcium sensor reveals that nodulation gene expression is regulated by calcium spike number and the developmental status of the cell. *Plant J* 48:883–894.

Monshausen, G.B., Bibikova, T.N., Messerli, M.A., Shi, C., and Gilroy, S. 2007. Oscillations in extracellular pH and reactive oxygen species modulate tip growth of *Arabidopsis* root hairs. *Proc Natl Acad Sci USA* 104:20996–21001.

Mori, I.C., Murata, Y., Yang YZ, Munemasa, S., Wang, Y.F., Andreoli, S., Tiriac, H., Alonso, J.M., Harper, J.F., Ecker, J.R., Kwak, J.M., and Schroeder, J.I. 2006. CDPKs CPK6 and CPK3 function in ABA regulation of guard cell S-type anion- and Ca^{2+}-permeable channels and stomatal closure. *PLoS Biol* 4:1749–1762.

Mori, I.C., Uozumi, N., and Muto, S. 2000. Phosphorylation of the inward-rectifying potassium channel KAT1 by ABR kinase in *Vicia* guard cells. *Plant Cell Physiol* 41:850–856.

Muir, S.R., and Sanders, D. 1997. Inositol 1,4,5-trisphosphate-sensitive Ca^{2+} release across non-vacuolar membranes in cauliflower. *Plant Physiol* 114:1511–1521.

Murata, Y., Pei, Z.M., Mori, I.C., and Schroeder, J. 2001. Abscisic acid activation of plasma membrane Ca^{2+} channels in guard cells requires cytosolic NADPH and is differentially disrupted upstream and downstream of reactive oxygen species production in *abi1–1* and *abi2–1* protein phosphatase 2C mutants. *Plant Cell* 13:2513–2523.

Mustilli, A.C., Merlot, S., Vavasseur, A., Fenzi, F., and Giraudat, J. 2002. *Arabidopsis* OST1 protein kinase mediates the regulation of stomatal aperture by abscisic acid and acts upstream of reactive oxygen species production. *Plant Cell* 14:3089–3099.

Nagata, T., Iizumi, S., Satoh, K., Ooka, H., Kawai, J., Carninci, P., Hayashizaki, Y., Otomo, Y., Murakami, K., Matsubara, K., and Kikuchi, S. 2004. Comparative analysis of plant and animal calcium signal transduction element using plant full-length cDNA data. *Mol Biol Evol* 10:1855–1870.

Navazio, L., Bewell, M.A., Siddiqua, A., Dickinson, G.D., Galione, A., and Sanders, D. 2000. Calcium release from the endoplasmic reticulum of higher plants elicited by the NADP metabolite nicotinic acid adenine dinucleotide phosphate. *Proc Natl Acad Sci USA* 97:8693–8698.

Navazio, L., Mariani, P., and Sanders, D. 2001. Mobilization of Ca^{2+} by cyclic ADP-ribose from the endoplasmic reticulum of cauliflower florets. *Plant Physiol* 125:2129–2138.

Navazio, L., Moscatiello, R., Genre A, Novero, M., Baldan, B., Bonfante, P., and Mariani P. 2007. A diffusible signal from arbuscular mycorrhizal fungi elicits a transient cytosolic calcium elevation in host plant cells. *Plant Physiol* 144:673–681.

Negi, J., Matsuda, O., Nagasawa, T., Oba, Y., Takahashi, H., Kawai-Yamada, M., Uchimiya, H., Hashimoto, M., and Iba K. 2008. CO_2 regulator SLAC1 and its homologues are essential for anion homeostasis in plant cells. *Nature* 452:483–486.

Ng, C.K.Y., Carr, K., McAinsh, M.R., Powell, B., and Hetherington, A.M. 2001. Drought-induced guard cell signal transduction involves sphingosine–1-phosphate. *Nature* 410:596–599.

Ng, C.K.Y., and McAinsh, M.R. 2003. Encoding specificity in plant calcium signalling: hot-spotting the ups and downs and waves. *Ann Botany* 92:477–485.

Nurnberger, T., Nennstiel, D., Jabs, T., Sacks, W.R., Hahlbrock, K., and Scheel, D. 1994. High affinity binding of a fungal oligopeptide elicitor to parsley plasma membranes triggers multiple defense responses. *Cell* 78:449–460.

Nurnberger, T., and Scheel, D. 2001. Signal transmission in the plant immune response. *Trends Plant Sci* 6:372–379.

Obermeyer, G., and Weisenseel, M.H. 1991. Calcium channel blocker and calmodulin antagonists affect the gradient of free calcium ions in lily pollen tubes. *Eur J Cell Biol* 56:319–327.

Okada, M., Takezawa, D., Tachibanaki, S., Kawamura, S., Tokumitsu, H., and Kobayashi, R. 2003. Neuronal calcium sensor proteins are direct targets of the insulinotropic agent repaglinide. *Biochem J* 375:87–97.

Oldroyd, G.E.D., and Downie, J.A. 2004. Calcium, kinases and nodulation signalling in legumes. *NatureNatl Rev Mol Cell Biol* 5:566–576.

Oldroyd, G.E.D., and Downie, J.A. 2006. Nuclear calcium changes at the core of symbiosis signaling. *Curr Opin Plant Sci* 9:351–357.

Patil, S., Takezawa, D., and Poovaiah, B.W. 1995. Chimeric plant calcium/calmodulin-dependent protein-kinase gene with a neural visinin-like calcium-binding domain. *Proc Natl Acad Sci USA* 92:4897–4901.

Pauly, N., Knight, M.R., Thuleau, P., Van Der Luit, A.H., Moreau, M., Trewavas, A.J., Ranjeva, R., and Mazars, C. 2000. Cell signalling—control of free calcium in plant cell nuclei. *Nature* 405:754–755.

Pei, Z.M., Murata, Y., Benning, G., Thomine, S., Klusener, B., Allen, G.J., Grill, E., and Schroeder, J.I. 2000. Calcium channels activated by hydrogen peroxide mediate abscisic acid signalling in guard cells. *Nature* 406:731–734.

Pei, Z.M., Ward, J.M., Harper, J.F., and Schroeder, J.I. 1996. A novel chloride channel in *Vicia faba* guard cell vacuoles activated by the serine/threonine kinase, CDPK. *EMBO J* 15:6564–6574.

Pei, Z.M., Ward, J.M., and Schroeder, J.I. 1999. Magnesium sensitizes slow vacuolar channels to physiological cytosolic calcium and inhibits fast vacuolar channels in fava bean guard cell vacuoles. *Plant Physiol* 121:977–986.

Peiter, E., Maathuis, F.J.M., Mills, L.N., Knight, H., Pelloux, M., Hetherington, A.M., and Sanders, D. 2005. The vacuolar Ca^{2+}-activated channel TPC1 regulates germination and stomatal movement. *Nature* 434:404–408.

Peiter, E., Sun, J., Heckmann, A.B., Venkateshwaran, M., Riely, B.K., Otegui, M.S., Edwards, A., Freshour, G., Hahn, M.G., Cook, D.R., Sanders, D., Oldroyd, G.E.D., Downie, J.A., and Ane, J.M. 2007. The *Medicago truncatula* DMI1 protein modulates cytosolic calcium signaling. *Plant Physiol* 145:192–203.

Perera, I.Y., Heilmann, I., and Boss, W.F. 1999. Transient and sustained increases in inositol 1,4,5-trisphosphate precede the differential growth response in gravistimulated maize pulvini. *Proc Natl Acad Sci USA* 96:5838–5843.

Pierson, E.S., Miller, D.D., Callaham, D.A., Shipley, A.M., Rivers, B.A., Cresti, M., and Hepler, P.K. 1994. Pollen tube growth is coupled to the extracellular calcium ion-flux and the intracellular calcium gradient: effect of BAPTA-type buffers and hypertonic media. *Plant Cell* 6:1815–1828.

Pierson, E.S., Miller, D.D., Callaham, D.A., van Aken, J., Hackett, G., and Hepler, P.K. 1996. Tip-localized entry fluctuates during pollen tube growth. *Dev Biol* 174:160–173.

Pittman, J.K., and Hirschi, K.D. 2001. Regulation of CAX1, an *Arabidopsis* Ca^{2+}/H^+ antiporter. Identification of an N-terminal autoinhibitory domain. *Plant Physiol* 127:1020–1029.

Pittman, J.K., and Hirschi, K.D. 2003. Don't shoot the second messenger: endomembrane transporters and binding proteins modulate cytosolic Ca^{2+} levels. *Curr Opin Plant Sci* 6:257–262.

Pittman, J.K., Mills, R.F., O'Connor, C.D., and Williams, L.E. 1999. Two additional type IIA Ca^{2+}-ATPases are expressed in *Arabidopsis thaliana*: evidence that type IIA sub-groups exist. *Gene* 236:137–147.

Pittman, J.K., Shigaki, T., Marshall, J.L., Morris, J.L., Cheng, N.H., and Hirschi, K.D. 2004. Functional and regulatory analysis of the *Arabidopsis thaliana* CAX2 cation transporter. *Plant Mol Biol* 56:959–971.

Poinssot, B., Vandelle, E., Bentejac, M., Adrian, M., Levis, C., Brygoo, Y., Garin, J., Sicilia, F., Coutos-Thevenot, P., and Pugin A. 2003. The endopolygalacturonase 1 from *Botrytis cinerea* activates grapevine defense reactions unrelated to its enzymatic activity. *Mol Plant-Micr Interact* 16:553–564.

Poovaiah, B.W., Xia, M., Liu, Z.H., Wang, W.Y., Yang, T.B., Sathyanarayanan, P.V., and Franceschi, V.R. 1999. Developmental regulation of the gene for chimeric calcium/calmodulin-dependent protein kinase in anthers. *Planta* 209:161–171.

Pottosin, I.I., and Schonknecht, G. 2007. Vacuolar calcium channels. *J Exp Botany* 58:1559–1569.

Pottosin, I.I., Tikhonova, L.I., Hedrich, R., and Schonknecht, G. 1997. Slowly activating vacuolar channels can not mediate Ca^{2+}-induced Ca^{2+} release. *Plant J* 12:1387–1398.

Price, A.H., Taylor, A., Ripley, S.J., Griffiths, A., Trewavas, A.J., and Knight, M.R. 1994. Oxidative signals in tobacco increase cytosolic calcium. *Plant Cell* 6:1301–1310.

Qi, Z., Stephens, N.R., and Spalding, E.P. 2006. Calcium entry mediated by GLR3.3, an *Arabidopsis* glutamate receptor with a broad agonist profile. *Plant Physiol* 142:963–971.

Ranf, S., Wunnenberg, P., Lee, J., Becker, D., Dunkel, M., Hedrich, R., Scheel, D., and Dietrich P. 2008. Loss of the vacuolar cation channel, AtTPC1, does not impair Ca^{2+} signals induced by abiotic and biotic stresses. *Plant J* 53:287–299.

Rathore, K.S., Cork, R.J., and Robinson, K.R. 1991. A cytoplasmic gradient of Ca^{2+} is correlated with the growth of lily pollen tubes. *Dev Biol* 148:612–619.

Romano, L.A., Jacob, T., Gilroy, S., and Assmann, S.M. 2000. Increases in cytosolic Ca^{2+} are not required for abscisic acid-inhibition of inward K^+ currents in guard cells of *Vicia faba* L. *Planta* 211:209–217.

Romeis, T., Ludwig, A.A., Martin, R., and Jones, J.D. 2001. Calcium-dependent protein kinases play an essential role in a plant defence response. *EMBO J* 20:5556–5567.

Romeis, T., Piedras, P., and Jones, J.D.G. 2000. Resistance gene-dependent activation of a calcium-dependent protein kinase in the plant defense response. *Plant Cell* 12:803–815.

Rooney, E.K., Gross, J.D., and Satre, M. 1994. Characterization of an intracellular Ca^{2+} pump in *Dictyostelium*. *Cell Calcium* 16:509–522.

Rudd, J.J., and Franklin-Tong V. 2001. Unravelling response-specificity in Ca^{2+} signalling pathways in plant cells. *New Phytol* 151:7–33.

Sanders, D., Pelloux, J., Brownlee, C., and Harper, J.F. 2002. Calcium at the crossroads of signalling. *Plant Cell* 14:S401-S417.

Sathyanarayanan, P.V., Cremo, C.R., and Poovaiah, B.W. 2000. Plant chimeric Ca^{2+}/calmodulin-dependent protein kinase—role of the neural visinin-like domain in regulating autophosphorylation and calmodulin affinity. *J Biol Chem* 275:30417–30422.

Sathyanarayanan, P.V., Siems, W.F., Jones, J.P., and Poovaiah, B.W. 2001. Calcium-stimulated autophosphorylation site of plant chimeric calcium/calmodulin-dependent protein kinase. *J Biol Chem* 276:32940–32947.

Schroeder, J.I. 2006. Physiology—nitrate at the ion exchange. *Nature* 442:877–878.

Schroeder, J.I., Allen, G.J., Hugouvieux, V., Kwak, J.M., and Waner, D. 2001. Guard cell signal transduction. *Annu Rev Plant Physiol Plant Mol Biol* 52:627–658.

Schroeder, J.I., and Hagiwara, S. 1990. Repetitive increases in cytosolic Ca^{2+} of guard cells by abscisic acid activation of nonselective Ca^{2+} permeable channels. *Proc Natl Acad Sci USA* 87:9305–9309.

Schneggenburger, R., and Neher, E. 2005. Presynaptic calcium and control of vesicle fusion. *Curr Opin Neurobiol* 15:266–274.

Schulz-Lessdorf, B., and Hedrich, R. 1995. Proton and calcium modulate SV-channels in the vacuolar lysosomal compartment – channel interaction with calmodulin inhibitors. *Planta* 197:655–671.

Schumacher, K., Vafeados, D., McCarthy, M., Sze, H., Wilkins, T., and Chory, J. 1999. The *Arabidopsis det3* mutant reveals a central role for the vacuolar H^+-ATPase in plant growth and development. *Genes Dev* 13:3259–3270.

Schumaker, K.S., and Sze, H. 1986. Calcium-transport into the vacuole of oat roots – characterization of H^+/Ca^{2+} exchange activity. *J Biol Chem* 26:2172–2178.

Shacklock, P.S., Read, N.D., and Trewavas, A.J. 1992. Cytosolic free calcium mediates red light-induced photomorphogenesis. *Nature* 358:753–755.

Shaw, S.L., and Long, S.R. 2003. Nod factor elicits two separable calcium responses in *Medicago truncatula* root hairs. *Plant Physiol* 131:976–984.

Sheen, J. 1996. Ca^{2+}-dependent protein kinases and stress signal transduction in plants. *Science* 274: 1900–1902.

Shi, J., Kim K-.N., Ritz, O., Albrecht, V., Gupta, R., Harter, K., Luan, S., and Kudla, J. 1999. Novel protein kinases associated with calcineurin B-like calcium sensors in *Arabidopsis*. *Plant Cell* 11:2393–2405.

Shigaki, T., Rees, I., Nakhleh, L., and Hirschi, K.D. 2006. Identification of three distinct phylogenetic groups of CAX cation/proton antiporters. *J Mol Evol* 63:815–825.

Staxen, I., Pical, C., Montgomery, L.T., Gray, J.E., and Hetherington, A.M. 1999. Abscisic acid induces oscillations in guard-cell cytosolic free calcium that involve phosphoinositide-specific phospholipase C. *Proc Natl Acad Sci USA* 96:1779–1784.

Suh, S.J., Wang, Y.F., Frelet, A., Leonhardt, N., Klein, M., Forestier, C., Mueller-Roeber, B., Cho, M.H., Martinoia, E., and Schroeder, J.I. 2007. The ATP binding cassette transporter AtMRP5 modulates anion and calcium channel activities in *Arabidopsis* guard cells. *J Biol Chem* 282:1916–1924.

Sze, H., Li, X., and Palmgren, M.G. 1999. Energization of plant cell membranes by H^+-pumping ATPases. Regulation and biosynthesis. *Plant Cell* 11:677–690.

Sze, H., Liang, F., Hwang, I., Curran, A.C., and Harper, J.F. 2000. Diversity and regulation of plant Ca^{2+} pumps: Insights from expression in yeast. *Annu Rev Plant Physiol Plant Mol Biol* 51:433–462.

Tang, R.H., Han, S.C., Zheng, H.L., Cook, C.W., Choi, C.S., Woerner, T.E., Jackson, R.B., and Pei, Z.M. 2007. Coupling diurnal cytosolic Ca^{2+} oscillations to the CAS-IP3 pathway in *Arabidopsis*. *Science* 315:1423–1426.

Thion, L., Mazars, C., Nacry, P., Bouchez, D., Moreau, M., Ranjeva, R., and Thuleau P. 1998. Plasma membrane depolarization-activated calcium channels, stimulated by microtubule-depolymerizing drugs in wild-type *Arabidopsis thaliana* protoplasts, display constitutively large activities and a longer half-life in ton 2 mutant cells affected in the organization of cortical microtubules. *Plant J* 13:603–610.

Thuleau, P., Ward, J.M., Ranjeva, R., and Schroeder, J.I. 1994a. Voltage-dependent calcium-permeable channels in the plasma-membrane of a higher-plant cell. *EMBO J* 13:2970–2975.

Thuleau, P., Moreau, M., Schroeder, J.I., and Ranjeva, R. 1994b. Recruitment of plasma-membrane voltage-dependent calcium-permeable channels in carrot cells. *EMBO J* 13:5843–5847.

Tirichine, L., Imaizumi-Anraku, H., Yoshida, S., Murakami, Y., Madsen, L.H., Miwa, H., Nakagawa, T., Sandal, N., Albrektsen, A.S., Kawaguchi, M., Downie, A., Sato, S., Tabata, S., Kouchi, H., Parniske, M., Kawasaki, S., and Stougaard, J. 2006. Deregulation of a Ca^{2+}/calmodulin-dependent kinase leads to spontaneous nodule development. *Nature* 441:1153–1156.

Tsien, R.W., and Tsien, R.Y. 1990. Calcium channels, stores, and oscillations. *Annu Rev Cell Biol* 6:715–760.

Vahisalu, T., Kollist, H., Wang, Y.F., Nishimura, N., Chan, W.Y., Valerio, G., Lamminmaki, A., Brosche, M., Moldau, H., Desikan, R., Schroeder, J.I., and Kangasjarvi, J. 2008. SLAC1 is required for plant guard cell S-type anion channel function in stomatal signalling. *Nature* 452:487–491.

Vandelle, E., Poinssot, B., Wendehenne, D., Bentejac, M., and Pugin A. 2006. Integrated signaling network involving calcium, nitric oxide, and active oxygen species but not mitogen-activated protein kin in BcPG1-elicited grapevine defenses. *Mol Plant Micr Interact* 19:429–440.

van den Wijngaard, P.W.J., Bunney, T.D., Roobeek, I., Schonknecht, G., and de Boer, A.H. 2001. Slow vacuolar channels from barley mesophyll cells are regulated by 14–3–3 proteins. *FEBS Lett* 488:100–104.

Very, A.A, and Davies, J.M. 2000. Hyperpolarization-activated calcium influx channels at the tip of *Arabidopsis* root hairs. *Proc Natl Acad Sci USA* 97:9801–9806.

Wais, R.J., Galera, C., Oldroyd, G., Catoira, R., Penmetsa, R.V., Cook, D., Gough, C., Denarie, J., and Long, S.R. 2000. Genetic analysis of calcium spiking responses in nodulation mutants of *Medicago truncatula*. *Proc Natl Acad Sci USA* 97:13407–13412.

Walker, S.A., Viprey, V., and Downie, J.A. 2000. Dissection of nodulation signaling using pea mutants defective for calcium spiking induced by Nod factors and chitin oligomers. *Proc Natl Acad Sci USA* 97:13413–13418.

Ward, J.M., Pei, Z.M., and Schroeder, J.I. 1995. Roles of ion channels in initiation of signal transduction in higher plants. *Plant Cell* 7:833–844.

Ward, J.M., and Schroeder, J.I. 1994. Calcium-activated K^+ channels and calcium-induced calcium-release by slow vacuolar ion channels in guard-cell vacuoles implicated in the control of stomatal closure. *Plant Cell* 6:669–683.

Webb, A.A.R., McAinsh, M.R., Mansfield, T.A., and Hetherington, A.M. 1996. Carbon dioxide induces increases in guard cell cytosolic free calcium. *Plant J* 9:297–304.

Wu, Y., Kuzma, J., Marechal, E., Graeff, R., Lee, H.C., Foster, R., and Chua N-H. 1997. Abscisic acid signalling through cyclic ADP-ribose in plants. *Science* 278:2126–2130.

Wu, Z.Y., Liang, F., Hong, B.M., Young, J.C., Sussman, M.R., Harper, J.F., and Sze, H. 2002. An endoplasmic reticulum-bound Ca^{2+}/Mn^{2+} pump, ECA1, supports plant growth and confers tolerance to Mn^{2+} stress. *Plant Physiol* 130:128–137.

Xie, Y., Coukell, M.B., and Gombos, Z. 1996. Antisense RNA inhibition of the putative vacuolar H^+-ATPase proteolipid of *Dictyostelium* reduces intracellular Ca^{2+} transport and cell viability. *J Cell Sci* 109:489–497.

Xu, H., Heath, M.C. 1998. Role of calcium in signal transduction during the hypersensitive response caused by basidiospore-derived infection of the cowpea frust fungus. *Plant Cell* 10:585–598.

Xu, J., Li, H.D., Chen, L.Q., Wang, Y., Liu, L.L., He, L., and Wu, W.H. 2006. A protein kinase, interacting with two calcineurin B-like proteins, regulates K^+ transporter AKT1 in *Arabidopsis*. *Cell* 7:1347–1360.

Yang, T.B., Du, L.Q., and Poovaiah, B.W. 2007. Concept of redesigning proteins by manipulating calcium/calmodulin-binding domains to engineer plants with altered traits. *Funct Plant Biol* 34:343–352.

Yang, Y., Costa, A., Leonhardt, N., Siegel, R.S., and Schroeder, J.I. 2008. Isolation of a strong *Arabidopsis* guard cell promoter and its potential as a research tool. *Plant Meth* 4:6.

Yang, T., and Poovaiah, B.W. 2002. A calmodulin-binding/CGCG box DNA-binding protein family involved in multiple signalling pathways in plants. *J Biol Chem* 277:45049–45058.

Yoo, B.C., and Harmon, A.C. 1996. Intramolecular binding contributes to the activation of CDPK, a protein kinase with a calmodulin-like domain. *Biochemistry* 35:12029–12037.

Yoon, G.M., Dowd, P.E., Gilroy, S., and McCubbin, A.G. 2006. Calcium-dependent protein kinase isoforms in Petunia have distinct functions in pollen tube growth, including regulating polarity. *Plant Cell* 18:867–878.

Yoshioka, K., Moeder, W., Kang, H.G., Kachroo, P., Masmoudi, K., Berkowitz, G., and Klessig DF. 2006. The chimeric *Arabidopsis* CYCLIC NUCLEOTIDE-GATED ION CHANNEL11/12 activates multiple pathogen resistance responses. *Plant Cell* 18:747–763.

Young, J.J., Mehta, S., Israelsson, M., Godoski, J., Grill, E., and Schroeder, J.I. 2006. CO_2 signaling in guard cells: calcium sensitivity response modulation, a Ca^{2+}-independent phase, and CO_2 insensitivity of the *gca2* mutant. *Proc Natl Acad Sci USA* 103:7506–7511.

Zhang, X., Zhang, L., Dong, F.C., Gao, J.F., Galbraith, D.W., and Song, C.P. 2001. Hydrogen peroxide is involved in abscisic acid-induced stomatal closure in *Vicia faba*. *Plant Physiol* 126:1438–1448.

Zhu, J.K. 2003. Regulation of ion homeostasis under salt stress. *Curr Opin Plant Biol* 6:441–445.

Zhu, S.Y., Yu, X.C., Wang, X.J., Zhao, R., Li, Y., Fan, R.C., Shang, Y., Du, S.Y., Wang, X.F., Wu, F.Q., Xu, Y.H., Zhang, X.Y., and Zhang, D.P. 2007. Two calcium-dependent protein kinases, CPK4 and CPK11, regulate abscisic acid signal transduction in *Arabidopsis*. *Plant Cell* 19:3019–3036.

Zimmermann, S., Nurnberger, T., Frachisse, J.M., Wirtz, W., Guern, J., Hedrich, R., and Scheel, D. 1997. Receptor-mediated activation of a plant Ca^{2+}-permeable ion channel involved in pathogen defense. *Proc Natl Acad Sci USA* 94:2751–2755.

5   Crosstalk in Pathogen and Hormonal Regulation of Guard Cell Signaling

Elena Bray Speth, Maeli Melotto, Wei Zhang, Sarah M. Assmann, and Sheng Yang He

Introduction

Stomata are microscopic pores on the surface of leaves, each surrounded by a pair of specialized epidermal cells referred to as guard cells. The size of the stomatal pore can be regulated through variations in the guard cells' turgor pressure. Fine tuning of the stomatal aperture is critical for regulating the exchange of oxygen and carbon dioxide and for controlling transpiration at the interface between plant and environment. Such fine control is achieved through a complex network of signal transduction events triggered by endogenous and environmental stimuli and culminates in the activation and/or inactivation of specific ion channels and transporters localized in the tonoplast and the plasma membrane, which alter the solute balance and consequently the turgor pressure in stomatal guard cells.

Changes in turgor pressure affect the shape and size of guard cells, resulting in closure or opening of the stomatal pore (Hetherington 2001; Schroeder et al. 2001; Nilson and Assmann 2007). For stomatal closure, the guard cell outwardly rectifying potassium channel GORK1 mediates K^+ efflux from the guard cells (Hosy et al. 2003). Opening of the voltage-regulated GORK channels is promoted by membrane depolarization resulting from inhibition of proton-extruding adenosine triphosphate hydrolases (H^+-ATPases) and activation of anion efflux. Efflux of the anions malate^{2-}, Cl^-, and NO_3^- is mediated by two types of anion channels, "slow" (S-type) and "rapid" (R-type). The slow anion channels are encoded by *SLOW ANION CHANNEL-ASSOCIATED 1*, which is distantly homologous to genes for dicarboxylate/malic acid transport proteins of microorganisms (Vahisalu et al. 2008); the R-type channel remains to be identified at the molecular level.

The efflux of ions from the guard cells promotes water loss, causing a decrease in guard cell turgor that results in closure of the stomatal pore (Fan et al. 2004; Israelsson et al. 2006). Stomatal opening, conversely, requires activation of the plasma membrane-localized H^+-ATPases, which pump protons out of the guard cells and cause plasma membrane hyperpolarization that drives K^+ influx (Schroeder et al. 2001). K^+ influx occurs through inwardly rectifying K^+ channels that are voltage-regulated and activated by membrane hyperpolarization. Inward K^+ channel genes that are expressed in guard cells include KAT1, KAT2, and AKT1 (Szyroki et al. 2001). Stomatal opening, caused by increased turgor pressure of guard cells, is also driven by metabolic changes leading to starch conversion into osmotically active malate (Schroeder et al. 2001) and sugar accumulation (Poffenroth et al. 1992; Stadler et al. 2003).

Besides playing a critical role in carbon dioxide and water vapor exchange between the plant and its environment, stomata represent gates through which some plant-pathogenic microorganisms gain access into the foliar apoplastic space, where they proliferate and cause disease. A variety of microbial elicitors could promote stomatal closure and/or inhibit stomatal opening (Underwood et al. 2007). In the case of bacterial infection, stomatal closure provides an initial line of defense mounted by plants to prevent infection (Melotto et al. 2006; W. Zhang et al. 2008). Certain plant-pathogenic bacteria, such as *Pseudomonas syringae*, produce virulence factors (e.g., the phytotoxin coronatine) that can counteract these processes and promote opening of the stomata, facilitating pathogen access into the leaf apoplast (Melotto et al. 2006; W. Zhang et al. 2008).

Much is known about signal transduction in stomatal guard cells in response to abiotic stimuli and some phytohormones. In contrast, the mechanisms leading to promotion of stomatal closure and inhibition of stomatal opening in response to pathogens, and stomata reopening in response to pathogen virulence factors, are only beginning to be elucidated. In this chapter, we review the current knowledge of stomatal regulation in response to pathogens and discuss how pathogen-derived signals might be integrated into dynamic hormonal regulation of the stomatal signal transduction network.

Hormonal Regulation of Stomatal Movements

It is well known that various abiotic stimuli, such as drought stress, atmospheric humidity, CO_2 concentration, and light, have profound effects on stomatal closure and opening (Hetherington and Woodward 2003). Responses to such stimuli are often hormonally mediated. Here, we briefly summarize current knowledge concerning the roles of various plant hormones in mediating stomatal movements.

Abscisic Acid

Abscisic acid (ABA) is the best characterized hormonal regulator of stomatal movements. A great deal of what is known about guard cell biology derives from studies on ABA, and understanding of the ABA-dependent regulatory pathways in stomata is far more advanced than the current knowledge of how other phytohormones influence stomatal movements.

ABA plays a critical role in mediating stomatal closure in response to water deficit. Leaf cells synthesize ABA when dehydrated, but ABA also serves as a long-distance drought stress signal, being produced in root cells and transported in the xylem sap under conditions of low soil water availability (Luan 2002; Wilkinson and Davies 2002). The cellular events leading from ABA perception by the guard cells to stomatal closure have been extensively investigated and remain a subject of intense ongoing research. Such events involve changes (oscillations) in cytosolic Ca^{2+} concentrations, accompanied by an increase in cytoplasmic pH; production of signal compounds such as nitric oxide (NO) and reactive oxygen species (ROS) such as hydrogen peroxide (H_2O_2); phosphorylation and dephosphorylation events (mediated by enzymes such as the guard cell-specific OST1 protein kinase and the ABI1 and ABI2 protein phosphatases); production of lipid-based signaling molecules such as inositol 1,4,5-triphosphate, sphingosine-1-phosphate, and phosphatidic acid; and malate catabolism and efflux through anion channels (Hetherington 2001; Schroeder et al. 2001; Mustilli et al. 2002; Fan et al. 2004; Israelsson et al. 2006; Kwak et al. 2006; Neill et al. 2008).

These signaling events together lead to inhibition of H^+-ATPase activity and promotion of outward K^+ channel and anion channel opening, which eventually leads to a net decrease in solute and water content and a narrowing of the stomatal aperture.

Salicylic acid

Salicylic acid (SA) is a plant signaling molecule known to play a critical role in plant resistance to many pathogens (Durrant and Dong 2004). In addition to its role in pathogen defense in infected tissues, SA is required for establishing systemic acquired resistance (SAR) throughout the infected plant, including uninfected parts (Gaffney et al. 1993; Delaney et al. 1994). The action of SA in inducing plant defense responses is mediated by cellular production of ROS (Chen et al. 1993). Interestingly, SA is also a potent inducer of stomatal closure (Manthe et al. 1992; J.-S. Lee 1998), and this effect was shown to be mediated, in epidermal peels of *Vicia faba*, by ROS, specifically peroxidase-generated superoxide ions (Mori et al. 2001).

Ethylene

Ethylene is an important regulator of plant growth and development, involved in many aspects of plants' life cycle, as well as in responses to biotic and abiotic stresses (Wang et al. 2002). Desikan and colleagues (2006) investigated ethylene's effects on stomatal movements. They demonstrated that application of either ethylene gas or its precursor 1-aminocyclopropane-1-carboxylic acid (ACC) to intact *Arabidopsis* leaves promoted stomatal closure. This effect was dependent on the ETR1 ethylene receptor (Schaller et al. 1995; K.L.C. Wang et al. 2002) and on H_2O_2 production, generated in the guard cells by the nicotinamide adenine diphosphate (NADPH) oxidase AtrbohF (Desikan et al. 2006). The same group had previously shown that ETR1 is expressed in guard cells and that different *etr1* mutant alleles produced varying degrees of stomatal closure in response to H_2O_2, ranging from complete insensitivity (*etr1-7*, *etr1-1*) to near-wildtype response (*etr1-3*). Because these mutants are similarly insensitive to ethylene, the ETR1 role in H_2O_2 response could be dissociated from its role in ethylene perception (Desikan et al. 2005).

Surprisingly, treatment with exogenous ethylene or with its precursor ACC inhibited ABA-induced stomatal closure in wildtype *Arabidopsis* epidermal peels and leaves (Tanaka et al. 2005; Desikan et al. 2006). Inhibition of ABA-mediated stomatal closure was also observed in epidermal peels of the ethylene-overproducing mutant *eto1-1* (Tanaka et al. 2005). These results indicate an antagonistic crosstalk between ABA and ethylene in guard cells, whereby either hormone alone promotes stomatal closure, but addition of both hormones together has an inhibitory effect on closure.

Jasmonic Acid

Jasmonic acid (JA) and its metabolically active derivatives (jasmonates) are important signaling molecules involved in many plant responses to biotic and abiotic stresses (Devoto and Turner 2003). Many aspects of the JA signaling pathway have been characterized, including its extensive crosstalk with other signaling networks, particularly the ABA, SA, and ethylene pathways, at the whole plant/tissue level (Lorenzo and Solano 2005; Fujita et al. 2006). Jasmonates are particularly well known as positive regulators of plant responses to herbivores and necrotrophic pathogens.

Exogenously applied methyl-JA (MeJA) triggers stomatal closure through a mechanism involving cytoplasmic pH changes and production of ROS (Gehring et al. 1997; Suhita et al. 2004). In patch-clamp experiments with guard cell protoplasts, low concentrations of exogenous MeJA (100 nM) inhibited K^+ influx channels and activated K^+ efflux channels, consistent with stomatal closure, whereas high MeJA concentrations (50 µM) inhibited the outward K^+ channels (Evans 2003). Stomatal closure at high concentrations may be driven via MeJA hyperactivation of anion efflux and calcium influx channels (Munemasa et al. 2007). The potential interconnections between ABA and MeJA pathways in *Arabidopsis* guard cells were explored using the ABA-insensitive *ost1-2* and the MeJA-insensitive *jar1-1* mutants of *Arabidopsis*, as well as the *gork1* and *atrbohD/F* mutants (Suhita et al. 2004). The results of this study provide evidence that distinct initial perception and signaling events for ABA and MeJA are followed by some overlapping mechanisms requiring the same players: in particular, the NADPH oxidases AtrbohD/F and the GORK1 K^+ efflux channel are required in both ABA and MeJA signal transduction pathways leading to stomatal closure (Suhita et al. 2004).

Munemasa and colleagues (2007) used the MeJA-insensitive *Arabidopsis* F-box protein mutant, *coi1* (Feys et al. 1994; D.X. Xie et al. 1998), to investigate the interplay between MeJA and ABA signaling in guard cells (Munemasa et al. 2007). They reported that in the *coi1* background, MeJA could not activate slow anion channels or Ca^{2+}-permeable channels of guard cells, induce an increase in NO and ROS, or induce stomatal closure, whereas ABA could still elicit all of these responses. The ABA-insensitive mutant *ABA-insensitive 2* (*abi2-1*) is blocked in the ABA signal transduction cascade at a step subsequent to production of NO and ROS (Murata et al. 2001). Neither ABA nor MeJA could induce stomatal closure in the *abi2-1* mutant, but they both caused NO and ROS production, suggesting that COI1 is required at an early step of MeJA signal transduction, upstream of a convergence point between the ABA and JA signaling pathways (Munemasa et al. 2007).

Although both ABA and MeJA individually promote stomatal closure, we recently observed that low concentrations of MeJA can actually counteract ABA-mediated stomatal closure. We found that high concentrations (20 µM) of JA and MeJA caused significant stomatal closure, as previously reported (Suhita et al. 2004); lower concentrations (e.g., 2 µM) did not cause stomatal closure but, rather, prevented ABA from closing stomata (Figure 5.1; M. Melotto and S.Y. He, unpublished data). Also, there seems to be an additive effect in promoting stomatal closure when a high concentration of MeJA is combined with ABA (Figure 5.1). These data, together with the bimodal effects of MeJA concentrations on outward K^+ channels (Evans 2003), indicate that there is a concentration threshold above and below which MeJA exerts different effects on stomatal response.

Auxin and Cytokinin

Auxins and cytokinins can induce stomatal opening. Endogenous and synthetic cytokinins counteracted CO_2-induced stomatal closure in maize (Blackman and Davies

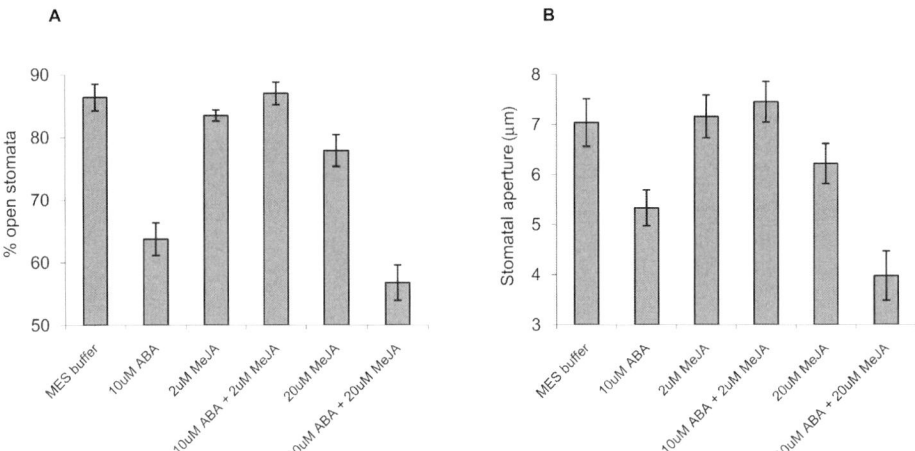

Fig. 5.1 Concentration-dependent effects of jasmonic acid (JA) and methyl-JA (MeJA) on abscisic acid (ABA)-induced stomatal movements. To maximize stomatal opening, *Arabidopsis* epidermes were peeled from light-conditioned, fully expanded leaves and placed on glass slides with the cuticle side in contact with MES buffer (25 mM MES-KOH, pH 6.15, and 10 mM KCl) for 2 hours under light. The purified chemicals ABA (Sigma) and MeJA (Sigma) were added to the MES buffer at the indicated concentrations. After 2 hours of incubation, photographs were taken of random regions. Stomatal aperture was scored as open (fully or partially) or closed to calculate the percentage of open stomata per treatment (panel A). The width of the stomatal aperture (panel B) was measured using the software Image-Pro, version 4.5 for Windows (Cybernetics, Silver Spring, Maryland, USA). Data points are average ± standard error ($n = 60$).

1984). In broad bean leaves, both auxins and cytokinins induced stomatal opening in darkness by reducing NO (She and Song 2006) and H_2O_2 (Song et al. 2006) levels in guard cells. In broad bean leaves, there is an optimal auxin concentration for promotion of stomatal opening; higher concentrations inhibit opening (Marten et al. 1991). Treatment with auxins and cytokinins could also counteract ABA-induced stomatal closure. Tanaka and colleagues used *Arabidopsis* mutants to investigate potential crosstalk between auxin, cytokinin, and ABA signaling in guard cells (Tanaka et al. 2006). They found that the antagonistic effects of auxin and cytokinin on ABA-induced stomatal closure were not observed in the ethylene-insensitive *ein3-1* mutant. Furthermore, treatment with 1-methylcyclopropene, a competitive inhibitor of ethylene-receptor binding, opposed the effect of auxin and cytokinin on ABA-induced stomatal closure. These findings led the group to suggest that auxin and cytokinin may act indirectly (possibly by enhancing ethylene biosynthesis) to antagonize ABA-mediated stomatal closure (Tanaka et al. 2006).

Brassinosteroids

There has been relatively little research on brassinosteroid modulation of stomatal movements. In *Arabidopsis*, the brassinosteroid synthetic mutant, *sax1*, has been reported to exhibit enhanced ABA-induced stomatal closure, suggesting that brassinosteroids oppose ABA responses (Ephritikhine et al. 1999). However, in broad bean, application of brassinolide, like ABA, enhances stomatal closure and inhibits stomatal opening and K^+ influx currents; furthermore, brassinolide coapplication does not oppose ABA action (Haubrick et al. 2006). In sorghum, brassinolide enhances the drought-protective effects of ABA (Xu et al. 1994). The contrasting results from *Arabidopsis* versus broad bean and sorghum may reflect species-specificity in brassinosteroid responses, method of perturbation (e.g., exogenous vs. endogenous), or developmental specificity of brassinosteroid responses, because the *sax1* mutant is a dwarf mutant.

Are ROS and NO Common Integrators of Hormonal Regulation of Stomatal Aperture?

From the foregoing discussion, we can draw three tentative conclusions regarding hormonal regulation of stomatal movements. First, it seems that ROS and/or NO production is associated with stomatal closure triggered by multiple plant hormones. It is possible that these molecules function as common integrators of hormonal regulation of stomatal movements (Figure 5.2). Second, current studies often use exogenously applied purified hormones or precursors to investigate the effects of hormones on

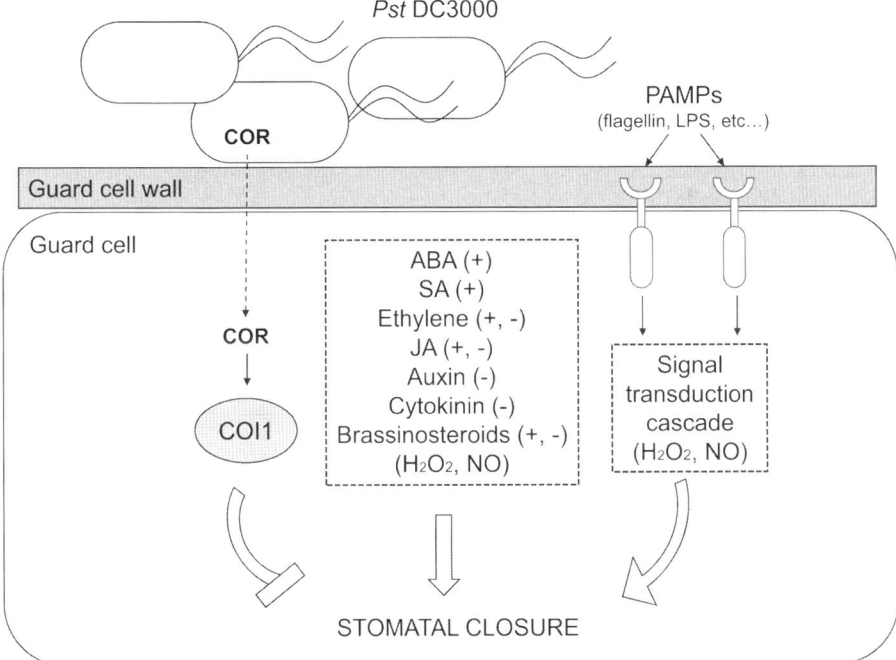

Fig. 5.2 A simplified diagram of guard cell signaling, indicating the effects of various hormonal and pathogen-derived stimuli on stomatal movements. Only relevant components discussed in this chapter are shown. Stomatal closure is triggered by pathogen/pathogen-associated molecular patterns (PAMPs) perception at the cell surface and counteracted by the bacterial toxin coronatine (COR), which acts through COI1, a key player in the jasmonic acid signaling pathway. The effects of various plant hormones are summarized in the middle box: the "+" and "−" symbols refer to the positive and negative effects, respectively, that each hormone exerts on stomatal closure. H_2O_2 and/or NO may be common modulators involved in pathogen and hormonal regulation of stomatal closure.

stomatal movements. A question that demands attention is that of the biological significance of these effects. So far, ABA has been thoroughly investigated in this respect, and its role in modulating stomatal aperture in response to abiotic stress (particularly water deficit) is relatively well understood (Schroeder et al. 2001; Fan et al. 2004; Israelsson et al. 2006). It is possible that other hormones modulate stomatal movements as well, but their physiological roles have not yet been clarified. Third, auxin, cytokinin, ethylene, and JA all seem to have the ability to counter ABA-induced stomatal closure, even though some of these hormones (e.g., ethylene and JA), when used alone, could close stomata under some conditions. In the case of JA and auxin, there seem to be concentration-dependent effects, as discussed earlier. It is possible that such different effects genuinely reflect dynamic crosstalk between endogenous hormone response pathways under different physiological conditions, but further study is necessary to solidify their biological relevance.

Pathogen Regulation of Stomatal Movements

Promotion of Stomatal Closure as Part of the Plant Defense (Innate Immune) Response

In addition to abiotic environmental cues, stomata have been found to respond to various microbe-derived compounds (Table 5.1). The fungal toxin fusicoccin has long been known to promote stomatal opening and to antagonize ABA-induced stomatal closure through activation of the plasma membrane H^+-ATPase (Marré 1979). Promotion of stomatal opening was also observed during *Sclerotinia sclerotiorum* infection of *Vicia faba* leaves (Guimaraes and Stotz 2004) and *Plasmopara viticola* infection of grapevine leaves (Allègre et al. 2007). Oligogalacturonic acid (OGA), an elicitor derived from the degradation of the plant cell wall by fungal cell wall-degrading enzymes, and chitosan, a component of the fungal cell wall, were both shown to affect stomatal movements in tomato (Lee et al. 1999). Both OGA and chitosan elicited H_2O_2 production in guard

Table 5.1 Biotic signals that stimulate stomatal movements.

Signals inducing stomatal closure	References
Oligogalacturonic acid (plant cell wall component)	(S. Lee et al. 1999)
Chitosan, yeast elicitor (fungal cell wall component)	(S. Lee et al. 1999; Klüsener et al. 2002)
Avr9 (fungal Avr protein)	(Blatt et al. 1999)
Live bacteria	(Melotto et al. 2006)
flg22, LPS (bacterial PAMPs)	(Melotto et al. 2006; Ali et al. 2007; Zhang et al., submitted)
Syringopeptins (bacterial toxins)	(Di Giorgio et al. 1996)
Signals inducing stomatal opening	
Fusicoccin (fungal toxin)	(Turner and Graniti 1969)
Oxalate (fungal virulence factor)	(Guimaraes and Stotz 2004)
Coronatine (bacterial virulence factor)	(Melotto et al. 2006; Zhang et al., 2008)

cells and inhibited light-induced opening of closed stomata. OGA also accelerates midday stomatal closure in tomato, and promotes closure and inhibits opening of stomata in *Commelina communis*. Moreover, chitosan and yeast elicitor derived from *Saccharomyces cerevisiae* were both found to induce stomatal closure and activate Ca^{2+} influx currents in a manner similar to ABA in *Arabidopsis* guard cells (Klüsener et al. 2002). However, the biological relevance of the reported responses of stomata to fungus-derived compounds with respect to plant defense or fungal invasion is not yet clear.

Stomata are particularly important entry portals for bacterial pathogens, which, unlike many fungal pathogens, cannot directly penetrate the plant epidermis. Study of *Arabidopsis* stomatal response to the bacterial pathogen *Pseudomonas syringae* pv. *tomato* DC3000 (*Pst* DC3000) provides evidence that guard cells perceive live bacterial cells and, within 1 hour, close the stomatal pore as an active innate immune response preventing the free passage of bacterial cells through the plant epidermis (Melotto et al. 2006).

Not only plant pathogenic bacteria but also human pathogens, such as *Escherichia coli* O157:H7, induce stomatal closure within the first hour of contact with plant tissue, suggesting that guard cells can sense and respond to bacterial molecules common between plant and human pathogens (Melotto et al. 2006). Pathogen-associated molecular patterns (PAMPs) are evolutionarily conserved microbial molecules that are perceived by eukaryotes as non-self, and consequently induce innate immune response in both plants and animals (Ausubel, 2005; Akira et al. 2006; He et al. 2007). Commonly used PAMPS include bacterial flagellin and its conserved 22-amino acid peptide flg22 (Felix et al. 1999), and bacterial lipopolysaccharide (LPS; Zeidler et al. 2004). Melotto et al. (2006) showed that, similar to live bacteria, flg22 and LPS could trigger stomatal closure, indicating that stomatal closure is part of the PAMP-induced innate immune response. flg22 also inhibits light-induced stomatal opening in *Arabidopsis*, suggesting that plants may keep their stomata more closed when exposed to pathogenic bacteria (W. Zhang et al. 2008).

Stomatal closure in response to bacteria, flg22, or LPS required, in addition to PAMP signaling, ABA biosynthesis and signal transduction (Melotto et al. 2006): neither flg22 nor LPS could induce stomatal closure in the ABA biosynthetic mutant *aba3-1* (Leon-Kloosterziel et al. 1996) or in the ABA signaling mutant *ost1-2* (Mustilli et al. 2002). Furthermore, neither bacteria nor LPS could elicit stomatal closure in *Arabidopsis* plants that are impaired in SA biosynthesis, such as the *eds16* mutant (Wildermuth et al. 2001), or in SA accumulation, such as the transgenic *NahG* plants (Delaney et al. 1994), thus revealing a key role for SA in stomatal response to pathogens (Melotto et al. 2006). As discussed earlier, SA is an important mediator of plant defenses and a strong inducer of stomatal closure; it is therefore very interesting that PAMP-triggered stomatal closure requires SA. These results provide strong evidence for the existence of interconnected networks regulating guard cell response to bacteria, which involve PAMP signaling, SA, and ABA biosynthesis and signal transduction. It is not yet understood how these different signaling networks merge in stomatal regulation in response to pathogens/PAMPs. However, PAMPs are known to induce ROS and NO production (Felix et al. 1999; Lee et al. 1999; Zeidler et al. 2004; Melotto et al. 2006), which, as discussed earlier, may function as common integrators of guard cell signaling leading to stomatal closure.

Promotion of Stomatal Opening as Part of a Pathogen Virulence Mechanism

PAMP-triggered stomatal closure would reduce the efficiency of bacterial entry through stomata. However, as in the case of response to other stimuli including ABA (Terashima et al. 1988; Mott and Buckley 2000), not all stomata in a leaf close in response to bacteria or PAMP treatments (Melotto et al. 2006). It is therefore possible that some pathogens may be highly virulent, and even a few bacteria that enter through nonresponsive open stomata would be sufficient to cause disease. Other bacterial pathogens, however, may need massive initial invasion to cause disease. In this case, pathogens must evolve mechanisms to overcome PAMP-triggered stomatal closure. Different pathogens likely employ different mechanisms to do so. One mechanism could be based on interplay between environmental and pathogen regulation of stomatal movements. At any given time, stomata must respond to a number of stimuli, including light, humidity, CO_2 concentration, microorganisms, and the circadian clock. How these inputs are prioritized by guard cells is not known. However, severe outbreaks of bacterial disease in crop fields are often associated with storms, heavy rain, or periods of high humidity (Hirano and Upper 2000). These conditions could create wounds at the leaf surface, which would facilitate bacterial entry independent of stomata. It is also possible that under such environmental conditions, PAMP-triggered stomatal defense is partially ineffective because high humidity inhibits stomatal closure (Talbott et al. 2003; Roelfsema and Hedrich 2005; X.D. Xie et al. 2006; Nejad and van Meeteren 2007), therefore allowing more bacteria to enter the leaf tissue through stomata to cause infection.

Pathogens may also have evolved specific virulence factors to actively overcome PAMP-triggered stomatal closure. This seems to be the case for *Pst* DC3000. Melotto and colleagues (2006) observed a significant difference between stomatal response to the human pathogen *E. coli* O157:H7 versus the plant pathogen *Pst* DC3000. After initial closure in response to both pathogens, stomata reopened after approximately 3 hours of incubation with *Pst* DC3000, but not with *E. coli* O157:H7. A likely explanation is that *Pst* DC3000, but not *E. coli* O157:H7, may have evolved a virulence mechanism that can counter PAMP-induced stomatal closure. This is an attractive hypothesis because, in the mid-1990s, Mittal and Davis (1995) suggested the possibility that the phytotoxin coronatine (COR), a virulence factor produced by *Pst* DC3000, suppresses an early defense response during *Pst* DC3000 infection of *Arabidopsis* and tomato. Mutant bacteria defective in COR biosynthesis (*cor*$^-$), inoculated by infiltration of bacterial suspension directly into *Arabidopsis* leaves, multiplied similarly to wildtype *Pst* DC3000 but caused fewer disease symptoms, including less chlorosis and necrosis. However, when leaves were surface-inoculated by dipping into a bacterial suspension, *cor*$^-$ mutant bacteria achieved significantly lower intraleaf population levels than wildtype *Pst* DC3000 and did not cause disease symptoms (Mittal and Davis 1995). Direct infiltration of bacteria in leaves is a widespread laboratory practice that bypasses stomata; the observation that *cor*$^-$ bacteria did not proliferate significantly or cause disease symptoms in surface-inoculated *Arabidopsis* leaves suggests a role for COR in bacterial entry through stomata into the leaf apoplast. Indeed, Melotto et al. (2006) found that COR is able to reverse bacteria- and PAMP-induced stomatal closure. Furthermore, COR can also reverse flg22 inhibition of light-induced stomatal opening (W. Zhang et al. 2008). These results suggest that *Pst* DC3000 has developed a strategy to overcome the stomata-based plant immune response and offer one explanation for the earlier observations on the role of COR in bacterial virulence.

Although COR is one of the most extensively studied bacterial phytotoxins, the precise molecular mechanisms through which COR acts to promote stomatal reopening and other virulence functions during infection are not yet clear. COR has long been hypothesized to be a structural and functional mimic of jasmonates: not only is COR structurally similar to JA conjugated to the amino acid isoleucine (JA-Ile; Staswick and Tiryaki 2004), but also COR and JA have been demonstrated to induce a number of similar plant responses, including induction of protease inhibitors and phytoalexins, as well as regulation of JA-responsive transcripts (Weiler et al. 1994; Palmer and Bender 1995; Tamogami and Kodama 2000; Zhao et al. 2003; Thilmony et al. 2006; Uppalapati et al. 2007). Furthermore, the *Arabidopsis* COI1 F-box protein is required for responses to both JA and COR at the whole plant level (Xie et al. 1998; Kloek et al. 2001), and it is also required for COR-mediated inhibition of stomatal closure (Melotto et al. 2006). COR is therefore likely to function as a molecular mimic of JA-Ile in the plant cell.

How COR activation of JA signaling elements would lead to inhibition of PAMP-induced stomatal closure is not yet known. However, several studies have shown a remarkable degree of positive and/or negative crosstalk between JA-, ethylene-, SA-, or ABA-mediated signaling in plant response to pathogen infection (Spoel et al. 2003; Anderson et al. 2004; Brooks et al. 2005; Glazebrook 2005; Lorenzo and Solano 2005; Laurie-Berry et al. 2006). In general, it seems that JA- and ethylene-mediated signaling is antagonistic to SA-dependent signaling (Glazebrook 2005). For instance, the *Arabidopsis* JA signaling mutant *coi1* and its corresponding tomato mutant *jai1* exhibited marked resistance to *Pst* DC3000 infection, and such resistance was correlated with upregulation of transcripts associated with SA-mediated defenses (Kloek et al. 2001; Zhao et al. 2003). Furthermore, plants defective in SA accumulation such as *NahG*, and plants defective in SA biosynthesis such as the *eds16* mutant supported a significantly enhanced multiplication of *cor⁻* mutant bacteria, compared to wildtype plants (Brooks et al. 2005; Melotto et al. 2006). Thus, by activating signaling elements that normally function downstream of JA, COR suppresses SA-mediated defenses, including PAMP-triggered stomatal closure.

JA signaling could act antagonistically to ABA signaling as well. The *Arabidopsis AtMYC2* gene encodes a basic helix–loop–helix leucine-zipper transcription factor that is an important positive regulator of ABA responses in the plant (Abe et al. 2003; Anderson et al. 2004; Lorenzo et al. 2004; Yadav et al. 2005). Interestingly, this locus was found to be allelic to *Jasmonate Insensitive 1* (*JIN1*), thus representing a common element in the JA and ABA signaling pathways (Laurie-Berry et al. 2006). *Atmyc2* mutant plants displayed elevated levels of transcription from some JA- and ethylene-responsive defense genes but were, of course, impaired in their ABA response (Anderson et al. 2004). It is not known whether AtMYC2 is involved in the crosstalk between COR and ABA in stomatal guard cells, but *Arabidopsis* mutants defective in ABA biosynthesis (*aba3-1*) or signaling (*ost1*) allowed enhanced multiplication of *cor⁻* mutant bacteria after surface inoculation (Melotto et al. 2006). As discussed earlier, different concentrations of JA and MeJA appear to have different effects on stomatal movements. Whereas high concentrations of exogenous JA and MeJA promote stomatal closure, low concentrations counteract ABA-induced stomatal closure (see Figure 5.1).

Therefore, in addition to its antagonistic effect on SA-dependent stomatal closure, COR produced by bacteria during infection could have an effect similar to low concentrations of JA/MeJA (e.g., 2 µM; see Figure 5.1), to inhibit ABA-mediated closure of the stomatal pore.

Ion Channel Regulation Associated with Pathogen Modulation of Stomatal Movements

Ion channels, especially K^+ and anion channels, are very downstream regulators of stomatal movements because they are incharge of net ion flux into and out of the guard cells (Pandey et al. 2007). A number of reports have documented regulation of guard cell ion channel activities by pathogens and the factors they produce. Like ABA, the bacterial PAMP flg22 inhibits guard cell inward K^+ currents, indicating that both stimuli may use some of the same signaling components (Zhang et al. 2008). Consistent with COR reversal of flg22 inhibition of stomatal opening, COR also reverses flg22 inhibition of inward K^+ currents (Zhang et al. 2008). Regulation of K^+ channels has also been found in a plant-fungal interaction that involves a specific plant disease resistance gene and its cognate pathogen elicitor. Strains of the tomato leaf mold pathogen *Cladosporium fulvum* expressing the Avr9 protein activate gene-for-gene resistance in tomato cultivars that express the *Cf-9* disease resistance gene (Jones et al. 1994). Similar to the ABA regulation pattern for K^+ currents that is observed in tobacco guard cells (Blatt 1990; Blatt and Armstrong 1993), Avr9 inhibits inward K^+ currents and activates outward K^+ currents of *Cf-9* transgenic guard cells (Blatt et al. 1999).

H_2O_2 and NO are potentially involved in pathogen regulation of outward K^+ currents. It has been found that, besides inhibiting guard cell inward K^+ currents, flg22 gradually inhibits the guard cell outward K^+ currents (Zhang et al. 2008). Similar effects occur in H_2O_2 or NO regulation of these channels (Zhang et al. 2001; Kohler et al. 2003; Sokolovski et al. 2005; Garcia-Mata and Lamattina 2007). As discussed earlier, H_2O_2 and NO are also important components of ABA signaling. Thus, it is possible that H_2O_2- and NO-mediated ion channel regulation is a point of crosstalk between PAMPs and phytohormones.

The Arabidopsis Cyclic Nucleotide Gated Channel2 (CNGC2/DND1) is activated by the bacterial PAMP LPS and mediates a Ca^{2+} current into guard cells to induce NO production (Ali et al. 2007). Changes in cytosolic free Ca^{2+} constitute an important second messenger in stress responses including ABA signaling (Hetherington 2001), and Ca^{2+} is therefore another signal held in common by both pathogen and hormonal signaling pathways.

Heterotrimeric G proteins are also involved in guard cell responses to both ABA and pathogens. It is well known that G-proteins play important roles in whole plant ABA signaling and non-guard-cell-related resistance to some pathogens (X.Q. Wang et al. 2001; Coursol et al. 2003; Perfus-Barbeoch et al. 2004; Assmann, 2005; Llorente et al. 2005; Trusov et al. 2006, 2007; Pandey et al. 2007). In guard cells, the G-protein alpha subunit GPA1 is required for ABA inhibition of stomatal opening, and *gpa1* null mutants are impaired in ABA inhibition of inward K^+ currents and ABA activation of pH-independent anion currents (Wang et al. 2001). flg22 inhibition of stomatal opening and inward K^+ currents by flg22 are similarly impaired in these mutants

(Zhang et al. 2008), providing another example of the interplay between phytohormone and pathogen signaling in guard cells.

In summary, common signal elements, including NO, H_2O_2, Ca^{2+}, and G-proteins, are involved in regulation of ion channel activities and stomatal movements by both pathogens and phytohormones, suggesting widespread interaction between biotic and abiotic regulatory pathways.

Concluding Remarks

Guard cell signaling has traditionally been studied in the context of abiotic stresses, yet stomata are a major portal through which many foliar pathogens enter and exit plants, representing two key steps in disease cycle and epidemiology (Boureau et al. 2002; Guimaraes and Stotz 2004; Melotto et al. 2008). In this chapter, we hope to have conveyed a sense of the current understanding of some of the shared and distinct signaling mechanisms and potential interactions between signaling networks in the guard cell in response to pathogen and hormonal signals (Figure 5.2). Clearly, we have much to learn, and this is a fruitful area for further study of signal integration and prioritization in the guard cell. Future studies integrating abiotic and biotic signals may shed light on the important question of whether abiotic conditions that promote stomatal opening diminish the effectiveness of stomate-based restriction of pathogen entry into and exit from the plant, thereby influencing disease epidemics. Conversely, because plants in nature often harbor diverse microbial communities, sometimes at very high densities (Lindow and Brandl 2003), it would be important to understand how such microbial communities influence stomatal response to abiotic signals in natural ecosystems.

Acknowledgments

Research in the SYH laboratory is supported by grants from the U.S. Department of Energy and National Institutes of Health. Research in the SMA laboratory is supported by grants from the National Science Foundation and the U.S. Department of Agriculture.

References

Abe, H., Urao, T., Ito, T., Seki, M., Shinozaki, K., and Yamaguchi-Shinozaki, K. 2003. *Arabidopsis* AtMYC2 (bHLH) and AtMYB2 (MYB) function as transcriptional activators in abscisic acid signaling. *Plant Cell* 15:63–78.

Akira, S., Uematsu, S., and Takeuchi, O. 2006. Pathogen recognition and innate immunity. *Cell* 124:783–801.

Ali R., Ma, W., Lemtiri-Chlieh, F., Tsaltas, D., Leng, Q., von Bodman, S., and Berkowitz, G.A. 2007. Death don't have no mercy and neither does calcium: *Arabidopsis* CYCLIC NUCLEOTIDE GATED CHANNEL2 and innate immunity. *Plant Cell* 19:1081–1095.

Allègre, M., Daire, X., Heloir, M.C., Trouvelot, S., Mercier, L., Adrian, M., and Pugin, A. 2007. Stomatal deregulation in *Plasmopara viticola*–infected grapevine leaves. *New Phytol* 173:832–840.

Anderson, J.P., Badruzsaufari, E., Schenk, P.M., Manners, J.M., Desmond, O.J., Ehlert, C., Maclean, D.J., Ebert, P.R., and Kazan, K. 2004. Antagonistic interaction between abscisic acid and jasmonate-ethylene signaling pathways modulates defense gene expression and disease resistance in *Arabidopsis*. *Plant Cell* 16:3460–3479.

Assmann, S.M. 2005. G protein regulation of disease resistance during infection of rice with rice blast fungus. *Science STKE* 2005:cm13.

Ausubel, F.M. 2005. Are innate immune signaling pathways in plants and animals conserved? *Nat Immunol* 6:973–979.

Blackman, P.G., and Davies, W.J. 1984. Modification of the CO_2 responses of maize stomata by abscisic acid and by naturally occurring and synthetic cytokinins. *J Exp Botany* 35:174–179.

Blatt, M.R. 1990. Potassium channel currents in intact stomatal guard cells–Rapid enhancement by abscisic acid. *Planta* 180:445–455.

Blatt, M.R., and Armstrong, F. 1993. K^+ channels of stomatal guard cells–Abscisic acid-evoked control of the outward rectifier mediated by cytoplasmic pH. *Planta* 191:330–341.

Blatt, M.R., Grabov, A., Brearley, J., Hammond-Kosack, K., and Jones, J.D.G. 1999. K^+ channels of *Cf-9* transgenic tobacco guard cells as targets for *Cladosporium fulvum* Avr9 elicitor-dependent signal transduction. *Plant J* 19:453–462.

Boureau, T., Routtu, J., Roine, E., and Taira, S., Romantschuk, M. 2002. Localization of hrpA-induced *Pseudomonas syringae* pv. *tomato* DC3000 in infected tomato leaves. *Mol Plant Pathol* 3:451–460.

Brooks, D.M., Bender, C.L., and Kunkel, B.N. 2005. The *Pseudomonas syringae* phytotoxin coronatine promotes virulence by overcoming salicylic acid–dependent defences in *Arabidopsis thaliana*. *Mol Plant Pathol* 6:629–639.

Chen, Z.X., Silva, H., and Klessig, D.F. 1993. Active oxygen species in the induction of plant systemic acquired resistance by salicylic acid. *Science* 262:1883–1886.

Coursol, S., Fan, L.M., Le Stunff, H., Spiegel, S., Gilroy, S., and Assmann S.M. 2003. Sphingolipid signalling in *Arabidopsis* guard cells involves heterotrimeric G proteins. *Nature* 423:651–654.

Delaney, T.P., Uknes, S., Vernooij, B., Friedrich, L., Weymann, K., Negrotto, D., Gaffney, T., Gutrella, M., Kessmann, H., Ward, E., and Ryals, J. 1994. A central role of salicylic acid in plant disease resistance. *Science* 266:1247–1250.

Desikan, R., Hancock, J.T., Bright, J., Harrison, J., Weir, L., Hooley, R., and Neill, S.J. 2005. A role for ETR1 in hydrogen peroxide signaling in stomatal guard cells. *Plant Physiol* 137:831–834.

Desikan, R., Last, K., Harrett-Williams, R., Tagliavia C., Harter, K., Hooley, R., Hancock J.T., and Neill S.J. 2006. Ethylene-induced stomatal closure in *Arabidopsis* occurs via AtrbohF-mediated hydrogen peroxide synthesis. *Plant J* 47:907–916.

Devoto, A., and Turner, J.G. 2003. Regulation of jasmonate-mediated plant responses in *Arabidopsis*. *Ann Botany* 92:329–337.

Di Giorgio D., Camoni L., Mott K.A., Takemoto J.Y., and Ballio A. 1996. Syringopeptins, *Pseudomonas syringae* pv. *syringae* phytotoxins, resemble syringomycin in closing stomata. *Plant Pathol* 45:564–571.

Durrant, W.E., and Dong, X. 2004. Systemic acquired resistance. *Annu Rev Phytopathol* 42:185–209.

Ephritikhine, G., Fellner, M., Vannini C., Lapous, D., and Barbier-Brygoo, H. 1999. The *sax1* dwarf mutant of *Arabidopsis thaliana* shows altered sensitivity of growth responses to abscisic acid, auxin, gibberellins and ethylene and is partially rescued by exogenous brassinosteroid. *Plant J* 18:303–314.

Evans, N.H. 2003. Modulation of guard cell plasma membrane potassium currents by methyl jasmonate. *Plant Physiol* 131:8–11.

Fan, L.-M., Zhao Z., and Assmann, S.M. 2004. Guard cells: a dynamic signaling model. *Curr Opin Plant Biol* 7:537–546.

Felix, G., Duran, J.D., Volko, S., and Boller, T. 1999. Plants have a sensitive perception system for the most conserved domain of bacterial flagellin. *Plant J* 18:265–276.

Feys, B.J.F., Benedetti, C.E., Penfold, C.N., and Turner, J.G. 1994. *Arabidopsis* mutants selected for resistance to the phytotoxin coronatine are male-sterile, insensitive to methyl jasmonate, and resistant to a bacterial pathogen. *Plant Cell* 6:751–759.

Fujita, M., Fujita, Y., Noutoshi, Y., Takahashi, F., Narusaka, Y., Yamaguchi-Shinozaki, K., and Shinozaki K. 2006. Crosstalk between abiotic and biotic stress responses: a current view from the points of convergence in the stress signaling networks. *Curr Opin Plant Biol* 9:436–442.

Gaffney, T., Friedrich, L., Vernooij, B., Negrotto, D., Nye, G., Uknes, S., Ward, E., Kessmann, H., and Ryals, J. 1993. Requirement of salicylic acid for the induction of systemic acquired resistance. *Science* 261:754–756.

Garcia-Mata, C., and Lamattina, L. 2007. Abscisic acid (ABA) inhibits light-induced stomatal opening through calcium- and nitric oxide-mediated signaling pathways. *Nitric Oxide* 17:143–151.

Gehring, C.A., Irving, H.R., McConchie, R., and Parish, R.W. 1997. Jasmonates induce intracellular alkalinization and closure of *Paphiopedilum* guard cells. *Ann Botany* 80:485–489.

Glazebrook, J. 2005. Contrasting mechanisms of defense against biotrophic and necrotrophic pathogens. *Annu Rev Phytopathol* 43:205–227.

Guimaraes, R.L., and Stotz H.U. 2004. Oxalate production by *Sclerotinia sclerotiorum* deregulates guard cells during infection. *Plant Physiol* 136:3703–3711.

Haubrick, L.L., Torsethaugen, G., and Assmann, S.M. 2006. Effect of brassinolide, alone and in concert with abscisic acid, on control of stomatal aperture and potassium currents of *Vicia faba* guard cell protoplasts. *Physiologia Plantarum* 128:134–143.

He, P., Shan, L., and Sheen, J. 2007. Elicitation and suppression of microbe-associated molecular pattern-triggered immunity in plant-microbe interactions. *Cell Microbiol* 9:1385–1396.

Hetherington, A.M. 2001. Guard cell signaling. *Cell* 107:711–714.

Hetherington, A.M., and Woodward, F.I. 2003. The role of stomata in sensing and driving environmental change. *Nature* 424:901–908.

Hirano S.S., and Upper, C.D. 2000. Bacteria in the leaf ecosystem with emphasis on *Pseudomonas syringae*—a pathogen, ice nucleus, and epiphyte. *Microbiol Mol Biol Rev* 64:624–653.

Hosy, E., Vavasseur, A., Mouline, K., Dreyer I., Gaymard, F., Poree, F., Boucherez, J., Lebaudy, A., Bouchez, D., Very, A.A., Simonneau, T., Thibaud, J.B., and Sentenac, H. 2003. The *Arabidopsis* outward K^+ channel GORK is involved in regulation of stomatal movements and plant transpiration. *Proc Natl Acad Sci USA* 100:5549–5554.

Israelsson, M., Siegel, R.S., Young, J., Hashimoto, M., Iba, K., and Schroeder, J.I. 2006. Guard cell ABA and CO_2 signaling network updates and Ca^{2+} sensor priming hypothesis. *Curr Opin Plant Biol* 9:654–663.

Jones, D.A., Thomas, C.M., Hammond-Kosack, K.E., Balint-Kurti, P.J., and Jones, J.D. 1994. Isolation of the tomato Cf-9 gene for resistance to *Cladosporium fulvum* by transposon tagging. *Science* 266:789–793.

Kloek, A.P., Verbsky, M.L., Sharma, S.B., Schoelz, J.E., Vogel, J., Klessig, D.F., and Kunkel, B.N. 2001. Resistance to *Pseudomonas syringae* conferred by an *Arabidopsis thaliana* coronatine-insensitive (*coi1*) mutation occurs through two distinct mechanisms. *Plant J* 26:509–522.

Klüsener, B., Young, J.J., Murata, Y., Allen, G.J., Mori, I.C., Hugouvieux, V., and Schroeder, J.I. 2002. Convergence of calcium signaling pathways of pathogenic elicitors and abscisic acid in *Arabidopsis* guard cells. *Plant Physiol* 130:2152–2163.

Kohler, B., Hills, A., and Blatt, M.R. 2003. Control of guard cell ion channels by hydrogen peroxide and abscisic acid indicates their action through alternate signaling pathways. *Plant Physiol* 131:385–388.

Kwak, J.M., Nguyen, V., and Schroeder, J.I. 2006. The role of reactive oxygen species in hormonal responses. *Plant Physiol* 141:323–329.

Laurie-Berry, N., Joardar V., Street, I.H., and Kunkel, B.N. 2006. The *Arabidopsis thaliana JASMONATE IN-SENSITIVE 1* gene is required for suppression of salicylic acid-dependent defenses during infection by *Pseudomonas syringae*. *Mol Plant Micr Interact* 19:789–800.

Lee, J.-S. 1998. The mechanism of stomatal closing by salicylic acid in *Commelina communis* L. *J Plant Biol* 41:97–102.

Lee, S., Choi, H., Suh, S., Doo, I.S., Oh, K.Y., Choi, E.J., Taylor, A.T.S., Low, P.S., and Lee, Y. 1999. Oligogalacturonic acid and chitosan reduce stomatal aperture by inducing the evolution of reactive oxygen species from guard cells of tomato and *Commelina communis*. *Plant Physiol* 121:147–152.

Leon-Kloosterziel, K.M., Gil, M.A., Ruijs, G.J., Jacobsen, S.E., Olszewski, N.E., Schwartz, S.H., Zeevaart, J.A.D., and Koornneef, M. 1996. Isolation and characterization of abscisic acid-deficient *Arabidopsis* mutants at two new loci. *Plant J* 10:655–661.

Lindow, S.E., and Brandl, M.T. 2003. Microbiology of the phyllosphere. *Appl Environ Microbiol* 69:1875–1883.

Llorente, F., Alonso-Blanco, C., Sanchez-Rodriguez, C., Jorda, L., and Molina A. 2005. ERECTA receptor-like kinase and heterotrimeric G protein from *Arabidopsis* are required for resistance to the necrotrophic fungus *Plectosphaerella cucumerina*. *Plant J* 43:165–180.

Lorenzo, O., Chico, J.M., Sanchez-Serrano, J.J., and Solano, R. 2004. *Jasmonate-insensitive1* encodes a MYC transcription factor essential to discriminate between different jasmonate-regulated defense responses in *Arabidopsis*. *Plant Cell* 16:1938–1950.

Lorenzo, O., and Solano, R. 2005. Molecular players regulating the jasmonate signalling network. *Curr Opin Plant Biol* 8:532–540.

Luan, S. 2002. Signalling drought in guard cells. *Plant Cell Environ* 25:229–237.

Manthe, B., Schulz, M., and Schnabl, H. 1992. Effects of salicylic acid on growth and stomatal movements of *Vicia faba* L: evidence for salicylic acid metabolization. *J Chem Ecol* 18:1525–1539.

Marré, E. 1979. Fusicoccin—tool in plant physiology. *Annu Rev Plant Physiol Plant Mol Biol* 30:273–288.

Marten, I., Lohse, G., and Hedrich, R. 1991. Plant growth hormones control voltage-dependent activity of anion channels in plasma membrane of guard cells. *Nature* 353:758–762.

Melotto, M., Underwood, W., and He, S.Y. 2008. Role of stomata in plant innate immunity and foliar bacterial diseases. *Annu Rev Phytopathol* 46:101–122.

Melotto, M., Underwood, W., Koczan, J., Nomura, K., and He, S.Y. 2006. Plant stomata function in innate immunity against bacterial invasion. *Cell* 126:969–980.

Mittal, S., and Davis, K.R. 1995. Role of the phytotoxin coronatine in the infection of *Arabidopsis thaliana* by *Pseudomonas syringae* pv *tomato*. *Mol Plant Micr Interact* 8:165–171.

Mori, I.C., Pinontoan, R., Kawano, T., and Muto, S. 2001. Involvement of superoxide generation in salicylic acid-induced stomatal closure in *Vicia faba*. *Plant Cell Physiol* 42:1383–1388.

Mott, K.A., and Buckley, T.N. 2000. Patchy stomatal conductance: emergent collective behaviour of stomata. *Trends Plant Sci* 5:258–262.

Munemasa, S., Oda, K., Watanabe-Sugimoto, M., Nakamura, Y., Shimoishi, Y., and Murata Y. 2007. The *coronatine-insensitive 1* mutation reveals the hormonal signaling interaction between abscisic acid and methyl jasmonate in *Arabidopsis* guard cells. Specific impairment of ion channel activation and second messenger production. *Plant Physiol* 143:1398–1407.

Murata, Y., Pei, Z.-M., Mori, I.C., and Schroeder, J. 2001. Abscisic acid activation of plasma membrane Ca^{2+} channels in guard cells requires cytosolic NAD(P)H and is differentially disrupted upstream and downstream of reactive oxygen species production in *abi1-1* and *abi2-1* protein phosphatase 2C mutants. *Plant Cell* 13:2513–2523.

Mustilli, A.C., Merlot, S., Vavasseur, A., Fenzi, F., and Giraudat, J. 2002. *Arabidopsis* OST1 protein kinase mediates the regulation of stomatal aperture by abscisic acid and acts upstream of reactive oxygen species production. *Plant Cell* 14:3089–3099.

Neill, S., Barros, R., Bright, J., Desikan, R., Hancock, J., Harrison, J., Morris, P., Ribeiro, D., and Wilson I. 2008. Nitric oxide, stomatal closure, and abiotic stress. *J Exp Botany* 59:165–176.

Nejad, A.R., and van Meeteren, U. 2007. The role of abscisic acid in disturbed stomatal response characteristics of *Tradescantia virginiana* during growth at high relative air humidity. *J Exp Botany* 58:627–636.

Nilson, S.E., and Assmann, S.M. 2007. The control of transpiration. Insights from *Arabidopsis*. *Plant Physiol* 143:19–27.

Palmer, D.A., and Bender, C.L. 1995. Ultrastructure of tomato leaf tissue treated with the pseudomonad phytotoxin coronatine and comparison with methyl jasmonate. *Mol Plant Micr Interact* 8:683–692.

Pandey, S., Zhang, W., and Assmann, S.M. 2007. Roles of ion channels and transporters in guard cell signal transduction. *FEBS Lett* 581:2325–2336.

Perfus-Barbeoch, L., Jones, A.M., and Assmann, S.M. 2004. Plant heterotrimeric G protein function: insights from *Arabidopsis* and rice mutants. *Curr Opin Plant Biol* 7:719–731.

Poffenroth, M., Green, D.B., and Tallman, G. 1992. Sugar concentrations in guard cells of *Vicia faba* illuminated with red or blue light—analysis by high-performance liquid chromatography. *Plant Physiol* 98:1460–1471.

Roelfsema, M.R.G., and Hedrich, R. 2005. In the light of stomatal opening: new insights into "the Watergate." *New Phytologist* 167:665–691.

Schaller, G.E., Ladd, A.N., Lanahan, M.B., Spanbauer, J.M., and Bleecker, A.B. 1995. The ethylene response mediator ETR1 from *Arabidopsis* forms a disulfide-linked dimer. *J Biol Chem* 270:12526–12530.

Schroeder, J.I., Allen G.J., Hugouvieux, V., Kwak, J.M., and Waner, D. 2001. Guard cell signal transduction. *Annu Rev Plant Physiol Plant Mol Biol* 52:627–658.

She, X.-P., and Song, X.-G. 2006. Cytokinin- and auxin-induced stomatal opening is related to the change of nitric oxide levels in guard cells in broad bean. *Physiol Plantarum* 128:569–579.

Sokolovski, S., Hills, A., Gay, R., Garcia-Mata, C., Lamattina, L., and Blatt, M.R. 2005. Protein phosphorylation is a prerequisite for intracellular Ca^{2+} release and ion channel control by nitric oxide and abscisic acid in guard cells. *Plant J* 43:520–529.

Song, X.G., She, X.P., He, J.M., Huang, C., and Song, T.S. 2006. Cytokinin- and auxin-induced stomatal opening involves a decrease in levels of hydrogen peroxide in guard cells of *Vicia faba*. *Funct Plant Biol* 33:573–583.

Spoel, S.H., Koornneef, A., Claessens, S.M.C., Korzelius, J.P., Van Pelt, J.A., Mueller, M.J., Buchala, A.J., Metraux, J.P., Brown, R., Kazan, K., Van Loon, L.C., Dong, X.N., and Pieterse, C.M.J. 2003. NPR1 modulates cross-talk between salicylate- and jasmonate-dependent defense pathways through a novel function in the cytosol. *Plant Cell* 15:760–770.

Stadler, R., Buttner, M., Ache, P., Hedrich, R., Ivashikina, N., Melzer, M., Shearson S.M., Smith S.M., and Sauer, N. 2003. Diurnal and light-regulated expression of *AtSTP1* in guard cells of *Arabidopsis*. *Plant Physiol* 133:528–537.

Staswick, P.E., and Tiryaki, I. 2004. The oxylipin signal jasmonic acid is activated by an enzyme that conjugates it to isoleucine in *Arabidopsis*. *Plant Cell* 16:2117–2127.
Suhita, D., Raghavendra, A.S., Kwak, J.M., and Vavasseur, A. 2004. Cytoplasmic alkalization precedes reactive oxygen species production during methyl jasmonate- and abscisic acid-induced stomatal closure. *Plant Physiol* 134:1536–1545.
Szyroki, A., Ivashikina, N., Dietrich, P., Roelfsema, M.R.G., Ache, P., Reintanz, B., Deeken, R., Godde, M., Felle, H., Steinmeyer, R., Palme, K., and Hedrich, R. 2001. KAT1 is not essential for stomatal opening. *Proc Natl Acad Sci USA* 98:2917–2921.
Talbott, L.D., Rahveh, E., and Zeiger, E. 2003. Relative humidity is a key factor in the acclimation of the stomatal response to CO_2. *J Exp Botany* 54:2141–2147.
Tamogami, S., and Kodama, O. 2000. Coronatine elicits phytoalexin production in rice leaves (*Oryza sativa* L.) in the same manner as jasmonic acid. *Phytochemistry* 54:689–694.
Tanaka, Y., Sano, T., Tamaoki, M., Nakajima, N., Kondo, N., and Hasezawa, S. 2005. Ethylene inhibits abscisic acid-induced stomatal closure in *Arabidopsis*. *Plant Physiol* 138:2337–2343.
Tanaka, Y., Sano, T., Tamaoki, M., Nakajima, N., Kondo, N., and Hasezawa, S. 2006. Cytokinin and auxin inhibit abscisic acid-induced stomatal closure by enhancing ethylene production in *Arabidopsis*. *J Exp Botany* 57:2259–2266.
Terashima, I., Wong, S.C., Osmond, C.B., and Farquhar, G.D. 1988. Characterization of non-uniform photosynthesis induced by abscisic acid in leaves having different mesophyll anatomies. *Plant Cell Physiol* 29:385–394.
Thilmony, R., Underwood, W., He, S.Y. 2006. Genome-wide transcriptional analysis of the *Arabidopsis thaliana* interaction with the plant pathogen *Pseudomonas syringae* pv. *tomato* DC3000 and the human pathogen *Escherichia coli* O157:H7. *Plant J* 46:34–53.
Trusov, Y., Rookes, J.E., Chakravorty, D., Armour, D., Schenk, P.M., and Botella, J.R. 2006. Heterotrimeric G proteins facilitate *Arabidopsis* resistance to necrotrophic pathogens and are involved in jasmonate signaling. *Plant Physiol* 140:210–220.
Trusov, Y., Rookes, J.E., Tilbrook, K., Chakravorty, D., Mason, M.G., Anderson, D., Chen J.G., Jones A.M., and Botella J.R. 2007. Heterotrimeric G protein γ subunits provide functional selectivity in Gβ dimer signaling in *Arabidopsis*. *Plant Cell* 19:1235–1250.
Underwood, W., Melotto, M., and He S.Y. 2007. Role of plant stomata in bacterial invasion. *Cell Microbiol* 9:1621–1629.
Uppalapati, S.R., Ishiga, Y., Wangdi, T., Kunkel, B.N., Anand, A., Mysore, K.S., and Bender, C.L. 2007. The phytotoxin coronatine contributes to pathogen fitness and is required for suppression of salicylic acid accumulation in tomato inoculated with *Pseudomonas syringae* pv. *tomato* DC3000. *Mol Plant Micr Interact* 20:955–965.
Vahisalu, T., Kollist, H., Wang, Y.F., Nishimura, N., Chan, W.Y., Valerio, G., Lammimäki, A., Brosché, M., Moldau, H., Desikan, R., Schroeder J.I., and Kangasjärvi J. 2008. SLAC1 is required for plant guard cell S-type anion channel function in stomatal signalling. *Nature* 452:487–491.
Wang K.L.C., Li, H., and Ecker, J.R. 2002. Ethylene biosynthesis and signaling networks. *Plant Cell* 14:S131–151.
Wang X.Q., Ullah, H., Jones, A.M., and Assmann, S.M. 2001. G protein regulation of ion channels and abscisic acid signaling in *Arabidopsis* guard cells. *Science* 292(5524):2070–2072.
Weiler, E.W., Kutchan, T.M., Gorba, T., Brodschelm, W., Niesel, U., and Bublitz, F. 1994. The *Pseudomonas* phytotoxin coronatine mimics octadecanoid signaling molecules of higher plants. *FEBS Lett* 349:317.
Wildermuth, M.C., Dewdney, J., Wu, G., and Ausubel, F.M. 2001. Isochorismate synthase is required to synthesize salicylic acid for plant defence. *Nature* 414:562–565.
Wilkinson, S., and Davies, W.J. 2002. ABA-based chemical signalling: the co-ordination of responses to stress in plants. *Plant Cell Environ* 25:195–210.
Xie, D.X., Feys, B.F., James, S., Nieto-Rostro, M., Turner, J.G. 1998. *COI1*: an *Arabidopsis* gene required for jasmonate-regulated defense and fertility. *Science* 280:1091–1094.
Xie, X.D., Wang, Y.B., Williamson, L., Holroyd, G.H., Tagliavia, C., Murchie, E., Theobald, J., Knight, M.R., Davies, W.J., Leyser, H.M.O., and Hetherington, A.M. 2006. The identification of genes involved in the stomatal response to reduced atmospheric relative humidity. *Curr Biol* 16:882–887.
Xu, H.L., Shida, A., Futatsuya, F., and Kumura, A. 1994. Effects of epibrassinolide and abscisic acid on sorghum plants growing under soil water deficit. 1. Effects on growth and survival. *Japan J Crop Sci* 63:671–675.
Yadav, V., Mallappa, C., Gangappa, S.N., Bhatia, S., and Chattopadhyay, S. 2005. A basic helix-loop-helix transcription factor in *Arabidopsis*, MYC2, acts as a repressor of blue light-mediated photomorphogenic growth. *Plant Cell* 17:1953–1966.

Zeidler, D., Zahringer, U., Gerber, I., Dubery I., Hartung, T., Bors, W., Hutzler, P., and Durner J. 2004. Innate immunity in *Arabidopsis thaliana*: lipopolysaccharides activate nitric oxide synthase (NOS) and induce defense genes. *Proc Natl Acad Sci USA* 101:15811–15816.

Zhang, W., He, S.Y., and Assmann, S.M. 2008. The plant innate immunity response in stomatal guard cells invokes G-protein-dependent ion channel regulation. *Plant J* 56:984–996.

Zhang, X., Miao, Y.C., An, G.Y., Zhou, Y., Shangguan, Z.P., Gao, J.F., and Song, C.P. 2001. K^+ channels inhibited by hydrogen peroxide mediate abscisic acid signaling in *Vicia* guard cells. *Cell Res* 11:195–202.

Zhao, Y., Thilmony, R., Bender, C.L., Schaller, A., He, S.Y., and Howe, G.A. 2003. Virulence systems of *Pseudomonas syringae* pv. *tomato* promote bacterial speck disease in tomato by targeting the jasmonate signaling pathway. *Plant J* 36:485–499.

6 Environmental Sensitivity in Pathogen Resistant *Arabidopsis* Mutants

Wolfgang Moeder and Keiko Yoshioka

Introduction

Plants possess a large number of defense systems that can be deployed to prevent pathogen infection. Plants, unlike mammals, do not have mobile cells that are specifically developed as defender cells such as T cells or B cells, but have developed another type of immune system, which enables each individual cell to respond directly to pathogen attack. Furthermore, they acquired the ability to transmit a systemic signal from infected sites to uninfected leaves, making them more immune to subsequent infection (Ryals et al. 1996; Vlot et al. 2008). Whether plant defense responses are successful appears to be determined largely by the intensity and rapidity of their activation. It has been found that the rapidity and intensity is related to the type of recognition involved. There are largely two branches of the plant immune system: 1) PAMP (pathogen-associated molecular patterns)- or MAMP (microbe-associated molecular patterns)-triggered immunity and 2) effector-triggered immunity (Jones and Dangl 2006). The PAMP/MAMP-triggered immunity is triggered by recognition of molecules common to a wide range of microbes, including nonpathogenic microbes. Because the details of PAMP-triggered immunity are discussed in depth in other chapters in this book, it is not reviewed here. Effector-triggered immunity, however, is a strong and rapid defense response that is triggered by the direct or indirect interaction between the products of a plant *resistance* (*R*) gene and a pathogen *avirulence* (*avr*) gene (Flor 1971; Keen 1990). It has been proposed that effector-triggered immunity has evolved as a result of an "arms race" between plants and microorganisms (Jones and Dangl 2006). Following this effector recognition (gene-for-gene-specific recognition or *R* gene-mediated recognition), the inoculated leaves usually exhibit increased levels of salicylic acid (SA), the induction of several families of defense genes, including those encoding *the pathogenesis-related* (*PR*) genes, and the development of an hypersensitive response (HR) (Hammond-Kosack and Jones 1996). The HR is characterized by apoptosis-like programmed cell death at the site(s) of pathogen entry and is thought to contribute to the restriction of pathogens to the cells/area within and immediately surrounding the necrotic lesion(s). However, because there are several studies showing pathogen resistance without HR formation or HR without pathogen resistance (Hunt et al. 1997; Yu et al. 1998; Cole et al. 2001; Gassmann 2005), the direct role of HR in resistance is still not unequivocal. At later times after infection, elevated SA levels and *PR* gene expression are also detected in the uninoculated portions of the plant, concurrent with development of systemic acquired resistance (SAR), a long-lasting, broad-range resistance to subsequent infection (Ryals et al. 1996; Dempsey et al. 1999). It has been suggested that SA is a critical signaling molecule in the pathways leading to local and systemic resistance; however, it seems unlikely to be the mobile signal triggering SAR (Vernooij et al. 1994). Several candidates for the actual mobile signal that travels through the phloem from the site of infection to establish systemic resistance have emerged in the past two years, including the methylated derivative of SA (methyl salicylate), the defense hormone jasmonic acid, an undefined glycerolipid-derived factor, and a group of peptides that is involved in cell-to-cell basal defense signaling (Park et al. 2007; Truman et al. 2007; Chaturvedi et al. 2008; Ryan et al. 2007; Vlot et al. 2008). The precise transmission mechanism of the systemic signal is expected to be revealed in the near future.

To date, a substantial number of *R* genes has been isolated and characterized (Jones and Dangl 2006; Ellis et al. 2000). The majority of *R* genes cloned at this point possess a nucleotide-binding site (NBS) and a leucine-rich repeat (LRR) region. It is interesting to note that resistance mediated by NBS-LRR type *R* genes is only effective against biotrophic or hemi-biotrophic pathogens, not against necrotrophic pathogens (Glazebrook 2005). This may be related to the activation of HR cell death by this type of *R* genes. The NBS-LRR-containing *R* genes are further classified into two major subclasses on the basis of the structure of their amino terminal domain, one with the Toll/interleukin-1 receptor (TIR) domain, which shows similarity to the cytoplasmic signaling domain of the Drosophila Toll protein, the mammalian interleukin receptor, and a family of mammalian Toll-like receptors (Martin et al. 2003). Another type of NBS-LRR genes has a coiled-coil domain at the N terminus. On the basis of the *Arabidopsis* genome sequence, approximately 125 NBS-LRR type genes exist in *Arabidopsis*, suggesting the importance of this class of *R* genes (Martin et al. 2003). Some other *R* genes possess a serine/threonine protein kinase domain. The *Pto* gene, which confers resistance against bacterial speck disease in tomato, encodes a serine/threonine protein kinase. Interestingly, this *R* gene does not have a LRR region; however, it requires the presence of the corresponding NBS-LRR gene, *Prf*, for resistance activity (Salmeron et al. 1996). The rice *Xa21* gene that confers resistance against bacterial blight disease caused by *Xanthomonas oryzae* also posses a serine/threonine protein kinase domain (Song et al. 1995). These findings strongly suggest that protein-kinase mediated signal transduction plays an important role in resistance responses. *Xa21* further possess an extracellular LRR domain. This type of extracellular LRR domain was also found in the *Cf* genes of tomato conferring resistance against the fungal pathogen *Cladosporium fulvum* that causes leaf mold disease on tomato (Jones et al. 1994; Colwyn et al. 1998). However, the *Cf* genes lack a significant intracellular region that could constitute a signaling component, suggesting that defense response activation is mediated through the interaction with other partners (Rivas and Thomas 2005).

Environmental Effects on *R* Gene-Mediated Pathogen Resistance

Environmental effects on Cf/Avr *conferred resistance against* C. fulvum

The interaction between *C. fulvum* and tomato has been well studied. The availability of near-isogenic tomato lines carrying different *Cf* genes and race specific elicitors from *C. fulvum* carrying the corresponding *Avr* genes made this pathosystem an excellent experimental model to elucidate the *R* gene mediated resistance signal transduction pathway. It has been reported that the recognition of the cognate *Avr* products by *Cf* genes induces an oxidative burst, lipid peroxidation, production of glutathione, an alteration of the cellular redox state, and production of SA and ethylene. Furthermore, in the case of the *Cf-9/Avr9* interaction, rapid opening of stomata was observed (Hammond-Kosack et al. 1996; May et al. 1996). Curiously, these responses and subsequently induced HR cell death were found to be environmentally sensitive. Elevated relative humidity (98%) either suppresses or delays these responses induced by *Cf-9/Avr9* and *Cf-2/Avr2* (Hammond-Kosack et al. 1996; May et al. 1996). This phenomenon was also reported for the *Cf-4/Avr4* interaction (Wang et al. 2005). Furthermore, in the case of the *Cf-9/Avr9* and *Cf-4/Avr4* interactions, relatively high temperature works synergistically with high humidity on HR suppression (Wang et al. 2005). The precise mechanism of this suppression has not been elucidated yet. However, de Jong et al. (2002) have reported that elevated temperature attenuates binding of Avr9 to a high-affinity binding site, which is present in plasma membrane. This high-affinity binding site is thought to be a coreceptor for Avr9. The amount of binding is reduced by 80% at 33°C compared with that observed at 20°C. This could be the direct cause of suppression of the subsequent activation of resistance responses by elevated temperatures, but it does not explain the effect of humidity. Another interesting aspect in *Cf/Avr* conferred resistance is light sensitivity. HR-like cell death formation and SA accumulation in the *Cf-9/Avr9* and *Cf-4/Avr4* interactions was significantly accelerated under dark conditions (Hammond-Kosack et al. 1996).

N *gene–mediated HR and its temperature sensitivity*

The tobacco *N* (necrosis) gene that confers resistance to TMV is one of the first cloned *R genes* from the TIR-NBS-LRR class of *R* genes (Whitham et al. 1994). The *N* gene-TMV interaction has been a classical model system for the study of HR and SAR. Plants carrying the *N* gene mount a rapid HR after infection with TMV, and the virus is thought to be restricted at the region immediately surrounding the HR (Goodman et al. 1986). In contrast, tobacco cultivars that lack the *N* gene do not induce HR and allow TMV to spread systemically throughout the plant and develop mosaic symptoms (susceptible). Interestingly, it has long been known that the HR does not develop in tobacco after TMV infection above 28°C. Under these conditions, plants allow TMV to propagate well and spread systemically even in plants carrying the *N* gene (Samuel 1931). When these systemically infected plants are shifted to a permissive temperature (21°C), they develop systemic HR throughout the plant. This manipulation of HR development by temperature shift has been widely used in various experiments to induce HR cell death in a more synchronized manner (Guo et al. 1998; Hatsugai et al. 2004).

This temperature shift not only affects HR cell death but also SA accumulation and the expression of defense-related genes (Niki et al. 1998; Ohtsubo et al. 1999; Yamakawa et al. 2001). Furthermore, the development of HR cell death induced by temperature shift is reversible. When the plants are shifted back to 30°C, cell death formation is suppressed within 4 hours (Ohtsubo et al. 1999).

These two examples of environmentally sensitive pathogen resistance responses are intriguing. However, the precise molecular mechanism and biological significance of these phenomena has not been explored in detail. Recently, a similar environmental sensitivity in pathogen resistance responses, especially HR induction, has been reported in a number of lesion mimic mutants that exhibit HR-like spontaneous cell death development and also for *R* gene–related transgenic plants.

Lesion Mimic Mutants and Their Environmental Sensitivity

Although the direct contribution of HR in the execution of resistance against pathogens is still elusive, induction of HR is one of the most frequently observed resistance responses in many plant species. To study the process of pathogen resistance responses, mutants that display HR-like spontaneous cell death have been isolated by many researchers (Lorrain et al. 2003; Moeder and Yoshioka 2008). These mutants are called lesion mimic mutants (LMMs). Over the past decade a substantial number of LMMs have been identified mostly in *Arabidopsis*, rice, maize, and barley, and a growing number of the genes have been cloned. Two classes of LMMs have been defined: initiation- and propagation mutants (Lorrain et al. 2003). Initiation mutants show constitutive formation of lesions of variable sizes. On the other hand, the propagation mutants are unable to control cell death once it has started. The overwhelming number of LMMs that were isolated so far belong to the initiation group. They usually develop spontaneous cell death (necrotic, chlorotic, or HR-like) without pathogen infection. Many LMMs display overproduction of SA and constitutive expression of resistance related genes, including *PR* genes. Consequently, these mutants often display enhanced resistance against various pathogens, mainly biotrophic pathogens. Interestingly, some of the initiation type LMMs were reported to be environmentally sensitive. Table 6.1 summarizes *Arabidopsis* LMMs and defense-related mutants that are known to be environmentally sensitive. After the *lsd6* (lesion simulating disease 6) LMM was initially reported to be humidity sensitive (Weymann et al. 1995), a detailed characterization of environmental sensitivity has been reported in two *Arabidopsis* LMMs: *cpr22* (constitutive expresser of *PR* genes 22, Yoshioka et al. 2001) and *cpn1/bon1* (copine1/bonzai1; Jambunathan et al. 2001; Hua et al. 2001).

cpr22 was isolated through a screen for plants displaying constitutive expression of the *PR-1* gene. Heterozygous *cpr22* mutant plants exhibit elevated level of SA, stunted growth with curly leaves, and heightened resistance against various virulent pathogens, such as the bacterial pathogen *Pseudomonas syringae* and the oomycete pathogen, *Hyaloperonospora parasitica*. Additionally, constitutive expression of the defensin gene *PDF 1.2*, which is a marker gene for ethylene and jasmonic acid signal activation, is also observed in *cpr22* (Yoshioka et al. 2001). Recently, the mutation was identified as a 3-kb deletion that fuses two cyclic nucleotide-gated ion channel (CNGC)-encoding genes, *AtCNGC11* and *AtCNGC12*, to generate a novel chimeric gene, *AtCNGC11/12* (Yoshioka et al. 2006). Genetic, molecular, and complementation analyses suggest that AtCNGC11/12, as well as AtCNGC11 and AtCNGC12, form functional cyclic adenosine monophosphate (cAMP)-activated CNGCs and that the phenotype conferred by *cpr22* is attributable to the expression of *AtCNGC11/12*, not the absence of either *AtCNGC11* or *AtCNGC12* (Yoshioka et al. 2006). An interesting aspect of the *cpr22* mutant is that the enhanced accumulation of SA and all the SA-dependent defense responses, as well as morphological phenotypes including homozygous lethality are suppressed by relatively high temperature and high humidity conditions, whereas the SA-independent responses are unaffected (Yoshioka et al. 2001; Mosher et al. 2008).

Therefore, it has been hypothesized that there is an environmentally sensitive factor upstream of SA biosynthesis or accumulation in the *cpr22*-mediated signal transduction pathway (Yoshioka et al. 2001).

Before *cpr22*, two *Arabidopsis* mutants that have a mutation in other members of the CNGC family, which has 20 members in *Arabidopsis* (Mäser et al. 2001), have been reported: *dnd1 (defense, no death 1)* is defective in HR in response to avirulent *Pseudomonas syringae*; however, it maintains *R* gene–mediated restriction of pathogen growth. It also exhibits constitutive activation of various resistance responses, such as increased expression of *PR* genes or enhanced SA accumulation (Yu et al. 1998). Subsequently, it was reveled that *DND1* encodes *AtCNGC2* and that *dnd1*-mediated heightened resistance is attributable to SA accumulation (Clough et al. 2000). With respect to environmental sensitivity, this mutant has a curious story. According to Clough et al. (2000), when uninfected *dnd1* mutant plants were grown in a variety of environments in the United States and United Kingdom for their early work, no substantial leaf flecking, necrotic spotting, or other lesion mimic phenotypes were observed upon evaluation by naked eye or in microscopic studies using autofluorescence or trypan blue staining (Yu et al. 1998). However, at the University of Wisconsin in the United States, leaf lesions in some batches of noninoculated *dnd1* plants started to be observed; thus, this mutant was later categorized as a conditional lesion mimic mutant (Clough et al. 2000). One plausible explanation of this story is the influence of environmental factors that were different at the University of Wisconsin. However, there has been no further report regarding a possible factor that affected lesion development in *dnd1*. Subsequently, another *CNGC* mutant, *hlm1*, has been isolated as a LMM (Balague et al. 2003). *HLM1* encodes *AtCNGC4*, which shares the highest similarity to *AtCNGC2* among the 20 members of the AtCNGC family (Mäser et al. 2001). The *hlm1* mutant, a null mutant due to a T-DNA insertion in *AtCNGC4*, shows various phenotypes of a typical LMM, such as enhanced SA accumulation, heightened pathogen resistance, and constitutive *PR* gene expression (Balague et al. 2003). In contrast, *dnd2* has been isolated also as a null mutant of *AtCNGC4*, but it did not exhibit a lesion mimic phenotype (Jurkowski et al. 2004). Considering that both mutants are null mutants for *AtCNGC4*, the contradicting results regarding lesion mimic phenotypes are intriguing. Whereas Balague et al. (2003) reported the length of the photo-periods, light intensity and humidity conditions did not cause significant alterations in lesion formation in *hlm1*, given the story of *dnd1*, unknown environmental conditions could still contribute the differences between *hlm1* and *dnd2* that have mutations in the same gene.

The *cpn1-1* mutant was identified as a humidity-dependent LMM (Jambunathan et al. 2001). *cpn1-1* plants grown in low humidity (35%–45%) conditions are smaller than wildtype plants, and their leaves were curly, thicker, and irregular in shape with short petioles. Similar to *cpr22*, low-humidity grown *cpn1-1* plants show constitutive expression of *PR* genes and heightened resistance to different types of pathogens. All these *cpn1*-related phenotypes were found to be humidity and temperature sensitive (Jambunathan et al. 2001; Jambunathan and McNellis 2003). High relative humidity (85%–95%) and modestly elevated temperature (24°C) can suppress all these phenotypes. Genetic and molecular analyses revealed that *CPN1* encodes a copine, a calcium-dependent phospholipid binding protein that is found in a wide range of organisms. Although it was initially hypothesized that *CPN1* is involved in developmental regulation, it was later found that the *CPN1* transcript was induced rapidly

Table 6.1 Environmentally sensitive *Arabidopsis* lesion mimic and defense-related mutants

Mutant name	Gene	Function	Mutant phenotypes[a]	Humidity[b]	Temperature[b]	Light and other conditions[c]	References
lsd1*	*LSD1* (At4g20380)	Zinc finger (may have transcription factor activity)	SA, PR, LM SA dependent	H-suppresses**	ND	Shifting SD to LD initiates runaway cell death	Dietrich et al. 1994; Jabs et al. 1996
lsd6	ND	ND	SA, PR, LM SA dependent	H-suppresses	ND	ND	Weymann et al. 1995
dnd1	*AtCNGC2* (At5g15410)	Ion channel	SA, PR SA dependent	L-enhances	ND	ND	Clough et al. 2000 Yoshioka, unpublished
cpn1/bon1	Copine/Bonzai (At5g61900)	Phospholipid binding protein	SA, PR, LM, DF	H-suppresses	H-suppresses	ND	Jambunathan et al. 2001; Hua et al. 2001; Jambunathan et al. 2003
cpr22	*AtCNGC11/12* (At2g46440, At2g46450)	Ion channel	SA, PR, LM, DF SA dependent	H-suppresses L-enhances	H-suppresses L-enhances	SD may enhance	Yoshioka et al. 2001; Yoshioka, unpublished
cpr5	*CPR5/OLD1* (At5g64930)	Type IIIa transmembrane protein	SA, PR, LM, DF SA dependent	H-suppresses	ND	High light suppresses dwarf phenotype	Bowling et al. 1997; Kirik et al. 2001; Mateo et al 2006
cep1	*CPR20 CPR21*	ND	SA, PR, LM, DF	H-suppresses	H-suppresses	ND	Silva et al., 1999; Yoshioka and Shirano, unpublished
ssi1	ND	ND	SA, PR, LM, DF	H-suppresses	H-suppresses	ND	Shah et al. 1999; Greenberg 2000; Yoshioka and Shirano; unpublished

Mutant	Gene	Protein	Phenotype	Effect 1	Effect 2	Light effect	Reference
ssi4	SSI4 (AY179750)	TIR-NBS-LRR type resistance gene	SA, PR, LM, DF	H-suppresses L-enhances	H-suppresses L-enhances	SD may enhance	Shirano et al. 2002; Zhou et al. 2004
nsl1	NSL1 (At1g28380)	ND	SA, PR, LM SA dependent	H- suppresses only growth arrest, not lesion	No effect	ND	Noutoshi et al. 2006
slh1	SLH1 (At5g45260)	TIR-NBS-LRR type resistance gene	SA, PR, LM SA dependent	H-suppresses L-enhances	H-suppresses	ND	Noutoshi et al. 2005
mekk1	MEKK1 (At4g08500)	MAPK kinase kinase	PR, LM, DF	No effect	H-suppresses	ND	Ichimura et al. 2006
mkk1/mkk2 double mutant (not LMM)	MKK1 MKK2 (At4g26070) (At4g29810)	MAPK kinase	SA, PR, DF partially SA dependent	ND	H-suppresses	ND	Qiu et al. 2008
bap1 (not LMM)	BAP1 (At3g61190)	Unknown lipid binding protein (BON1/Copine binding protein)	SA, PR, DF SA dependent	ND	H-suppresses	Light sensitive Constant light suppresses growth arrest	Yang et al. 2006
cpr1 (not LMM)	CPR1	ND	SA, PR, SA dependent	ND	H-suppresses pathogen induced HR	High light suppresses dwarf phenotype	Mateo et al. 2006; Yoshioka and Shirano, unpublished

[a] SA = enhanced accumulation of SA; PR = constitutive expression of *PR* genes; LM = lesion mimic phenotype; DF = dwarf, SA dependent; major phenotypes are SA dependent.
[b] H = high; L = low.
[c] SD = short day; LD = long day.
[*] Propagation type of lesion mimic mutants (see text).
[**] High-humidity suppression of lesion formation has been observed with other *lsd* mutants (Hunt and Weymann, unpublished data).
ND = not determined.

by infection with an avirulent pathogen or exogenous treatment with SA, suggesting an important role of this gene in plant disease resistance responses (Jambunathan and McNellis 2003).

Independently, the *bon1* (*bonzai1*) mutant was isolated as a temperature-sensitive growth defect mutant, which turned out to have a mutation in the same *CPN1* gene (Hua et al. 2001). In the initial study, the biological role of this gene was also suggested, mostly for developmental aspects, particularly maintenance of normal plant size at low temperature. However, later it was discovered to be a negative regulator of an *R* gene, *SNC1* (suppressor of *npr1-1*, constitutive 1), and this negative regulation appears to suppress the activation of defense responses (Yang and Hua 2004). Interestingly, a gain of function mutation in *SNC1* results in constitutive activation of various resistance responses including constitutive expression of *PR* genes (Li et al. 2001). Furthermore, the *snc1* mutant accumulates SA in the absence of pathogen infection; thus, all these phenotypes are extremely similar to LMMs. *SNC1* (also known as *BAL*) is a TIR-NB-LRR type *R* gene residing in the *RPP5* gene cluster that is very divergent among different accessions in *Arabidopsis* (Noël et al. 1999; Stokes et al. 2002; Zhang et al. 2003). It has been reported that *SNC1* is regulated at the transcript level by temperature and SA. The expression of *SNC1* at 28°C is lower than at 22°C, and SA application induces the level of its transcript (Yang and Hua 2004). There is an intriguing similarity between *SNC1* and another *R* gene, *SSI4*, which belongs to the same TIR-NB-LRR type *R* gene group. A gain of function mutation in *SSI4* induces almost identical phenotypes as seen in *snc1*, including heightened pathogen resistance, lesion mimic phenotype, and environmental sensitivity (Shirano et al. 2002; Zhou et al. 2004). Again, high relative humidity and modestly high temperature (28°C) suppress *ssi4* conferred HR-like spontaneous cell death, as well as constitutive resistance responses including spontaneous cell death (Zhou et al. 2004).

Transgenic plants carrying multiple copies of a genomic fragment that contains *RPW8.1* and *RPW8.2* also display characteristics of LMMs. *Arabidopsis RPW8.1* and *RPW 8.2* are two naturally polymorphic *R* genes, which individually control resistance to a broad range of powdery mildew pathogens (Xiao et al. 2001). Transgenic *Arabidopsis* plants that have high expression of *RPW8* genes under the control of their native promoter due to insertion of multiple copies were found to display a lesion mimic phenotype in the absence of pathogen infection. Furthermore, this spontaneous HR-like lesion development was suppressed on agar medium, by low light, and by high temperature and humidity (Xiao et al. 2003).

Recently, Noutoshi and colleagues reported two environmentally sensitive LMMs, *slh1* (*sensitive to low humidity 1*; Noutoshi et al. 2005) and *nsl1* (*necrotic spotted lesions 1*: Noutoshi et al. 2006). *SLH1* encodes an *R* gene–like protein, which consists of a TIR-NBS-LRR domain plus a WRKY transcription factor domain at the C-terminal region. *slh1* also shows SA-dependent lesion mimic phenotypes, including heightened pathogen resistance. It was reported that *slh1*-conferred constitutive pathogen resistance phenotypes are also suppressed by relatively high temperature and humidity.

Additionally, the tissue-specific cell death phenotypes observed in the *mekk1* mutant were also revealed to be environmentally sensitive (Ichimura et al. 2006). It has been reported that mitogen-activated protein kinase (MAPK) cascades are activated by pathogen recognition and play a crucial role to mediate the downstream signaling after pathogen recognition in several plant species including *Arabidopsis* (Pedley and Martin

2005). The activation of MAP kinases is thought to be an early event after pathogen recognition and occurs much before visual development of HR. Interestingly, Ichimura et al. (2006) reported that an insertion mutant of *MEKK1* (MAPK kinase kinase 1), *mekk1*, exhibits a LMM type phenotype, including dwarfism, elevated expression of *PR* genes, tissue-specific cell death, and H_2O_2 accumulation, and that these *mekk1*-related phenotypes are partially SA dependent (Ichimura et al. 2006). Interestingly, these partial SA-dependent phenotypes are suppressed by higher temperature (28°C), but not by high humidity in *mekk1* (Ichimura et al. 2006). Although such tissue specific lesion formation was not observed, a knockout mutant of *MPK4* (MAP kinase 4), *mpk4*, also exhibits constitutive activation of pathogen resistance phenotypes, including SA accumulation, constitutive expression of *PR* genes, and dwarfism, which is extremely similar to *mekk1* (Petersen et al. 2000). Su et al. (2007) have tested the effect of temperature on *mpk4* mutant plants. Strikingly, the *mpk4* dwarf phenotype is also rescued by elevated temperature conditions, indicating that this characteristic is shared by both the *mekk1* and *mpk4* mutants (Su et al. 2007). Because of this phenotypic similarity and other data, MPK4 has been shown to act downstream of MEKK1 in the regulation of defense-response pathways.

Collectively, environmental sensitivities appear to be the norm rather than an exception for various LMMs/transgenic plants with alterations in pathogen resistance responses. These observations strongly suggest that plant resistance responses to pathogens are tightly related to various environmental conditions.

What Are the Common Aspects in the Described Examples?

Assuming that there is an environmentally sensitive factor that affects the plant pathogen resistance response, what could it be? Various models have been proposed (Yoshioka et al. 2001; Xiao et al. 2003; Zhou et al. 2004; Noutoshi et al. 2005). However because they were made to explain each particular case, it is difficult to speculate about a shared mechanism underlying all of these phenotypes associated with a wide range of LMMs. However, because the number of examples is continuously growing, finding common aspects in these already-reported examples may help to generate a hypothesis to test. First, almost all mutants/transgenic plants show similar pathogen resistance phenotypes, such as spontaneous lesion formation, *PR* gene expression, and enhanced pathogen resistance. These phenotypes are either fully or partially SA-dependent in most cases. Although the effect of light conditions is not consistent or not reported for all cases, it is safe to say that high humidity and moderately high temperature can suppress these SA-dependent pleiotropic phenotypes. In some cases, the endogenous SA levels of mutants grown under high temperature or humidity condition were directly measured and found to be suppressed. It is not necessary that the effect of humidity and temperature act through the same mechanism, although the phenomena seem quite similar. As it is still debatable whether lesion formation induces SA accumulation or vice versa (Weymann et al. 1995), it is difficult to conclude which one of these precedes the other. Nevertheless, overwhelming data suggest that suppression of SA accumulation is a key for this phenomenon, and no LMMs maintain their high SA levels when they are kept at high humidity or temperature. However, relatively lower temperature and lower humidity enhanced phenotypes of these environmentally sensitive LMMs

(Xiao et al. 2003; Zhou et al. 2004; Mosher et al. 2008). Therefore, an environmental factor that modulates the LMM phenotypes appears to regulate SA accumulation both positively and negatively.

Furthermore, this environmental effect was also observed in aforementioned *Cf* and *N* gene-mediated pathogen resistance responses in wildtype tomato and tobacco plants, as well as for resistance of *Arabidopsis* to *Pseudomonas syringae* (*AvrRpt2*; W. Urquhart and K. Yoshioka, unpublished data). Taken together, it raises the possibility that a humidity/temperature sensitive factor is present upstream of SA accumulation including the steps in SA biosynthesis, as Yoshioka et al. (2001) hypothesized. Alternatively, it could be somewhere in the SA positive feedback loop (a self-amplifying loop), which has been suggested by many researchers (Draper 1997; Shirasu et al. 1997; Van Camp et al. 1998). The components in this SA positive feedback loop are not completely clear yet; however, it is possible that this feedback system is required to cross a threshold level of SA to induce HR and SA-related phenotypes. Interruption of the SA feedback loop by certain environmental conditions could be a main cause for the suppression of the SA-dependent LMM phenotypes.

Possible Mechanisms for Environmental Sensitivity

Is R-*Gene–Mediated Pathogen Recognition Environmentally Sensitive?*

The starting point of all pathogen resistance responses is recognition of pathogen infection directly or indirectly. Therefore, one can speculate that this pathogen recognition step might be environmentally sensitive. In this scenario, this sensitivity could lead to the observed environmentally sensitive resistance responses. As mentioned earlier, de Jong et al. (2002) have reported that elevated temperature suppressed the binding of Avr9 with a high-affinity binding site at the plasma membrane, which is a putative coreceptor for Avr9. This reduction in binding consequently interrupts the recognition of Avr9 by the cognate resistance protein, Cf9. This suggests that *R* gene-mediated recognition of pathogens can be altered under different temperature conditions by an unknown mechanism.

Alternatively, the amount of *R* gene transcript or protein may be altered by different environmental conditions. The expression of the NBS-LRR-type *R* gene *SNC1* was lower at 28°C than at 22°C (Yang and Hua 2004), indicating that the amount of some *R* genes can be altered at the transcriptional level by environmental effects. Similarly, the expression of the tobacco *N* gene is also reported to be temperature sensitive. Takabatake et al. (2006) showed that the tobacco *N* gene generates two alternative transcripts; N (S) encodes the full length of *N*-gene, whereas N (L) encodes a truncated version. Interestingly, they found that both transcripts accumulate at 20°C but not at 30°C, indicating that the *N* gene transcript level is also regulated by temperature. Furthermore, expression of the *N* gene was significantly induced after shifting the infected plants from 30° to 20°C to trigger cell death formation.

Bieri et al. (2004) studied the steady-state levels of the polymorphic barley MLA R proteins that confer allelic race-specific resistance to the powdery mildew fungus *Blumeria graminis* f. sp. *hordei*. In this study, they proposed the existence of a conserved mechanism to reach threshold levels of NBS-LRR type R proteins that are required to trigger effective resistance responses ("the threshold model"; Bieri et al. 2004). Through this study, they showed that RAR1, which is required for several *R* gene-mediated resistance responses in a broad range of plant species (Shirasu and Schulze-Lefert 2003), controls R protein steady-state levels. Concerning the environmental effect, Bieri et al. (2004) accidentally found that the MLA protein abundance was temperature sensitive. They monitored the levels of MLA1 and MLA6 protein after shifting temperature from 18°C to 37°C and found a gradual reduction of the amount of both proteins. However, contrary to *Arabidopsis SNC1* and the tobacco *N* gene, the reduction in protein levels of MLA1 and MLA6 at higher temperatures did not correlate with changes in their transcript levels, suggesting the alteration of protein abundance occurred on the posttranscriptional or posttranslational level. Additionally, it has been shown that this reduction of protein abundance was not observed in other components of the R protein complex, such as RAR1, SGT1, and HSP90, suggesting that MLA proteins, and possibly some other R proteins, might be temperature sensitive. It would be interesting to know whether this is the case for other *R* gene products, such as SSI4.

This observation leads to further questions regarding the enhancement of resistance responses at lower temperature in some cases. Because the aforementioned threshold model hypothesized that NBS-LRR type R proteins are quantitative regulators, enhancement of resistance responses should be due to increased amounts of R proteins. Does this mean the expression of some *R* gene is higher at lower temperature or the degradation/turnover of R proteins is slowed down at lower temperature? These questions seem straightforward to address. Alternatively, if the physical binding of an R protein and a target ligand is environmentally sensitive, as in the case of Avr9 and Cf9, what kind of effect by lower temperature and humidity can be hypothesized? Does this mean that the R protein/ligand binding will be more stable or that the binding frequency will be higher? It will be interesting to see further progress for this aspect.

Is the SA Positive Feedback Loop Environmentally Sensitive?

Because *cpr22* encodes cyclic nucleotide gated ion channel, the model by Yoshioka et al. (2001) suggested that the environmental-sensitive factor is located downstream of *R* gene activation. However, if this is the case, then how can we explain the sensitivity in *R* gene related cases? In this respect, Xiao et al. (2003) and Zhou et al. (2004) have proposed interesting models in which the environmentally sensitive factor acts in the SA positive feedback loop. As already noted, many researchers have suggested a self-amplifying SA positive feedback loop. If there is a threshold level of SA to induce resistance responses and this positive feedback loop is required for SA to reach to this threshold level, then the factors in this loop play a crucial role to activate pathogen resistance. Regarding the SA feedback loop, Shirano et al. (2002) showed interesting results before Zhou's model (2004). A gain of function LMM, *ssi4*, has a mutation in a NBS-LRR type *R* gene, resulting in a conditional lesion-mimic phenotype with extremely enhanced SA accumulation. In this mutant, they found that the TIR subclass NBS-LRR *R* genes, *RPP1* and *RPS4* but not the CC subclass *R* genes, *RPM1* and *RPS2*, were overexpressed in *ssi4*. Because these overexpressed genes are SA inducible, it suggested that the accumulation of SA in *ssi4* induced the expression of these genes. Additionally, Zhou et al. (2004) observed overexpression of *RPW8* genes in *ssi4* and showed that this expression was suppressed under high-humidity conditions. Taken together, these data suggest that SA accumulation induces increased expression of a certain group of *R* genes, and this may be part of the SA positive feedback loop to amplify resistance response signals. Although the models proposed by Xiao et al. (2003) and Zhou et al. (2004) are different in details, the principle of both models is that the environmental-sensitive factor is located in this SA positive feedback loop. This can explain the environmental sensitivity in various mutants including those cases where *R* genes are affected.

COPINE may be one of the components in this SA positive feedback loop and could be important for the regulation of environmental sensitivity in pathogen resistance responses. The *copine* gene in wildtype plants is strongly induced by exogenous treatment with SA as well as during the *R* gene-mediated resistance response, but not after infection with a nonpathogenic microbe, suggesting that *copine* is a crucial player during the pathogen resistance response and may be involved in the SA feedback loop (Jambunathan and McNellis 2003). Furthermore, the expression of *copine* in wildtype plants is induced by low temperature and low humidity, indicating the transcriptional regulation of this gene by environmental conditions.

With respect to SA synthesis, Strawn et al. (2007) recently reported interesting properties of isochorismate synthase 1 (ICS1), which plays a crucial role for pathogen-induced SA accumulation (Wildermuth et al. 2001). ICS catalyzes the reversible conversion of chorismate to isochorismate, thereby controlling chorismate partitioning to isochorismate-derived products, such as SA. In *Arabidopsis*, null mutations in *AtICS1* suppress the induction of SA accumulation and pathogen defense responses, thus indicating that this gene is responsible for SA biosynthesis after pathogen infection (Wildermuth et al. 2001). Interestingly, a detailed analysis of the enzyme properties of AtICS1 reveled that AtICS1 has a broad temperature range for its activity (Strawn et al. 2007). It shows >90% of maximal activity from 4°C to 37°C and maintains >50% of maximal activity even at 50°C. This means that at least the activity of the key component for SA biosynthesis, ICS1, is not the factor for suppression of SA accumulation in relatively high temperature.

Recently, another biological role for SA in temperature stress responses was suggested (Scott et al. 2004; Larkindale et al. 2005). Scott et al. (2004) reported that *Arabidopsis* Columbia-0 wildtype plants accumulate SA, particularly in its glucosylated form in response to extended growth at chilling temperature (5°C) and that a SA-deficient mutant, *eds5*, and transgenic *NahG* plants displayed a greater growth rate at 5°C. On the contrary, the *cpr1* mutant with elevated levels of SA (Bowling et al. 1994) showed extreme inhibition of growth at this temperature. Because AtICS1 is active at this temperature, Strawn et al. (2007) suggested that SA made via AtICS1 plays a role in cold acclimation or cold-tolerant growth. Collectively, these data suggested that LMMs that have elevated levels of SA compared with wildtype plants accumulate even greater SA levels under cold temperature and consequently exhibit enhanced phenotypes under colder temperature conditions. This scenario is likely to explain the enhancement of LMMs at low temperature. However, again the nature of the factor that triggers the enhanced production of SA under low temperature is still not clear.

What Other Components Can Be Candidates for the Environmentally Sensitive Factor?

Receptor-Like Protein Kinases. A differential screening to find genes that are involved in HR formation in TMV-infected tobacco plants was conducted (Seo et al. 1995). Through this screening, a receptor-like protein kinase, *WRK* (*wound-induced receptor like protein kinase*), was identified (Ito et al. 2002). A high level of transcript accumulation was observed in TMV-infected tobacco plants, and this accumulation was *N* gene dependent. The accumulation was observed before the formation of HR, indicating that *WRK* may be involved in the initiation of HR formation. Interestingly, although the kinase activity itself was not affected, the expression of the *WRK* gene was induced when TMV-inoculated plants were shifted from 30°C to 20°C. This result indicated that the transcriptional regulation of *WRK* is temperature sensitive. Furthermore, Ito et al. (2002) suggested that expression of *WRK* was also altered by changes in osmotic pressure.

WRK is an interesting candidate for the environmentally sensitive factor for defense responses; however, it is suggested to be involved in JA signaling rather than SA signaling. Thus, more work is required to establish the role of WRK in disease resistance.

MAP Kinases. Mitogen-activate protein kinase (MAPK) cascades are used in all eukaryotes including plants to transmit signals induced by various stimuli (Tena et al. 2001). The MAPK cascade module consists of three protein kinases, a MAPK kinase kinase (MAPKKK), a MAPK kinase (MAPKK), and a MAP kinase (MAPK). Upon signal perception, a MAPKKK phosphorylates a MAPKK. This phosphorylation activates MAPKK, which in turn phosphorylates a MAPK. Finally, the terminal kinase phosphorylates various downstream targets, resulting in the activation of cellular responses, including gene expression. Recently it has become apparent that MAPK cascades play a crucial role in a wide range of signaling pathways from development to abiotic/biotic stress responses in plants (Tena et al. 2001; Pedley and Martin 2005). Thus, MAPK cascades can be pivotal nods for signaling pathways upon various stresses.

Recent studies have begun to uncover the role of MEKK1 (MAPK kinase kinase) in both abiotic and biotic stress responses. For example, the MEKK1-MAPKK2-MPK4/MPK6 cascade has been reported to be involved in abiotic stress responses, such as cold and salt signaling (Teige et al. 2004). MPK4 and MPK6 were also reported to be activated transiently by various environmental stresses, such as low temperature, low humidity, and hyperosmolarity (Ichimura et al. 2000). The activation of the cascade seems to be regulated at the phosphorylation level, not at the transcriptional or translational levels. Interestingly, Asai et al. (2002) have proposed the first complete plant MAPK cascade consisting of MEKK1-MKK4/MKK5-MPK3/MPK6 as the downstream signaling cascade of the bacterial flagellin receptor FLS2 for induction of flg22-tiggered immunity using a protoplast transient expression system, suggesting the involvement of MEKK1 in the pathogen resistance response as well. However, it was later suggested that the activation of MEKK1 was not required for flagellin-triggered MPK3/MPK6 activation using *mekk1* knockout mutant plants (Ichimura et al. 2006), thereby challenging the results by Asai et al. (2002). On the basis of the results of Ichimura et al. (2006), MEKK1 is required for flagellin-triggered activation of MPK4, a negative regulator of systemic acquired resistance. This is further supported by the observation that both *mekk1* and *mpk4* null mutants exhibit similar activation of pathogen resistance responses, such as elevated *PR* gene expression and callose deposition (Ichimura et al. 2006; Petersen et al. 2000). Furthermore, as already mentioned, *mekk1* exhibits LMM-like phenotypes, and these are suppressed by high temperature (Ichimura et al. 2006). This is of particular interest because these phenotypes are partially dependent on SID2, which encodes aforementioned AtICS1 (isochorismate synthase), involved in SA biosynthesis upon pathogen infection. Additionally, the double loss of function mutant for MKK1 and MKK2, MAP kinase kinases, which have been implicated in biotic and abiotic stress responses as a part of a signaling cascade including MEKK1 and MPK4, showed marked phenotypes in disease resistance similar to *mekk1* and *mpk4* (Qiu et al. 2008). Interestingly, these phenotypes in the *mkk1/mkk2* double mutant also were suppressed under high-temperature conditions. Furthermore, global transcriptome profiling of the *mkk1/mkk2* double mutant revealed that the signatures of its transcriptional phenotype strongly intersects with those of the *mekk1*, *mpk4*, and, surprisingly *cpr5*, which is an environmentally sensitive LMM (Qiu et al. 2008; Table 6.1). These findings strongly argue the involvement of MEKK1, MKK1, MKK2, and MPK4 for the crosstalk of abiotic and biotic stress responses and also the environmental sensitivity in LMMs.

As mentioned earlier, the *Arabidopsis ssi4* mutant, which contains a gain-of function mutation in a TIR-NBS-LRR type *R* gene, exhibits environmentally sensitive LMM phenotypes (Shirano et al. 2002; Zhou et al. 2004). Zhou et al. (2004) demonstrated that in moderate relative humidity, *ssi4* displays constitutive activation of the MAP kinases AtMPK6 and AtMPK3, but not AtMPK4, and this activation was suppressed under high humidity. Collectively, MAP kinase cascades seem deeply involved in environmental sensitivity in pathogen resistance responses. For further details regarding MAP kinase cascades in abiotic/biotic crosstalk, refer to Chapter 2 in this book.

Reactive Oxygen Species and Related Components. One of the earliest events activated during pathogen resistance responses is the accumulation of hydrogen peroxide (H_2O_2) and other reactive oxygen species (ROS), known as the oxidative burst (OB; Lamb and Dixon 1997). This increase in ROS appears to play several roles in establishing plant resistance. It may directly inhibit pathogen colonization by serving as an antimicrobial agent and by promoting the cross-linking of cell wall glycoproteins (Bradley et al. 1992). In addition, elevated levels of H_2O_2 stimulate multiple defense responses, including *PR* gene expression (Chamnongpol et al. 1998; Neuenschwander et al. 1995), phytoalexin production (Lamb and Dixon 1997), HR development (Levine et al. 1994), and enhanced resistance to pathogen infection (Chamnongpol et al. 1998; Wu et al. 1995). ROS and H_2O_2 are thought to signal many defense responses through a positive feedback loop that also involves SA and cell death (Draper 1997; Shirasu et al. 1997; Van Camp et al. 1998). Supporting this hypothesis, H_2O_2 treatment induces SA accumulation in tobacco (Leon et al. 1995; Neuenschwander et al. 1995) and cell death in soybean cells (Levine et al. 1994; Shirasu et al. 1997), additionally SA potentiates both H_2O_2 accumulation and cell death in pathogen-infected soybean suspension cells (Shirasu et al. 1997).

Consistent with this observation, many LMMs that show pathogen resistance activation were found to accumulate ROS/H_2O_2 (Zhou et al. 2004; K. Yoshioka et al. unpublished data).

It has been reported that the oxidative burst after pathogen infection induces the accumulation of glutathione and lipid peroxidation products (Vera-Estrella et al. 1992; May and Leaver 1993). Interestingly, May et al. (1996) reported that the accumulation of ROS, glutathione, and lipid peroxidation products was suppressed in high humidity (RH 98%) in both *Cf-2/Avr2* and *Cf-9/Avr9* interactions, suggesting a suppression of the oxidative burst under high-humidity conditions.

Studies during the past several years have suggested that a plasma membrane-localized nicotinamide adenine dinucleotide phosphate (NADPH) oxidase is a major source of apoplastic ROS production in plants after pathogen infection (Lamb and Dixon 1997; Grant and Loake 2000; Torres et al. 2002). NADPH oxidase, also known as the respiratory burst oxidase (RBO), was initially described in mammalian neutrophils, where it forms a multicomponent complex. The *Arabidopsis* RBO homologs *AtrbohD* and *AtrbohF* have been reported to be involved in the HR after infection with avirulent bacteria (Torres et al. 2002).

Recently, Király et al. (2008) reported that the suppression of HR in tobacco plants carrying the *N* gene after TMV infection under 30°C condition was rescued by exogenous treatment with ROS-producing agents. Furthermore, they reported a suppression of *NtrbohD* expression at 30°C, implicating that less ROS production at this temperature may lead to the suppression of HR and that this is attributable to the expression level of *NtrbohD*. Considering that ROS were suggested to be involved in the SA positive feedback loop, it is of interest to know how temperature and humidity affect the regulation of ROS production. Indeed, the functional study of rboh begins to provide insight into abiotic stress responses (Miller et al. 2008; see also Chapter 7 in this book). Further studies regarding the role of rboh will shed a more light on the role of ROS in both biotic and biotic stress responses and may reveal its role as a crossroad of abiotic and biotic stress responses.

Ion Homeostasis and Ca^{2+} Signaling. To enhance the photosynthetic capacity of higher plants, Mittler et al. (1995) generated transgenic tobacco plants expressing a gene encoding the bacterio-opsin (bO) protein from *Halobacterium halobium*. This protein functions as a light-driven proton pump that uses a light spectrum different from that used by the photosynthetic apparatus of higher plants. Unexpectedly, the transgenic tobacco plants exhibited spontaneous lesion formation resembling LMMs. Further characterization revealed that the transgenic plants indeed activated various pathogen resistance responses, including enhanced SA accumulation and *PR* gene expression. These phenotypes were attributable to an active proton pumping function of the inserted gene because a mutated version of *bO*, which lost its pumping function, did not induce such phenotypes. Interestingly, they found that these pathogen resistance responses including the lesion mimic phenotype were suppressed at 30°C, just like the HR of TMV-infected tobacco plants carrying the *N* gene. The inhibition of bO-related phenotypes at this temperature was not due to a decrease in *bO* gene expression or protein accumulation, because there was no difference in transcript and protein levels between plants grown at 22°C and 30°C. Furthermore, to test whether this phenomenon was related to the *N* gene, they generated another transgenic tobacco line in the SR1 background that does not possess the *N* gene. Strikingly, these plants exhibit the same temperature sensitive lesions, indicating that the temperature-sensitive component(s) is not the *N* gene itself, nor the pathogen recognition step. These data indicate that the temperature-sensitive factor locates downstream of pathogen recognition. Additionally, they observed the same lesion mimic phenotypes in transgenic *Arabidopsis* expressing *bO*, suggesting that *bO* expression activates a common pathway to trigger cell death in various plant species. Mittler et al. (1995) suggested the ectopic expression of *bO* may result in an artificial increase in proton–pumping activity that may perturb the natural ionic homeostasis of the plant cells. It has been known for some time that triggering of HR by certain pathogens requires the activation of a plasma membrane K^+ efflux/H^+ influx (Atkinson and Baker 1989). Furthermore, Ca^{2+} influx coincides with the onset of K^+ efflux/H^+ influx after pathogen infection, suggesting a connection of the two events (Atkinson et al. 1990). Later, Ca^{2+} influx was reported to be downstream of the oxidative burst (ROS production) in the activation of cell death using soybean culture cells (Levine et al. 1996).

Cytosolic Ca^{2+} changes are apparent during the signal transduction upon various abiotic and biotic stimuli (Sanders et al. 2002). Rapid changes in cytosolic Ca^{2+} were reported as a response to infection with various pathogens and after treatment with phytotoxins (Reddy 2001; White and Broadley 2003). It is also evident that cytosolic Ca^{2+} changes play a crucial role for programmed cell death including HR development (Heath 2000; Lam 2004). Coincidentally, the environmentally sensitive LMMs, *cpr22* and *cpn1/bon1* have mutations in a cyclic nucleotide-gated ion channel that is a Ca^{2+} permeable channel and a calcium-dependent, phospholipids binding protein, respectively (Hua et al. 2001; Jambunathan et al. 2001; Yoshioka et al. 2001; Urquhart et al. 2007). Additionally, a number of studies have linked cyclic nucleotides with Ca^{2+} signaling (Moutinho et al. 2001; Frietsch et al. 2007; Ali et al. 2007). Collectively, it is possible that Ca^{2+} signaling or ion homeostasis play a role for environmental sensitivity for certain pathogen resistance responses.

The involvement of Ca^{2+} signaling in abiotic stress responses such as light, ultraviolet irradiation, touch, low and high temperature, hyperosmotic stress, and oxidative stress are well established (Knight and Knight 2001; Sanders et al. 2002). Some data suggest that the production of ROS is involved in ABA-induced stomatal closure and activation of Ca^{2+} permeable channels play essential roles in this response (Pei et al. 2000; Zhang et al. 2001). Considering that ABA plays an important role for temperature and drought stress, it is of interest to consider the involvement of ABA in environmentally sensitive pathogen resistance responses. Indeed, Yasuda et al. (2008) recently reported antagonistic crosstalk between SA-mediated SAR signaling and ABA-mediated abiotic stress responses using a combination of *Arabidopsis* mutants and SAR activators. Additionally, aforementioned LMM mutant *cpr22* exhibited partial ABA insensitivity (Mosher et al. 2008). Therefore, the connection of Ca^{2+} signaling and ABA could be a key to understand the environmental effects on pathogen resistance responses. Further investigations on this aspect will shed more light on ABA involvement in pathogen resistance responses (see also Chapter 1 in this book).

Concluding Remarks

In this chapter, we have introduced the currently known environmentally sensitive *Arabidopsis* LMMs. A summary of the common elements of these examples and the comparison with similar phenomena observed in wildtype plants clearly argues that this is a general response and not an artifact due to cellular perturbation and therefore is likely to have biological significance.

During their lifetime, plants encounter a large number of microbial attacks. Additionally, as sessile organisms, plants are constantly exposed to numerous abiotic stress conditions, such as heat, cold, and drought stress. To survive under natural conditions, plants must have sophisticated mechanisms to sense and react to various stresses and prioritize their responses in the most effective way to protect themselves.

It has long been known that, under higher temperature or humidity, more frequent disease outbreak is observed for many plant pathogens (Agrios, 2005). This could be due to some growth advantage for microbes at these conditions; however, it is also possible to be a breakdown or suppression of the pathogen resistance activation of plants under certain temperature and humidity conditions. Interestingly, as mentioned already, Bieri et al. (2004) observed a marked and rapid decrease of the protein levels of the *R* genes *MLA1* and *MLA6* upon elevation of temperature. However, they could not test whether this reduction of *R* gene product leads to a loss of resistance against concomitant infection with the pathogen *Blumeria graminis*, because this fungus is not able to colonize barley at temperature above 30°C. This makes us wonder whether, if this pathogen would be able to infect at temperatures greater than 30°C, host plants might have evolved to be able to respond to infections at such temperatures. In other words, they might have evolved to maintain the *R* gene product above 30°C. Recently, the crosstalk between abiotic and biotic stress responses has been an important topic and now we see a growing number of reports using many different pathosystems beyond the typical laboratory systems. Therefore, we anticipate that the biological significance of environmental sensitivity of pathogen resistance responses will be revealed in the near future.

Because of space limitations, several aspects could not be covered in this chapter. Among these, light conditions are the most important. As seen in Table 6.1, light conditions certainly influence the development of LMM phenotypes as well as pathogen resistance responses in general. Moreover, the influence of SA on short-term acclimation to high light stress in *Arabidopsis* has recently been reported (Mateo et al. 2006). However, the effect of light is not as consistent as the effect of humidity and temperature on LMMs. The reasons for this inconsistency are not clear, but one possible explanation could be that the effect of light conditions can be variable depending on other environmental factors or unknown factors in particular genotypes. To address this question, it is necessary to await more examples and studies that address the influence of light conditions.

In contrast, the effect of temperature and humidity seems very consistent across the examples presented here. High temperature and humidity tend to suppress SA-dependent pathogen resistance responses, and the opposite was observed under lower temperature and humidity. Compelling evidence is yet to be shown as to whether the molecular basis of the effect of these two factors is the same. However, as the number of potential components for this environmental sensitivity grows, a possible mechanism will be revealed soon, which will show the similarities and differences between temperature and humidity sensitivity in plant defense responses.

Finally, as described in the other chapters, it is now clear that there is an intense crosstalk between plant hormones. The importance of SA and ABA crosstalk was recently highlighted, connecting abiotic and biotic stress responses (Yasuda et al.

2008). However, the crosstalk of other phytohormones also should be considered to decipher the complex signal transduction network (Robert-Seilaniantz et al. 2007). To understand the complexity of plant defense mechanisms against abiotic and biotic stresses, the future challenge is to uncover the components that play the role of nods in this crosstalk. To this end, a systems biology approach and the wealth of *Arabidopsis* mutants will facilitate to create new hypotheses to be tested in future studies.

Acknowledgments

We thank Dr. Hong-Gu Kang and Mr. Steve Mosher for critical reading and fruitful comments on this manuscript. This work was supported by a Discovery grant of the Natural Science and Engineering Research Council of Canada to K.Y.

References

Agrios, G.N. 2005. *Plant Pathology*, 5th ed. Ames: Academic Press, London.

Ali, R., Ma, W., Lemtiri-Chlieh, F., Tsaltas, D., Leng, Q., von Bodman, S., and Berkowitz, G.A. 2007. Death don't have no mercy and neither does calcium: *Arabidopsis* CYCLIC NUCLEOTIDE GATED CHANNEL2 and innate immunity. *Plant Cell* 19;1081–1095.

Asai, T., Tena, G., Plotnikova, J., Willmann, M.R., Chiu, W.-L. Gomez-Gomez, L., Boller, T., Ausubel, F.M., and Sheen, J. 2002. MAP kinase signalling cascade in *Arabidopsis* innate immunity. *Nature* 415;977–983.

Atkinson, M.M., and Baker, C.J. 1989. Role of the plasmalemma H^+-ATPase in *Pseudomonas syringae*-induced K^+/H^+ exchange in suspension-cultured tobacco cells. *Plant Physiol* 91:298–303.

Atkinson, M.M., Keppler, L.D., Orlandi, E.W., Baker, C.J., and Mischke, C.F. 1990. Involvement of plasma membrane calcium influx in bacterial induction of the K^+/H^+ and hypersensitive responses in tobacco. *Plant Physiol.* 92:215–221.

Balague, C., Lin B, Alcon, C., Flottes, G., Malmstrom, S., Köhler, C., Neuhaus, G., Pelletier, G., Gaymard, F., and Roby, D. 2003. HLM1, an essential signaling component in the hypersensitive response, is a member of the cyclic nucleotide-gated channel ion channel family. *Plant Cell* 15:365–379.

Bieri, S., Mauch, S., Shen, Q.-H., Peart, J., Devoto, A., Casais, C., Ceron, F., Schulze, S., Steinbi, H.-H., Shirasu, K., and Schulze-Lefert, P. 2004. RAR1 positively controls steady state levels of barley MLA resistance proteins and enables sufficient MLA6 accumulation for effective resistance. *Plant Cell* 16;3480–3495.

Bowling, S.A., Guo, A., Cao, H., Gordon, A.S., Klessig, D.F., and Dong, X. 1994. A mutation in *Arabidopsis* that leads to constitutive expression of systemic acquired resistance. *Plant Cell* 6:1845–1857.

Bradley, D.J., Kjellbom, P., and Lamb, C.J. 1992. Elicitor- and wound-induced oxidative cross-linking of a proline-rich plant cell wall protein: a novel, rapid defense response. *Cell* 70:21–30.

Chamnongpol, S., Willekens, H., Moeder, W., Langebartels, C., Sanderman, H.J., Van Montagu, M., Inze, D., and Van Camp, W. 1998. Defense activation and enhanced pathogen tolerance induced by H2O2 in transgenic tobacco. *Proc Natl Acad Sci USA* 95:5818–5823.

Chaturvedi, R., Krothapalli, K., Makandar, R., Nandi, A., Sparks, A.A., Roth, M.R., Welti, R., and Shah, J. 2008. Plastid ω3-fatty acid desaturase-dependent accumulation of a systemic acquired resistance inducing activity in petiole exudates of *Arabidopsis thaliana* is independent of jasmonic acid. *Plant J* 54:106–117.

Clough, S.J., Fengler, K.A., Yu, I.C., Lippok, B., Smith, R.K. Jr., Bent, A.F. 2000. The *Arabidopsis dnd1* "defense, no death" gene encodes a mutated cyclic nculeotide-gated ion channel. *Proc Natl Acad Sci USA* 97:9323–9328.

Cole, A.B., Király, L., Ross, K., and Schoelz, J.E. 2001. Uncoupling resistance from cell death in the hypersensitive response of *Nicotiana* species to cauliflower mosaic virus infection. *Mol Plant Microbe Interact* 14:31–41.

Colwyn, M.T., Dixon, M.S., Parniske, M., Golstein, C., and Jones, J.D.G. 1998. Genetic and molecular analysis of tomato Cf genes for resistance to *Cladosporium fulvum*. *Philos Trans R Soc Lond B Biol Sci* 353:1413–1424.

Dempsey, D., Shah, J., and Klessig, D.F. (1999). Salicylic acid and disease resistance in plants. *Crit Rev Plant Sci* 18:547–575.

Dietrich, R.A., Delaney, T.P., Uknes, S.J., Ward, E.R., Ryals, J.A., and Dangl, J.L. 1994. *Arabidopsis* mutants simulating disease resistance response. *Cell* 77:565–567.

Dietrich, R.A., Richberg, M.H., Schmidt, R., Dean, C., and Dangl, J.L. 1997. A novel zinc finger protein is encoded by the *Arabidopsis* LSD1 gene and functions as a negative regulator of plant cell death. *Cell* 88:685–694.

de Jong, C.-F., Takken, F.L.W., Cai, X., de Wit, P.J.G.M., and Joosten, M.H.A.J. 2002. Attenuation of Cf-mediated defense responses at elevated temperatures correlates with a decrease in elicitor-binding sites. *Mol Plant Microbe Interact* 15;1040–1049.

Draper, J. 1997. Salicylate, superoxide synthesis and cell suicide in plant defense. *Trends Plant Sci* 2:162–165.

Ellis, J., Dodds, P., and Pryor, T. 2000. Structure, function and evolution of plant disease resistance genes. *Curr Opin Plant Biol* 3:278–284.

Flor, H. 1971. Current status of gene-for-gene concept. *Annu Rev Phytopathol* 9:275–296.

Frietsch, S., Wang, Y.-F., Sladek, C., Poulsen, L.R., Romanowsky, S.M., Schroeder, J.I., and Harper, J.F. 2007. A cyclic nucleotide-gated channel is essential for polarized tip growth of pollen. *Proc Natl Acad Sci USA* 104;14531–14536.

Gassmann, W. 2005. Natural variation in the *Arabidopsis* response to the avirulence gene *hoppsya* uncouples the hypersensitive response from disease resistance. *Mol Plant Microbe Interact* 18:1054–1060.

Glazebrook, J. 2005. Contrasting mechanisms of defense against biotrophic and necrotrophic pathogens. *Annu Rev Phytopathol* 43:205–227.

Goodman, R.N., Kiraly, Z., and Wook, K.R. 1986. *The Biochemistry and Physiology of Plant Disease*. Columbia: University of Missouri Press.

Grant, J.J., and Loake, G.J. 2000. Role of reactive oxygen intermediates and cognate redox signaling in disease resistance. *Plant Physiol* 124;21–30.

Greenberg, J.T. 2000. Positive and negative regulation of salicylic acid-dependent cell death and pathogen resistance in *Arabidopsis lsd6* and *ssi1* mutants. *Mol Plant-Microbe Interact* 13;877–881.

Guo, A., Durner, J., and Klessig, D.F. 1998. Characterization of a tobacco epoxide hydrolase gene induced during the resistance response to TMV. *Plant J* 15:647–656.

Hammond-Kosack, K.E., and Jones, J.D.G. 1996. Resistance gene-dependent plant defense responses. *Plant Cell* 8:1773–1791.

Hammond-Kosack, K.E., Silverman, P., Raskin, I., and Jones, J.D.G. 1996. Race-specific elicitors of *Cladosporium fulvum* induce changes in cell morphology and the synthesis of ethylene and salicylic acid in tomato plants carrying the corresponding Cf disease resistance gene. *Plant Physiol* 110;1381–1394.

Hatsugai, N., Kuroyanagi, M., Yamada, K., Meshi, T., Tsuda, S., Kondo, M., Nishimura, M., and Hara-Nishimura, I. 2004. A plant vacuolar protease, VPE, mediates virus-induced hypersensitive cell death. *Science* 305:855–858.

Heath, M.C. 2000. Hypersensitive response-related death. *Plant Mol Biol* 44:321–334.

Hua, J., Grisafi, P., Cheng, S.-H., and Fink, G.R. 2001. Plant growth homeostasis is controlled by the *Arabidopsis BON1* and *BAP1* genes. *Genes Dev* 15;2263–2272.

Hunt, M.D., Delaney, T.P., Dietrich, R.A., Weymann, K.B., Dangl, J.L., and Ryals, J.A. 1997. Salicylate-independent lesion formation in *Arabidopsis* lsd mutants. *Mol Plant Microbe Interact* 10:531–536.

Ichimura, K., Casais, C., Peck, C.C., Kazuo Shinozaki, K., and Shirasu, K. 2006. MEKK1 Is required for MPK4 activation and regulates tissue-specific and temperature-dependent cell death in *Arabidopsis*. *J Biol Chem* 281:36969–36976.

Ito, N., Takabatake, R., Seo, S., Hiraga, S., Mitsuhara, I., and Ohashi, Y. 2002. Induced expression of a temperature-sensitive leucine-rich repeat receptor-like protein kinase gene by hypersensitive cell death and wounding in tobacco plant carrying the n resistance gene. *Plant Cell Physiol* 43:266–274.

Jabs, T., Dietrich, R.A., and Dangl, J.L. 1996. Initiation of runaway cell death in an *Arabidopsis* mutant by extracellular superoxide. *Science* 273:1853–1856.

Jambunathan, N., and McNellis, T.W. 2003. Regulation of *Arabidopsis* COPINE 1 gene expression in response to pathogens and abiotic stimuli. *Plant Physiol* 132:1370–1381.

Jambunathan, N., Siani, J.M., and McNellis, T.W. 2001. A humidity-sensitive *Arabidopsis* copine mutant exhibits precocious cell death and increased disease resistance. *Plant Cell* 13:2225–2240.

Jones, D.A., Thomas, C.M., Hammond-Kosack, K.E., Balint-Kurti, P.J., and Jones J.D. 1994. Isolation of the tomato Cf-9 gene for resistance to *Cladosporium fulvum* by transposon tagging. *Science* 266:789–793.

Jones, J.D., and Dangl, J.L. 2006. The plant immune system. *Nature* 444:323–329.

Jurkowski, G.I., Smith, R.K., Jr., Yu, I.-C., Ham, J.H., Sharma, S.B., Klessig, D.F., Fengler, K.A., and Bent, A.F. 2004. *Arabidopsis DND2*, a second cyclic nucleotide-gated ion channel gene for which mutation causes the "*Defense, No Death*" phenotype. *Mol Plant-Microbe Interact* 17:511–520.

Keen, N.T. 1990. Gene-for-gene complementarity in plant-pathogen interactions. *Annu Rev Genet* 24:447–463.

Király, L., Hafez, Y.M., Fodor, J., and Király, Z. 2008. Suppression of tobacco mosaic virus-induced hypersensitive-type necrotization in tobacco at high temperature is associated with downregulation of NADPH oxidase and superoxide and stimulation of dehydroascorbate reductase. *J Gen Virol* 89:799–808.

Kirik, V., Bouyer, D., Schöbinger, U., Bechtold, N., Herzog, M., Bonneville, J.-M., and Hülskamp, M. 2001. *CPR5* is involved in cell proliferation and cell death control and encodes a novel transmembrane protein. *Current Biol* 23:1891–1895.

Knight, H., and Knight, M.R. 2001. biotic stress signaling pathways: specificity and cross-talk. *Trends Plant Sci* 6:262–267.

Lam, E. 2004. Controlled cell death, plant survival and development. *Nature Rev Mol Cell Biol* 5:305–315.

Lamb, C., and Dixon, R.A. 1997. The oxidative burst in plant disease resistance. *Annu Rev Plant Physiol Plant Mol Biol* 48:251–275.

Larkindale, J., Hall, J.D., Knight, M.R., and Vierling, E. 2005. Heat stress phenotypes of *Arabidopsis* mutants implicate multiple signaling pathways in the acquisition of thermotolerance. *Plant Physiol* 138:882–897.

Leon, J., Lawton, M.A., and Raskin, I. 1995. Hydrogen peroxide stimulates salicylic acid biosynthesis in tobacco. *Plant Physiol* 108:1673–1678.

Levine, A., Tenhaken, R., Dixon, R., and Lamb, C. 1994. H_2O_2 from the oxidative burst orchestrates the plant hypersensitive disease resistance response. *Cell* 79:585–593.

Levine, A., Pennell, R.I., Alvarez, M.E., Palmer, R., and Lamb, C. 1996. Calcium-mediated apoptosis in a plant hypersensitive disease resistance response. *Curr Biol* 6:427–437.

Li, X., Clarke, J.D., Zhang, Y., and Dong, X. 2001. Activation of an EDS1-mediated *R*-gene pathway in the *snc1* mutant leads to constitutive, NPR1-independent pathogen resistance. *Mol Plant Microbe Interact* 14;1131–1139.

Lorrain, S., Vailleau, F., Balague, C., and Roby, D. 2003. Lesion mimic mutants: keys for deciphering cell death and defense pathways in plants? *Trends Plant Sci* 8:263–271.

Martin, G.B., Bogdanove, A.J., and Sessa, G. 2003. Understanding the functions of plant disease resistance proteins. *Annu Rev Plant Biol* 54:23–61.

Mäser, P., Thomine, S., Schroeder, J.I., Ward, J.M., Hirschi, K., Sze, H., Talke, I.N., Amtmann, A., Maathuis, F.J., Sanders, D., Harper, J.F., Tchieu, J., Gribskov, M., Persans, M.W., Salt, D.E., Kim, S.A., and Guerinot, M.L. 2001. Phylogenetic relationships within cation transporter families of *Arabidopsis*. *Plant Physiol* 126:1646–1667.

Mateo, A., Funck, D., Mühlenbock, P., Kular, B., Mullineaux, P.M., and Karpinski, S. 2006. Controlled levels of salicylic acid are required for optimal photosynthesis and redox homeostasis. *J Exp Bot* 57:1795–1807.

May, M.J., and Leaver, C.J. 1993. Oxidative stimulation of glutathione synthesis in *Arabidopsis thaliana* suspension cultures. *Plant Physiol* 103:621–627.

May, M.J., Hammond-Kosack, K.E., and Jones, J.D.G. 1996. Involvement of reactive oxygen species, glutathione metabolism, and lipid peroxidation in the Cf-gene-dependent defense response of tomato cotyledons induced by race-specific elicitors of *Cladosporium fulvum*. *Plant Physiol* 110:1367–1379.

Miller, G., Shulaev, V., and Mittler, R. 2008. Reactive oxygen signaling and abiotic stress. *Physiol Plant* 133:481–489.

Mittler, R., Shulaev, V., and Lam, E. 1995. Coordinated activation of programmed cell death and defense mechanisms in transgenic tobacco plants expressing a bacterial proton pump. *Plant Cell* 7:29–42.

Moeder, W., and Yoshioka, K. 2008. Lesion mimic mutants. *Plant Signal Behav* 3:764–767.

Mosher, S., Moeder, W., and Yoshioka, K. 2008. Investigation of abiotic stress responses in the pathogen resistance mutant *cpr22*. Presented at the *19th International Conference on* Arabidopsis *Research*, Montreal, Canada, July 23–27.

Moutinho, A., Hussey, P.J., Trewavas, A.J., and Malhó, R. 2001. cAMP acts as a second messenger in pollen tube growth and reorientation. *Proc Natl Acad Sci USA* 98:10481–10486.

Niki, T., Mitsuhara, I., Seo, S., Ohtsubo, N., and Ohashi, Y. 1998. Antagonistic effect of salicylic acid and jasmonic acid on the expression of pathogenesis-related (PR) protein genes in wounded mature tobacco leaves. *Plant Cell Physiol* 39:500–507.

Noël, L., Moores, T.L., Van Der Biezen, E.A., Parniske, M., Daniels, M.J., Parker, J.E., and Jones, J.D. 1999. Pronounced intraspecific haplotype divergence at the RPP5 complex disease resistance locus of *Arabidopsis*. *Plant Cell* 11:2099–2112.

Neuenschwander, U., Vernooij, B., Friedrich, L., Uknes, S., Kessmann, H., and Ryals, J. 1995. Is hydrogen peroxide a second messenger of salicylic acid in systemic acquired resistance? *Plant J* 8:227–233.

Noutoshi, Y., Ito, T., Seki1, M., Nakashita, H., Yoshida, S., Marco, Y., Shirasu, K., and Shinozaki, K. 2005. A single amino acid insertion in the WRKY domain of the Arabidopsis TIR–NBS–LRR–WRKY-type disease resistance protein SLH1 (sensitive to low humidity 1) causes activation of defense responses and hypersensitive cell death. *Plant J* 43:873–888.

Noutoshi, Y., Kuromori, T., Wada, T., Hirayama, T., Kamiya, A., Imura, Y., Yasuda, M., Nakashita, H., Shirasu, K., Shinozaki, K. 2006. Loss of NECROTIC SPOTTED LESIONS 1 associates with cell death and defense responses in Arabidopsis thaliana. *Plant Mol Biol* 62:29–42.

Ohtsubo, N., Mitsuhara, I., Koga, M., Seo, S., and Ohashi, Y. 1999. Ethylene promotes the necrotic lesion formation and basic PR gene expression in TMV-infected tobacco. *Plant Cell Physiol* 40:808–817.

Park, S.-W., Kaimoyo, E., Kumar, D., Mosher, S., and Klessig, D.F. 2007. Methyl salicylate is a critical mobile signal for plant systemic acquired resistance. *Science* 318:113–116.

Pedley, K.F., and Martin, G.B. 2005. Role of mitogen-activated protein kinases in plant immunity. *Curr Opin Plant Biol* 8:541–547.

Pei, Z.M., Murata, Y., Benning, G., Thomine, S., Klüsener, B., Allen, G.J., Grill, E., and Schroeder, J.I. 2000. Calcium channels activated by hydrogen peroxide mediate abscisic acid signalling in guard cells. *Nature* 406:731–734.

Petersen, M., Brodersen, P., Naested, H., Andreasson, E., Lindhart, U., Johansen, B., Nielsen, H.K., Lacy, M., Austin, M.J., Parker, J.E., Sharma, S.B., Klessig, D.F., Martienssen, R., Mattsson, O., Jensen, A.B., and Mundy, J. 2000. *Arabidopsis* MAP kinase 4 negatively regulates systemic acquired resistance. *Cell* 103:1111–1120.

Qiu, J.-L., Zhou, L., Yun, B.-W., Nielsen, H.B., Fiil, B.K., Petersen, K., MacKinlay, J., Loake, G.J., Mundy, J., and Morris, P.C. 2008. *Arabidopsis* mitogen–activated protein kinase kinases MKK1 and MKK2 have overlapping functions in defense signaling mediated by MEKK1, MPK4, and MKS1. *Plant Physiol* 148:212–222.

Reddy, A.S.N. 2001. Calcium: silver bullet in signalling. *Plant Sci* 160:381–404.

Rivas, S., and Thomas, C.M. 2005. Molecular interactions between tomato and the leaf mold pathogen *Cladosporium fulvum*. *Ann Rev Phytopathol* 43:395–436.

Robert-Seilaniantz, A., Navarro, L., Bari, R., and Jones, J.D.G. 2007. Pathological hormone imbalances. *Curr Opin Plant Biol* 10:372–379.

Ryals, J.A., Neuenschwander, U.H., Willits, M.G., Molina, A., Steiner, H.-Y., and Hunt, M. D. 1996. Systemic acquired resistance. *Plant Cell* 8:1809–1819.

Ryan, C.A., Huffaker, A., and Yamaguchi, Y. 2007. New insights into innate immunity in Arabidopsis. *Cell Microbiol* 9:1902–1908.

Salmeron, J.M., Oldroyd, G.E.D., Rommens, C.M.T., Scofield, S.R., Kim, H.-S., Lavelle, D.T., Dahlbeck, D., and Staskawicz, B.J. 1996. Tomato Prf is a member of the leucine-rich repeat class of plant disease resistance genes and lies embedded within the Pto kinase gene cluster. *Cell* 86:123–133.

Samuel, G. 1931. Some experiments on inoculating methods with plant virus and on local lesions. *Ann Appl Biol* 18:494–507.

Sanders, D., Pelloux, J., Brownlee, C., and Harper, J.F. 2002. Calcium at the crossroads of signaling. *Plant Cell* 14 (Suppl.): s401–s417.

Scott, I.M., Clarke, S.M., Wood, J.E., and Mur, L.A.J. 2004. Salicylate accumulation inhibits growth at chilling temperature in *Arabidopsis*. *Plant Physiol* 135:1040–1049.

Seo, S., Okamoto, M., Seto, H., Ishizuka, K., Sano, H., and Ohashi, Y. 1995. Tobacco MAP kinase: a possible mediator in wound signal transduction pathways. *Science* 270:1988–1992.

Shah, J., Kachroo, P., and Klessig, D.F. 1999. The *Arabidopsis* ssi1 mutation restores pathogenesis-related gene expression in npr1 plants and renders defensin gene expression salicylic acid dependent. *Plant Cell* 11;191–206.

Shirano, Y., Kachroo, P., Shah, J., and Klessig, D.F. 2002. A gain-of-function mutation in an Arabidopsis Toll Interleukin1 receptor-nucleotide binding site-leucine-rich repeat type R gene triggers defense responses and results in enhanced disease resistance. *Plant Cell* 14:3149–3162.

Shirasu, K., Nakajima, H., and Krishnamachari, R.V. 1997. Salicylic acid potentiates an agonist-dependent gain control that amplifies signals in the activation of defense mechanisms. *Plant Cell* 9:261–270.

Shirasu, K., and Schulze-Lefert, P. 2003. Complex formation, promiscuity and multi-functionality: protein interactions in disease-resistance pathways. *Trends Plant Sci* 8:252–258.

Silva, H., Yoshioka, K., Dooner, H.K., and Klessig, D.F. 1999. Characterization of a new *Arabidopsis* mutant exhibiting enhanced disease resistance. *Mol Plant Microbe Interact* 12:1053–1063.

Song, W.Y., Wang, G.L., Chen, L.L., Kim, H.S., Pi, L.Y., Holsten, T., Gardner, J., Wang, B., Zhai, W.X., Zhu, L.H., Fauquet, C., and Ronald, P. 1995. A receptor kinase-like protein encoded by the rice disease resistance gene, Xa21. *Science* 15:1804–1806.

Stokes, T.L., Kunkel, B.N., and Richards, E.J. 2002. Epigenetic variation in *Arabidopsis* disease resistance. *Genes Dev* 16;171–182.

Strawn, M.A., Marr, S.K., Inoue, K., Inada, N., Zubieta, C., and Wildermuth, M.C. 2007. *Arabidopsis* isochorismate synthase functional in pathogen-induced salicylate biosynthesis exhibits properties consistent with a role in diverse stress responses. *J Biol Chem* 282:5919–5933.

Su, S.-H., Suarez-Rodriguez, M.C., and Krysan, P. 2007. Genetic interaction and phenotypic analysis of the *Arabidopsis* MAP kinase pathway mutations mekk1 and mpk4 suggests signaling pathway complexity. *FEBS Lett* 581:3171–3177.

Takabatake, R., Seo, S., Mitsuhara, I., Tsuda, S., and Ohashi, Y. 2006. Accumulation of the two transcripts of the N gene, conferring resistance to tobacco mosaic virus, is probably important for N gene-dependent hypersensitive cell death. *Plant Cell Physiol.* 47:254–261.

Teige, M., Scheikl, E., Eulgem, T., Dóczi, R., Ichimura, K., Shinozaki, K., Dangl, J.L., and Hirt, H. 2004. The MKK2 pathway mediates cold and salt stress signaling in *Arabidopsis*. *Mol Cell* 15:141–152.

Tena, G., Asai, T., Chiu, W.L., and Sheen, J. 2001. Plant mitogen-activated protein kinase signaling cascades. *Curr Opin Plant Biol* 4:392–400.

Torres, M.A., Dangl, J.L., and Jones, J.D.G. 2002. *Arabidopsis* gp91phox homologues AtrbohD and AtrbohF are required for accumulation of reactive oxygen intermediates in the plant defense response. *Proc Natl Acad Sci USA* 99:517–522.

Truman, W., Bennett, M.H., Kubigsteltig, I., Turnbull, C., and Grant, M. 2007. *Arabidopsis* systemic immunity uses conserved defense signaling pathways and is mediated by jasmonates. *Proc Natl Acad Sci USA* 16;1075–1080.

Urquhart, W., Gunawardena, A.H.L.A.N., Moeder, W., Ali, R., Berkowitz, G.A., and Yoshioka, K. 2007. The chimeric cyclic nucleotide-gated ion channel ATCNGC11/12 constitutively induces programmed cell death in a Ca^{2+} dependent manner. *Plant Mol Biol* 65:747–761.

Van Camp, W., Van Montagu, M., and Inze, D. 1998. H_2O_2 and NO: redox signals in disease resistance. *Trends Plant Sci.* 3:330–334.

Vera-Estrella, R., Higgins, V.J., and Blumwald, E. 1994. Plant defense response to fungal pathogens. II. G-protein mediated changes in host plasma membrane redox reactions. *Plant Physiol* 106:97–102.

Vernooij, B., Friedrich, L., Morse, A., Reist, R., Kolditz-Jawhar, R., Ward, E., Uknes, S., Kessmann, H., and Ryals, J. 1994. Salicylic acid is not the translocated signal responsible for inducing systemic acquired resistance but is required in signal transduction. *Plant Cell* 6:959–965.

Vlot, A.C., Klessig, D.F., and Park, S.W. 2008. Systemic acquired resistance: the elusive signal(s). *Curr Opin Plant Biol* 11:436–442.

Wang, C., Cai, X., and Zheng, Z. 2005. High humidity represses Cf-4/Avr4- and Cf-9/Avr9-dependent hypersensitive cell death and defense gene expression. *Planta* 222:947–956.

Weymann, K., Hunt, M., Uknes, S., Neuenschwander, U., Lawton, K., Steiner, H.-Y., and Ryals, J. 1995. Suppression and restoration of lesion formation in *Arabidopsis* lsd mutants. *Plant Cell* 7:2013–2022.

White, P.J., and Broadley, M.R. 2003. Calcium in plants. *Ann Bot* 92:487–511.

Whitham, S., Dinesh-Kumar, P., Choi, D., Hehl, R., Corr, C., and Baker, B. 1994. The product of the tobacco mosaic virus resistance gene Al: similarity to toll and the interleukin-1 receptor. *Cell* 79:1101–1115.

Wildermuth, M.C., Dewdney, J., Wu, G., and Ausubel, F.M. 2001. Isochorismate synthase is required to synthesize salicylic acid for plant defence. *Nature* 414:562–565.

Wu, G., Shortt, B.J., Lawrence, E.B., Levine, E.B., Fitzsimmons, K.C., and Shah, D.M. 1995. Disease resistance conferred by expression of a gene encoding H2O2-generating glucose oxidase in transgenic potato plants. *Plant Cell* 7:1357–1368.

Xiao, S., Brown, S., Patrick, E., Brearley, C., and Turner, J.G. 2003. Enhanced transcription of the *Arabidopsis* disease resistance genes *RPW81* and *RPW82* via a salicylic acid–dependent amplification circuit is required for hypersensitive cell death. *Plant Cell* 15:33–45.

Xiao, S., Ellwood, S., Calis, O., Patrick, E., Li, T., Coleman, M., and Turner, J.G. 2001. Broad-spectrum mildew resistance in *Arabidopsis thaliana* mediated by *RPW8*. *Science* 291:118–120.

Yamakawa, H., Mitsuhara, I., Ito, N., Seo, S., Kamada, H., and Ohashi, Y. 2001. Transcriptionally and post-transcriptionally regulated response of 13 calmodulin genes to tobacco mosaic virus-induced cell death and wounding in tobacco plant. *Eur J Biochem* 268;3916–3929.

Yang, H., Li, Y., and Hua, J. 2006. The C2 domain protein BAP1 negatively regulates defense responses in *Arabidopsis*. *Plant J* 48:238–248.

Yang, S., and Hua, J. 2004. A haplotype-specific resistance gene regulated by BONZAI1 mediates temperature-dependent growth control in *Arabidopsis*. *Plant Cell.* 16:1060–1071.

Yasuda, M., Ishikawa, A., Jikumaru, Y., Seki, M., Umezawa, M., Asami, T., Maruyama-Nakashita, A., Kudo, T., Shinozaki, K., Yoshida, S., and Nakashita, H. 2008. Antagonistic interaction between systemic acquired resistance and the abscisic acid–mediated abiotic stress response in *Arabidopsis*. *Plant Cell* 20:1678–1692.

Yoshioka, K., Kachroo, P., Tsui, F., Sharma, S.B., Shah, J., and Klessig, D.F. 2001. Environmentally-sensitive, SA-dependent defense response in the *cpr22* mutant of *Arabidopsis*. *Plant J* 26:447–459.

Yoshioka, K., Moeder, W., Kang, H.G., Kachroo, P., Masmoudi, K., Berkowitz, G., and Klessig, D.F. 2006. The chimeric Arabidopsis CYCLIC NUCLEOTIDE-GATED ION CHANNEL11/12 activates multiple pathogen resistance responses. *Plant Cell* 18:747–763.

Yu, I.C., Parker, J., and Bent, A.F. 1998. Gene-for-gene disease resistance without the hypersensitive response in *Arabidopsis dnd1* mutant. *Proc Natl Acad Sci USA* 95:7819–7824.

Zhang, X., Zhang, L., Dong, F.C., Gao, J.F., Galbraith, D.W., and Song, C.P. 2001. Hydrogen peroxide is involved in abscisic acid-induced stomatal closure in *Vicia faba*. *Plant Physiol* 126:1438–1448.

Zhang, Y., Goritschnig, S., Dong, X., and Li, X. 2003. A gain-of-function mutation in a plant disease resistance gene leads to constitutive activation of downstream signal transduction pathways in *suppressor of npr1-1, constitutive 1*. *Plant Cell* 15:2636–2646.

Zhou, F., Menke, F.L., Yoshioka, K., Moeder, W., Shirano, Y., and Klessig, D.F. 2004. High humidity suppresses ssi4-mediated cell death and disease resistance upstream of MAP kinase activation, H_2O_2 production and defense gene expression. *Plant J* 39:920–932.

7 Reactive Oxygen Species, Nitric Oxide and Signal Crosstalk

Steven J. Neill, John T. Hancock, and Ian D. Wilson

Introduction

Reactive Oxygen Species and Nitric Oxide

Originally thought of as toxic byproducts of metabolism or noxious air pollutants, reactive oxygen species (ROS) and nitric oxide (NO; Figure 7.1) are now firmly placed within the canon of signaling molecules that are made and recognized by plant cells. The term ROS is used to encompass a group of relatively reactive compounds derived from the reduction of molecular oxygen (Halliwell 2006). Similarly, reactive nitrogen species (RNS) are a group of reactive compounds, the most studied of which is NO. Usually NO is thought of as being the free radical, NO$^{\bullet}$. However, it can either gain or lose an electron to form the nitroxyl anion (NO$^-$) and nitrosonium cation (NO$^+$), respectively (Figure 7.1) and it must be recognized that these compounds have differences in their reactivity. There are also other related reactive molecules that can all be termed 'reactive species' (Halliwell 2006). However, because the most well studied are ROS and NO, which are often made and function concurrently in plant cells, for the sake of convenience, we here refer to ROS and NO collectively as RS where appropriate.

Plant Stress Signaling

Plants are unavoidably exposed to many abiotic and biotic stresses and in their natural environment, likely to be exposed simultaneously to more than one such stress. Moreover, these stresses will inevitably superimpose their downstream effects on the existing signaling processes inherent to everyday developmental and rhythmic cues. For example, plants following an underlying circadian rhythm may be subjected simultaneously to challenge by several potential pathogens or pests, while at the same time experiencing both cold and drought stress. Plants synthesize an abundance of signaling molecules and respond not only to internal chemical cues but also to external ones that initiate internal chemical signaling. Thus, the biological state and activity of an individual plant cell represents its ability to integrate a variety of signaling inputs into a specific output that may change as the spectrum of inputs is altered. Given the obvious complexity of signaling inputs and the apparently limited repertoire of proteins involved in signaling cascades, it is not surprising that plant cell signaling does not result from the activity of individual linear signaling pathways for each input but rather entails a web or network of signaling proteins on which multiple stimuli have a combined effect. In such signaling webs, there is ample opportunity for signal crosstalk in which different stimuli can access common components of signaling pathways and

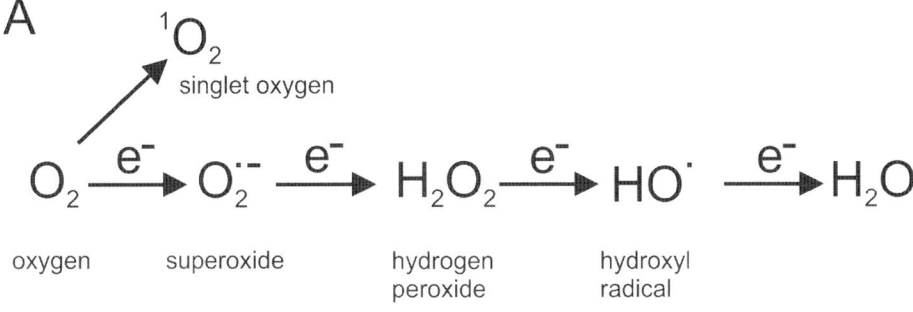

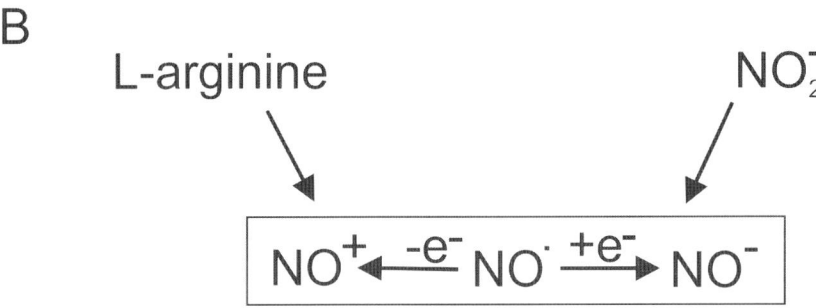

Fig. 7.1 The most common RS (reactive oxygen species and nitric oxide collectively) found in biological systems. (A) Dimolecular oxygen, itself a radical because it can possess two unpaired electrons, can be reduced sequentially to produce superoxide, hydrogen peroxide, the hydroxyl radical, and finally water. The protons are not shown stoichiometric here for simplicity. Electrons in dimolecular oxygen can also be elevated to other orbitals forming singlet oxygen, which will therefore possess different chemical properties. (B) Nitric oxide, although usually referred to as a free radical, can either gain or lose electrons, forming nitric oxide species with different chemical properties. Nitric oxide in biological systems is often formed from L-arginine or nitrite.

potentially elicit responses that, although different overall, may partly overlap. This is not to say that simple linear signal pathways do not exist or that particular signals suppress some signal transduction pathways while activating others and thus give the appearance of linear signaling. However, a growing amount of research demonstrates that signal crosstalk is a reality (Taylor and McAinsh 2004; Fujita et al. 2006).

RS as Signaling Intermediates

Increasing data demonstrate that RS are signaling intermediates that mediate responses to stress stimuli such as those elicited by pathogens, cellular dehydration, low temperatures, and the hormone abscisic acid (ABA). RS are also involved in various "nonstress" developmental and physiological processes driven by external and internal stimuli (Table 7.1). Consequently, RS can mediate crosstalk not only between various stresses but also with other signaling pathways. There has been an upsurge of research interest in RS in plants with several recent reviews (Lamattina et al. 2003; Apel and Hirt 2004; Bailey-Serres and Mittler 2006; Arasimowicz and Floryszak-Wieczorek 2007; Wilson et al. 2008).

Table 7.1 Some examples of RS and plant biology[a]

Stresses	RS signaling[b]
Pathogen challenge, elicitors, PAMPs	- Oxidative burst - ROS generation, mainly by NOX and peroxidases - ROS generation in chloroplasts - NO generation by NOS and NR - RS interaction with MAPK signaling cascades - ROS/NO interactions to induce defence responses such as programmed cell death/hypersensitive response (HR), defense gene expression, phytoalexin production
Temperature extremes (cold, heat)	- ROS generation from NOX - NO generation from NOS
Drought	- Turgor loss causes induction of both ROS and NO generation - ABA induces increases in levels of H_2O_2 and NO - Induction of antioxidant enzymes, modulating RS levels (and in other stresses[c])
Salt	- ROS generation from NOX - NO generation from NOS
Heavy metals	- ROS generation from NOX - NO generation from NOS
Ozone challenge	- ROS generation from NOX - NO generation from NOS
Wounding	- NO generation is involved in the wound-healing response - NOX-mediated ROS involved in wound responses - NO involved in jasmonic acid signaling
Hypoxia	- NO involved in increased aerenchyma formation (NR) - Increased ROS generation (NOX) - NO involved in sensing of low oxygen
Excess light	- ROS; chloroplast-mediated ROS generation
Ultraviolet light	- Increased ROS and NO generation - NO and ROS interactions - Involvement of MAPK signaling cascades
Other processes	
Seed germination	- NO stimulates germination - NO relieves salt reduced germination
Shoot growth/development	- NOX-mediated ROS signaling
Root initiation and growth	- NO generation from NOS and NR - NO partly mediates auxin signaling - NO effects mediated by MAPK signaling cascades
Root hair growth and development	- ROS, mediated by NOX (generation of superoxide and hydroxyl radicals) - ROS oscillations - ROS effects mediated by protein kinase signaling - NO generation involved in root hair development
Cell wall metabolism	- H_2O_2-mediated secondary wall formation, e.g., in cotton fibers - H_2O_2 effects mediated by protein kinases
Gravitropism	- ROS generation probably involving NOX - NO generated from NOS and/or NR - ROS and NO mediate auxin effects
Flowering	- NO delays flowering
Polarity initiation	- NO required in fern spores
Somatic embryogenesis	- NO mediation of auxin effects (generated from NOS)
Senescence	- NO as an anti-senescence factor - NO also reported as pro-senescence - Increased ROS generation - Altered antioxidant levels

Table 7.1 (*Continued*)

Other processes	RS signaling[b]
Stomatal closure/inhibition of opening	- Generation of ROS mediated by NOX, or other sources - Generation of NO, mediated by NOS and/or NR - H_2O_2 and NO mediate effects of other signals, e.g., ABA, blue light, ultraviolet light, CO_2, ethylene, elicitors

Hormone/signal molecule effects on RS metabolism	- Many hormones/signals have their responses mediated by RS, including
	    Abscisic acid
	    Auxin
	    Cytokinins
	    Jasmonate
	    Gibberellins
	    Polyamines
	    Salicylic acid

ABA = abscisic acid; MAPK = mitogen-activated protein kinase; NO = nitric oxide; NOX = NADPH oxidase or NADPH oxidase-like; NR = nitrate reductase; NOS = nitric oxide synthase-like enzyme; PAMPs = pathogen associated molecular patterns; ROS = reactive oxygen species; RS = ROS and NO collectively.

[a]Information pertinent to the above has been obtained from the following review articles: Lamattina et al. 2003; Apel and Hirt 2004; Gapper and Dolan 2006; Kwak et al. 2006; Torres et al. 2006; Van Breusegem and Dat 2006; Zaninotto et al. 2006; Arasimowicz and Floryszak-Wieczorek 2007; Hong et al. 2007; Neill et al. 2008; Wilson et al. 2008.

[b]Some experimental data were obtained using gene-specific knockouts, but many come from the use of enzyme inhibitors.

[c]Antioxidant enzymes induced when RS levels increased.

An exhaustive review of the many aspects of RS signaling in plants would be overwhelming and is not appropriate to attempt this here. Instead, we attempt to highlight what we think are the most important considerations concerning how RS might function as signaling intermediates in a range of scenarios in response to various stress signals and how RS might thereby represent a node of signaling crosstalk. Consequently, we provide some information on how RS are generated and removed in plants and how they might be perceived by plant cells and effect downstream signaling. Rather than discuss a long series of biological phenomena and experimental studies, we have tried to provide a generic overview and illustrate this with the specific example of stomatal closure for which there are many papers demonstrating roles for both ROS and NO. Consequently, where possible, we have restricted citations to relevant review articles.

RS as Nodes for Signaling Crosstalk

To understand and clarify the involvement of RS in stress signal crosstalk, it is necessary to have a clear view of the "RS landscape" within and between cells, something that is not yet fully available. As shown in Figure 7.1, there are several ROS and NO variants that can interact and affect separate and/or overlapping processes. Furthermore, there are several intra- and extracellular sites and biochemical routes of RS generation, and thus, the cellular effects of individual RS are likely to differ depending on their site of origin and the neighboring molecules with which they can interact. Such molecules include antioxidants, proteins, and enzymes that either react with or metabolize RS. Indeed, the interactions between RS and antioxidants are critical to cell signaling and function (Foyer and Noctor 2005). Intracellular movement of RS is possible and may represent a form of intracellular communication—for example, between either the chloroplast or mitochondrion and the nucleus. Some forms of RS such as H_2O_2, NO, and even superoxide, may diffuse between cells and, adding another level of complexity, along with others such as nitrite and S-nitrosoglutathione, may even be transported in the vascular system. RS could also occur exogenously. For example, they may either be generated by allelopathic, parasitic, and symbiotic organisms interacting with the plant or be present in the environment surrounding the plant as occurs when NO is released from the soil and other plants or is present as an air pollutant. Such exogenous RS are likely to elicit responses similar to a subset of endogenous RS and, depending on the intracellular concentrations achieved at various internal sites, may well disturb normal metabolism and elicit antioxidant and other responses.

There is no doubt that various stresses do indeed alter the RS landscape within plant cells (see Table 7.1). RS are continuously produced by cells as products of metabolism, and some abiotic stresses result directly in increased RS through their effects on photosynthesis, mitochondrial electron transport, and other metabolic processes. Often this can occur in an essentially nonregulated manner as opposed to resulting from any specific, regulated alteration in the activities of specific enzymes. If either the rate of RS generation exceeds that of its removal or if cells are exposed to exogenous RS, then either oxidative or nitrosative stress may result, perhaps ultimately inducing programmed cell death (PCD; Van Breusegm and Dat, 2006). It may have been the very ubiquity of RS that has led to them becoming useful signaling molecules. Excessive RS generation in organelles such as chloroplasts or mitochondria would result in spillover from these into the cytosol and other cell compartments. A failure of the protective antioxidant mechanisms to rapidly sequester or remove the excess RS would leave them free to react in various ways with a number of proteins, thereby initiating both altered enzyme activities and potential conformational changes that may, for example, trigger the enhanced transcription of genes encoding either antioxidant or NO-reactive proteins and other signaling proteins. If selectively beneficial, such processes might well have been evolutionarily maintained and adopted as useful components of signaling mechanisms ameliorating the adverse effects of stressful environments.

Specific, regulated, enzyme systems have evolved that generate RS. Clearly, if several signals affect RS generation, in the same or different intracellular sites, then some level of crosstalk is inevitable (Figure 7.2). One can envisage such crosstalk occurring through a number of mechanisms, but the end result would presumably be that the activity of particular signaling proteins at defined points in specific pathways would be either enhanced or repressed. This may arise directly through the interaction of specific proteins or indirectly either through the rerouting of second messengers or the depletion of second messenger pools. The elucidation of the underlying processes clearly requires large-scale holistic systems analyses approaches.

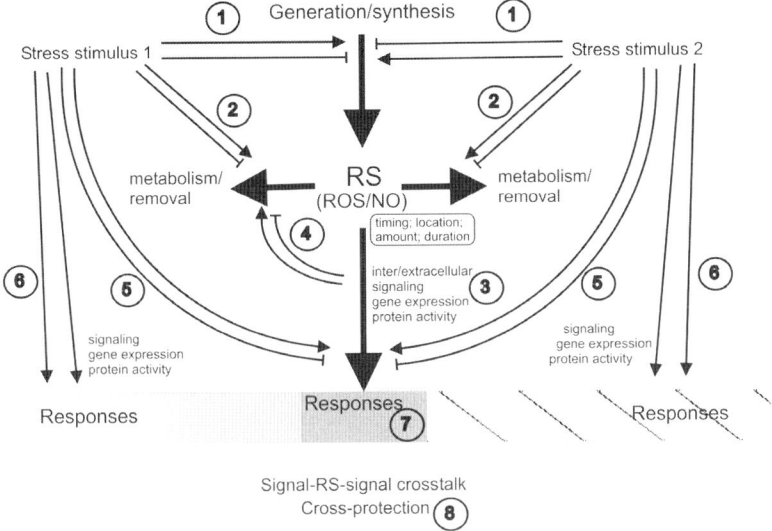

Fig. 7.2 RS (reactive oxygen species [ROS] and nitric oxide [NO] collectively) as mediators of crosstalk during stress signaling. **(1)** Stress stimuli could enhance or reduce RS generation/synthesis. **(2)** Stress stimuli could enhance or reduce RS metabolism/removal. **(3)** RS signaling results in various responses. **(4)** RS induction or repression of antioxidant defences. **(5)** Stress stimuli could amplify or reduce RS signaling. For example, RS signaling may affect turnover of second messengers, activation or inhibition of ion channels, modulation of transcription factor activity, and depletion of second messenger pools. **(6)** Stress stimuli induce responses via signaling pathways that are independent of RS but may overlap with each other in terms of signaling intermediates. **(7)** Cellular responses depend on the number and variety of stress and other stimuli. RS themselves induce certain responses depending on their amounts, molecular species and ratios, and subcellular locations. Stress stimuli also induce responses, some of which require RS signaling and some of which are RS-independent. Stress stimuli may also inhibit specific aspects of RS signaling, so that a particular suite of responses is dictated by the strength of stress stimulus 1 and 2. **(8)** RS-dependent stress responses represent the output of intracellular signal-RS-signal crosstalk. Cross-protection, the induction of stress tolerance by preexposure to another, is evident when different stresses induce a degree of oxidative stress which in turn activates antioxidant defence processes that afford protection against subsequent stresses.

Synthesis, Turnover, and Interactions of ROS and NO

ROS Generation

Many enzymatic routes exist for ROS generation in plants (Figure 7.3), with the most well studied of these being that involving the various differentially expressed, plasma-membrane-associated, nicotinamide adenine dinucleotide phosphate (NADPH) oxidase (often referred to as NOX) enzymes. Ten isoforms have been identified in *Arabidopsis* (Atrboh A–J; Sagi and Fluhr 2006) and gene-specific knockout mutagenesis studies have implicated them in a variety of individual roles (Table 7.1). *NOX* genes have been identified in other species and molecular, and biochemical studies have demonstrated their involvement in several stress responses (Table 7.1).

NOX enzymes contain EF-hands, protein domains that bind Ca^{2+} ions, and indeed, NOX activity is, in part, directly regulated by such a mechanism (Sagi and Fluhr 2006). However, Ca^{2+} effects may also occur indirectly. For example, calmodulin is

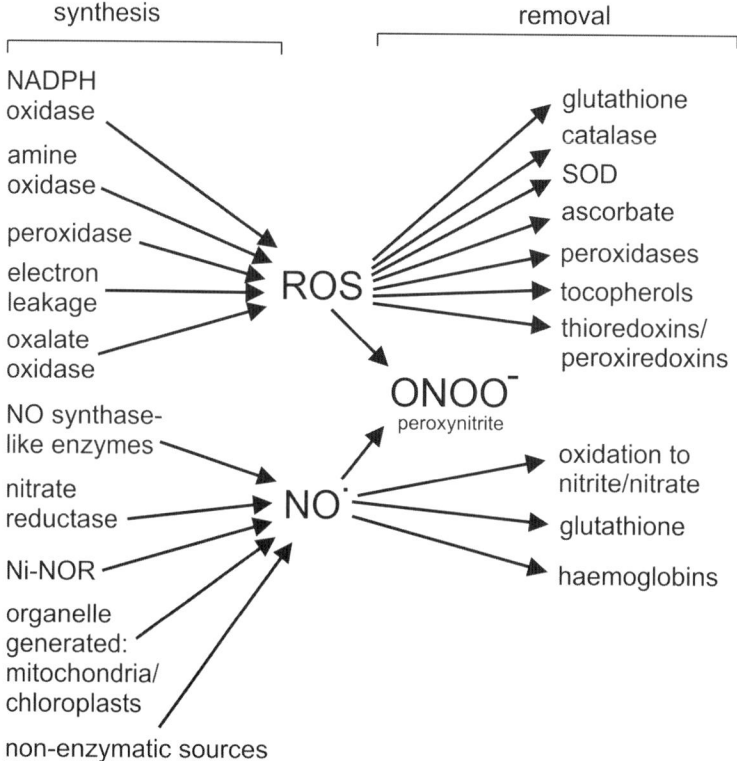

Fig. 7.3 Systems involved in the synthesis and removal of RS (reactive oxygen species [ROS] and nitric oxide [NO] collectively). Many enzyme systems are involved in the synthesis of ROS. Of critical importance are their activities under particular conditions, that is, their regulation so that RS are only produced when and where required. Electron leakage from redox pathways in organelles will mean that the associated ROS are located in a different place to those produced by either amine oxidase in the apoplast or nicotinamide adenine dinucleotide phosphate (NAPDH) oxidases in the plasma membrane; thus any signaling that ensues will be different. Similarly, different enzymes can generate NO in different cellular locations. The removal of both ROS and NO is also complex. Finally, ROS and NO can react together to form peroxynitrite.

required for ABA-induced ROS production (Hu et al. 2007), and Kobayashi et al. (2007) demonstrated that elevated Ca^{2+} induces the phosphorylation of a potato NOX on Ser-82 and Ser-97. Two calcium-dependent protein kinases (StCDPK4 and StCDPK5) were identified and shown to be responsible, and a specific mutagenesis approach implicated StCDPK5 in this mechanistic regulation of ROS production. NOX activity also depends on the activity of other kinases. The kinase open stomata 1 (OST1) has been implicated in the regulation of ABA-induced ROS production, and epistatic studies suggest that ABI1 and ROS generation are potentially linked by OST1 (Hirayama and Shinozaki 2007). Jiang et al. (2003) have also suggested that MEK1/2 and components of the MAP kinase signaling pathways mediate the increases in H_2O_2, which are induced following the addition of ABA. Small GTPases can also regulate NOX activity and ROS generation (Gapper and Dolan 2006).

An additional source of ROS arises from the action of the peroxidases, some of which generate superoxide anions and hydrogen peroxide. This activity may be particularly relevant in the extracellular space, especially with regard to ROS synthesis in response to pathogen invasion (Bindschedler et al. 2006). A number of other enzymes are also capable of generating ROS in plants. Both amine oxidase (An et al. 2008) and oxalate oxidase (Hu et al. 2003) can generate ROS, and there is some suggestion that, in this respect, these enzymes play a role in responses to ABA.

NO Generation

As with ROS, NO can be generated in several ways, in different cellular locations (Figure 7.3), and it is likely that the levels of bioactive NO reflect these various contributions. NO can simply be made as a product of nonenzymatic chemistry as with either the conversion of nitrite to NO or the reaction of nitrous acid with ascorbate to produce NO and dehydroascorbate (Wilson et al. 2008).

However, enzymatic sources of cellular NO also exist. In mammals, NO is made by a family of nitric oxide synthases (NOS) that catalyse a redox reaction in which L-arginine is converted through a nonreleased hydroxy-arginine intermediate to L-citrulline resulting in the release of NO. Many studies using inhibitors of animal NOS suggest similar enzyme activities in plants (Wilson et al. 2008). Comparative sequence searches indicate that there are no similar NOS sequences in the *Arabidopsis* genome. Using a snail sequence, Guo et al. (2003) found what they reported to be the first plant NOS equivalent, but recently, the actual function of this protein has been questioned (Neill et al. 2008). However, altering the expression of this protein has a profound effect on plant NO production, and on this basis, it has been suggested that it be named nitric oxide associated 1 (AtNOA1). Perhaps the protein forms a complex with other unidentified proteins to generate NO or regulate its synthesis.

So what is the plant NOS? It is interesting that whereas the human NOS polypeptide contains all the binding sites for all the prosthetic groups needed for full redox function, in reality, the redox pathway actually involves the transfer of electrons between two such polypeptides. Thus, it is possible that yet-to-be-identified polypeptides collude in such a way in plants. Whatever the identity of the plant NOS enzymes, increases in Ca^{2+} levels that can occur during stress responses have been implicated in the control of their activity (Besson-Bard et al. 2008).

NO synthesis can also result from the activity of nitrate reductases (NR) on nitrite (Wilson et al. 2008). *Arabidopsis* has two genes encoding NR, *NIA1* and *NIA2*, which are important for stomatal closure and of these it appears that the NIA1 isoform is specifically required (Bright et al. 2006). The control of NR-dependent NO production is clearly important if this signal is to be produced only when needed, and indeed, NR activity can be regulated by serine phosphorylation. The enzyme also interacts with 14-3-3 proteins, which may mediate its proteolysis (Wilson et al. 2008).

A number of other enzymes, including a root-specific plasma-membrane-bound protein, nitrite-NO-oxidoreductase (Ni-NOR) (Stohr and Stremlau 2006) and the enzyme xanthine oxidoreductase (XOR), may also contribute to the accumulation of NO under conditions of stress. However, more recent work has suggested that XOR does not have a role in plants (Planchet and Kaiser 2006). Others have noted that the addition of polyamines to plants can induce NO generation (Tun et al. 2006) and enzymes such as the polyamine oxidases may be directly involved in such activity (Yamasaki and Cohen 2006).

RS Removal

To be considered signals in the classical sense, there needs to be a mechanism for the removal of RS once a response to their perception has been initiated. Such removal mechanisms are also an important consideration with regard to the dynamic control of the cellular redox environment. As such, a plethora of mechanisms exist for removing RS (Figure 7.3). Dietary components such as vitamin C and E are known to have antioxidant properties and act similarly in planta. Tocopherol, for example, can remove various ROS species including singlet oxygen (Foyer and Noctor 2005).

Superoxide is quickly dismuted by members of the superoxide dismutase (SOD) family to H_2O_2, which can subsequently be removed by enzymes such as catalase. Indeed, loss of catalase function results in defective development and perturbed ROS signaling (Queval et al. 2007).

The ascorbate-glutathione (ASc-GSH) cycle also makes an important contribution to the removal of H_2O_2 (Foyer and Noctor 2005), and both ascorbate and reduced glutathione (GSH) are found at high (mM) concentrations in plant cells. The glutathione peroxidases (GPXs) are another family of enzymes that have also been shown to be involved in ROS removal. In *Arabidopsis*, it has been reported that AtGPX uses GSH as a reducing agent to remove H_2O_2, and as is described later, may act in the signaling pathway between ABA and H_2O_2 that leads to stomatal closure (Miao et al. 2006).

Peroxiredoxins are recognized as major participants in the removal of H_2O_2 in plants (Halliwell 2006). Their reactivity is a consequence of the cysteine groups that reside within their active sites that become oxidized to the sulphenic acid (–SOH) form by H_2O_2. A recent example of their importance in protecting the rice chloroplast from oxidative damage was reported by Pérez-Ruiz et al. (2006).

One of the major mechanisms for the removal of NO in plants could involve the hemoglobins. Nonsymbiotic hemoglobins (nsHbs) can form nitrate from NO in an NAD(P)H-dependent manner, and altering the level of *Ahb1* in *Arabidopsis* has been shown to have profound effects on the levels of cellular NO (Perazzolli et al. 2006).

Interactions between ROS and NO

RS can crossreact and therefore influence the bioavailability and function of each other. For example, the superoxide anion and NO radical can react to produce peroxynitrite (Figure 7.3). Thus, the production of superoxide may lower the level of NO in the cell and vice versa. Despite its instability and toxicity, peroxynitrite may itself have cell signaling effects (Klotz 2002) and when one considers that ROS and NO are often produced spatially and temporally together, such crossreactivity should be taken into account. Indeed, formation of peroxynitrite and interactions between H_2O_2 and NO are critical determinants of the PCD often associated with pathogen responses (Zaninotto et al. 2006).

RS can also modulate the synthesis of each other. Exogenous H_2O_2 or that induced by ABA has been shown to induce the production of NO in various tissues of a number of plant species (Neill et al. 2008). Although there is some suggestion that NO does not cause H_2O_2 synthesis in guard cells (Bright et al. 2006), others have suggested the opposite (He et al. 2005).

RS may also influence the removal of each other. For example, an increase in H_2O_2 levels may lead to an increase in the expression of antioxidant systems (de Pinto et al. 2006; Table 7.1). Increases in the level of NO can also lead to an increase in the level of antioxidants and the expression of enzymes that subsequently lower the ROS levels in the cell (Zhang et al. 2007; Table 7.1).

RS Signaling

It is clear that RS have biological activity in plants, but how this results is far from clear and the complexity of the underlying signaling mechanisms that operate remains to be fully described. Thus, the interest concerning their signaling must focus on the proteins involved in their perception and action and whether these can realistically constitute components of reversible signal transduction mechanisms. An overview of the mechanisms of RS signaling that may occur is shown in Figure 7.4.

Reversible Protein Phosphorylation

Reversible protein phosphorylation, in particular, that involving mitogen-activated protein kinase (MAPK) signaling, is a key component of RS signaling (Pitzscke and Hirt

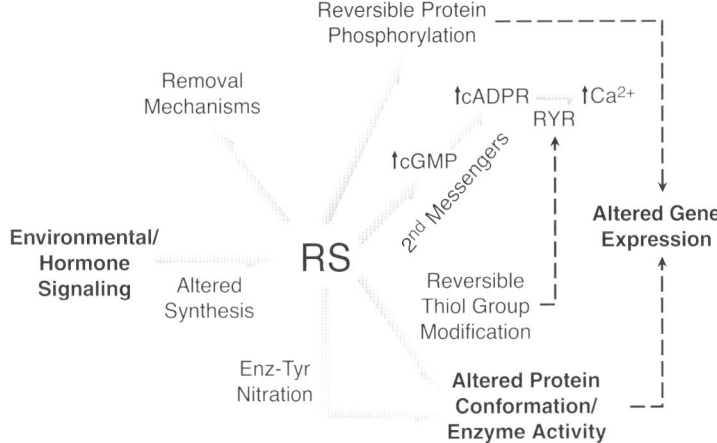

Fig. 7.4 Mechanisms of RS (reactive oxygen species [ROS] and nitric oxide [NO] collectively) signaling. Combinations of different stress stimuli and hormone signaling may result in specific levels of RS which subsequently signal through a number of mechanisms. RS may modify various amino acid residues in proteins either by thiol group oxidation and S-nitrosylation or by tyrosine nitration. Such modifications may either have direct effects on enzyme activities or may alter the conformation of proteins, modulating their susceptibility to other signaling-associated modifications such as phosphorylation, or enabling their interaction with other proteins. Altering the cellular redox state may engage specific suites of proteins in such signaling processes and act as an effective switch by which the cell can rapidly respond to environmental changes. RS signaling is known to involve kinase pathways, the generation of second messengers such as cyclic guanosine monophosphate (cGMP), cyclic adenosine-ribose (cADPR), and Ca^{2+}, and ultimately results in altered gene expression. Thus, RS can play a central role in mediating the overall downstream effects of multiple signaling inputs.

2006). The addition of H_2O_2 to *Arabidopsis* cell suspension cultures results in an increase in the activity of signaling kinases, and in particular, AtMPK3, AtMPK6, and AtMEK1 appear to be involved (Pitzscke and Hirt 2006; Xing et al. 2007). The downregulation of the expression of AtMPK6 by RNA interference results in plants that are hypersensitive to ozone (Miles et al. 2005). H_2O_2 also causes the activation of the MAPK kinase kinases (MAPKKKs) ANP1 and OMTK1 (Pitzscke and Hirt 2006). Similarly, NO has been shown to activate MAPK signaling pathways (Wilson et al. 2008). The signaling of singlet oxygen ($^1O_2$) may also involve MAP kinase signaling. Because $^1O_2$ signaling is often masked by that of other RS, this has proved difficult to demonstrate. However, recent studies that make use of the *Arabidopsis* conditional *fluorescent, flu*, mutant have begun to shed some light on the mechanisms that may play a part here (Laloi et al. 2007), and in *Arabidopsis* it has been shown that the protein kinase MAPK4 regulates SA-dependent singlet oxygen responses through the protein EDS1 (Brodersen et al. 2006).

Increases in MAPK activity involve elevated levels of phosphorylation. Not surprisingly, therefore, ROS also appear to decrease the activity of various phosphatases such as AtPTP1 and the protein phosphatase 2C enzymes ABI1 and ABI2 (Kalbina and Strid 2006; Kwak *et al.* 2006; Pitzscke and Hirt 2006). Thus, there appears to be a signaling mechanism by which the RS signal can be transduced. However, MAPKs such as those described above typically act in a cascade downstream of specific receptor molecules and thus, may not themselves be the direct targets of ROS. However, there is some suggestion that the activity of the phosphatases, ABI1 and ABI2, may be directly mediated by RS and this, therefore, may be one mechanism by which they exert their effects on the MAP kinase signaling cascade (Pitzscke and Hirt 2006).

A number of other kinases have also been implicated in RS signaling. The kinase oxidative signal-inducible 1 (OXI1) has been shown to be involved in H_2O_2 signaling and interestingly appears to act upstream of AtMPK3 and AtMPK6 (Rentel et al. 2004). In tobacco, the function of the kinase NtMPK4 is required for jasmonic acid signaling and appears to be involved in conferring ozone tolerance by influencing stomatal apertures (Gomi et al. 2005), perhaps by influencing the activity of the ABA signaling pathway. NO also signals through kinases other than MAPKs (Besson-Bard et al. 2008).

Second Messengers

In mammals, soluble guanylyl cyclase (sGC) plays a key role in NO signaling. NO binds to the haem domain in sGC and induces the production of cyclic guanosine monophosphate (cGMP), which in turn acts on a number of downstream targets. There is some evidence that NO similarly induces a transient rise in cGMP levels in plants (Durner et al. 1998) and the pharmacological use of sGC inhibitors has indicated that this may be required for normal ABA and NO signaling in guard cells (Neill et al. 2008). A key downstream signaling intermediate in cGMP-mediated signaling is cyclic adenosine (ADP)-ribose (cADPR; Wendehenne et al. 2001) which stimulates the release of Ca^{2+} through intracellular ryanodine receptor calcium channels (RYR). NO has been shown to cause elevated free Ca^{2+} levels (Wilson et al. 2008), and it is possible that this occurs through a cGMP- and cADPR-mediated signaling cascade. The use of RYR inhibitors prevents the normal cADPR-induced accumulation of transcripts encoding phenylalanine ammonia lysase (PAL) and the pathogenesis-related protein 1 (PR-1; Durner et al. 1998). NO, cGMP, and cADPR have all been shown to be involved in the normal responses of guard cells to ABA (Neill et al. 2008), and the scavenging of NO by 2-(4-carboxyphenyl)-4,4,5,5-tetramethylimidazoline-1-oxyl-3-oxide (cPTIO) during ABA-induced stomatal closure inhibits the associated inactivation of the Ca^{2+}-dependent inward rectifying K^+ channel and activation of the outward rectifying Cl^- channel (Garcia-Mata et al. 2003). Thus, there is a strong implication that NO may signal through a transduction cascade involving cGMP, cADPR, and Ca^{2+}. However, although animals and bacteria possess sGCs with haem domains capable of binding NO, plants do not appear to have a direct homolog of this protein, and the *Arabidopsis* guanylyl cyclase, AtGC1, seems not to be directly NO regulated (Ludidi and Gehring 2003). Thus, the mechanism by which NO stimulates a rise in cGMP remains unclear, and it is possible that the enzymes that generate cGMP in response to NO are significantly different from their mammalian counterparts.

ROS induce elevations in intracellular Ca^{2+} through direct and indirect effects on calcium channels (Kwak et al. 2006), and calmodulin is also a component of ROS signalling (Hu et al. 2007).

Thiol (-SH) Group Modification in RS Signaling

The detection and signaling of RS requires their initial perception by one or more proteins. It may be that this occurs through the model ligand and receptor mechanism, but to date no specific receptor molecules have been identified. In light of this, the reactivity of these compounds has led researchers to postulate that their recognition may instead occur through a chemical reaction. Various researchers have suggested that this may take the form of cysteine –SH group modification (Foyer and Noctor 2005), but other amino acids such as methionine, tyrosine, tryptophan, and histidine may be similarly oxidized. Whichever the case, because all proteins contain these amino acids, there must be a mechanism relating to amino acid sequence, protein conformation, localization, and interaction within complexes by which only a selected number are modified in this way.

One critical factor that allows cells to thrive is the reduction/oxidation (redox) potential of their contents. Many of the compounds crucial for enzyme activity, including a number of those used in metabolism such as nicotinamide adenine dinucleotid (NADH) and NADPH, must be maintained in the reduced state. RS are relatively oxidizing, and therefore, a breakdown in the balance between RS generation and removal could lead to a substantial increase in the oxidation of the cell cytoplasm. It has been argued that such unmanaged shifts in cellular redox state can lead to PCD (Schafer and Buettner 2001). However, peptides that possess groups such as thiols may be sensitive to more subtle changes in the redox state of their environment. Depending on the midpoint potentials of their thiol groups, RS-induced fluctuations in the "redox poise" of their environment may make individual proteins more or less susceptible to particular modifications. Mechanistically, this may allow particular cell responses to proceed only under certain conditions, for example, during a period of stress. Such conditional redox responses have, for example, been suggested to modulate day-length-dependent gene expression where the prevailing photoperiod plays a role in regulating H_2O_2-induced cell death (Queval et al. 2007).

One important way in which protein activity may be controlled is by alterations in structural conformation, and this potentially increases the number of modifications to which it is susceptible. For example, phosphorylation may alter its three-dimensional shape and result in the exposure of a thiol group for subsequent oxidation or nitrosylation. Conversely, oxidation or nitrosylation of a peptide may increase its susceptibility to the action of, for example, an associated kinase and, thus, confer a new level of regulation on its activity. Therefore, by altering the redox state, RS may not only act directly to regulate the function of particular proteins but may have an overall modulating effect that permits the activity of whole suites of proteins to be conditionally controlled during, for example, periods of stress. Such a mechanism would enable rapid and coordinated responses to occur and may be vital for cell and indeed whole plant survival under conditions in which levels of stress increase rapidly.

The ROS oxidation of –SH groups can yield a number of potential products. The oxidation of methionine yields methionine sulfoxide (MetSO) and, in the context of reversible signaling, this can be reduced by MetSO reductase (PMSR; Sadanandom et al. 2000). The oxidation of cysteine by H_2O_2 yields a number of potential products (Figure 7.5) and again, in the context of reversible signaling, the formation of some of these is readily reversible by thioredoxin, glutaredoxin, and sulphiredoxin-mediated reduction (Hancock et al. 2006). What is clear is that the degree of such oxidation of any –SH group would depend on its midpoint redox potential and its availability to the oxidant. Thus, it is likely that only a limited number of –SH groups within any one protein would be potentially modified. As part of a plant signaling mechanism, one could then envisage a scenario in which the de novo synthesis of an appropriate protein or the conformational change of an existing protein would make –SH groups available for modification by ROS.

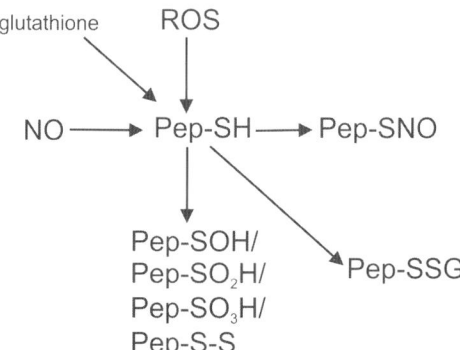

Fig. 7.5 The chemistry of thiol groups. Thiol groups can react with a variety of reactive compounds, resulting in either the loss of free RS (reactive oxygen species [ROS] and nitric oxide [NO] collectively) or the formation of a signal. The exact reaction that takes place on any particular thiol will be dependent on several factors, including the steric availability of the thiol to the reactive species, the concentration of the reactive species present, and the redox midpoint potential of the thiol. The latter will be influenced by the microenvironment in which the thiol resides and may be changed as the conformational topology of a peptide alters, perhaps as a result of other signaling modifications such as phosphorylation. A competition will result because glutathione, ROS, and NO are all likely to be present at the same place at the same time.

Growing evidence also suggests that NO may modify –SH and tyrosine groups in proteins and that this may constitute mechanisms by which it signals its increased presence. NO may either directly S-nitrosylate or indirectly trans-S-nitrosylate either the exposed cysteine residues in proteins or other low molecular weight compounds such as glutathione (Wilson et al. 2008). NO may also nitrate tyrosine residues through the action of peroxynitrite.

In animal systems, S-nitrosylation has become established as the key redox-based, posttranslational modification and has been shown to regulate a number of signaling events, structural protein conformations, and metabolic events (Wang et al. 2006). In *Arabidopsis*, a number of proteins involved in a wide range of cellular events have also been shown to be S-nitrosylated in vitro following the treatment of cell extracts with S-nitrosoglutathione (GSNO; Lindermayr et al. 2005) and in vivo during the hypersensitive response (Romero-Puertas et al. 2008). In the latter case, some of these proteins appear to be involved in biology consistent with this scenario. In vitro, the activity of one recombinant isoform of methionine adenosyl transferase appears to be altered by S-nitrosylation in a manner dependent on the presence or absence of Cys-114 (Lindermayr et al. 2006), and the activity of an *Arabidopsis* metacaspase seems to be similarly regulated (Belenghi et al. 2007).

Because S-nitrosylation may be achieved by reactive GSNO, formed by the reaction of NO with glutathione (GSH; Wang et al. 2006), it would seem likely that modulating the levels of this compound may also affect the signaling ability of NO. Indeed, loss of function mutations in the *Arabidopsis* gene *AtGSNOR1*, which encodes GSNO reductase, result in higher levels of S-nitrosylation and reduced resistance (R)-gene-related defense responses against microbial pathogenesis (Wang et al. 2006). Conversely, plants overexpressing AtGSNOR1 demonstrate enhanced pathogen resistance, and it has been shown that this protein positively regulates the salicylic signaling pathway that leads to plant immunity. In other plants such as peas, the expression of the gene encoding GSNO reductase and the activity of the protein are both reduced during cadmium stress, which is known to induce NO synthesis (Barroso et al. 2006).

In addition to oxidation and nitrosylation, thiol groups may also be modified by glutathionylation, and this would again be expected to have profound effects on both the topology and activity of a protein. Dixon et al. (2005) identified a range of proteins in *Arabidopsis*, which could be modified in this manner. Therefore, it is clear that there may be competition between the various mechanisms of thiol group modification that can occur, and this may have a direct bearing on the actual signaling (Figure 7.4) that results.

Changes in Gene Expression

Various experiments involving exogenous RS and manipulation of endogenous RS through genetic inhibition of ROS-removal mechanisms have shown that both ROS and NO alter substantially the transcriptome of plants (Gadjev et al. 2006; Zaninotto et al. 2006). Many of the transcripts that accumulate in response to RS encode proteins involved in antioxidative processes and responses to environmental stresses and pathogen challenge, including signaling proteins and transcription factors induced by hormonal stimuli such as ABA and ethylene. Not surprisingly, perhaps, although there are some genes with induction that is common to different ROS stimuli, perhaps representing general oxidative stress-response genes, the expression of others is restricted to specific ROS with differing reactivities and subcellular locations (Gadjev et al. 2006). However, during the cell death initiated by both NO and ROS, the dominant set of altered genes required both NO and ROS, indicating the substantial crosstalk between ROS and NO (Zago et al. 2006). Such data demonstrate that as abiotic and biotic stresses are likely to induce change in more than one RS, the resulting transcriptomic changes will reflect such complex interactions.

Assuming that such changes in gene expression do not all result from the operation of posttranscriptional silencing mechanisms, a significant portion of this alteration must result from the activity of signal transduction pathways and, as described later, subsequently involve changes in the activity of specific transcription factors that either repress or enhance gene promoter function.

For example, in yeast the oxidation of the glutathione peroxidase (GPx)-like enzyme, GPx3, by H_2O_2 causes Cys36 of Gpx3 to bridge to Cys598 on the transcription factor Yap1, which then activates the expression of genes encoding a number of antioxidant proteins (Delaunay et al. 2002). Suggesting that this protein may have a similar function in plants, mutation of the *Arabidopsis* homolog AtGPx3 results in enhanced ROS generation and altered stress responses (Miao et al. 2006).

In plants the acquisition of systemic acquired resistance is associated with the expression of particular pathogenesis related genes, much of which is modulated by the protein Non Expressor of Pathogenesis Related Genes 1 (NPR1) and a number of basic leucine zipper (bZIP) proteins, the TGA transcription factors. When inactive, NPR1 remains in an oligomeric form held together by S-S bonds. On activation, NPR1 becomes reduced resulting in the release of a monomeric form which accumulates in the nucleus and alters gene expression (Mou et al. 2003). Again, it would appear that –SH group modification plays an important role here. The removal of either Cys82 or Cys216 in NPR1 results in its constitutive activation and the DNA binding of TGA1 and TGA4, which is stimulated by their interaction with NPR1, also follows the reduction of S-S bonds within these bZIP proteins (Fobert and Despres 2005). Thus, NPR1 may act as a sensor of the increased levels of ROS known to occur during pathogenesis.

Interestingly, the protein glyceraldehyde-3-phosphate dehydrogenase (GAPDH), which is normally recognized for its vital glycolytic role, may also sense RS. This involves Cys modification by H_2O_2 and NO and subsequent translocation of the protein to the nucleus, where it is suggested it may affect gene expression (Hancock et al. 2005; Holtgrefe et al. 2008).

Two classes of transcription factors, the heat shock transcription factors (Hsfs) and those containing C2H2 zinc finger domains, have also been strongly implicated in RS signaling. There is growing evidence that Hsfs may directly sense ROS such as H_2O_2 and that there is a requirement for H_2O_2 synthesis during heat stress for expression of the normal complement of heat shock proteins to occur (Miller and Mittler 2006). Hsfs are a large gene family in plants and bind the consensus sequence nGAAnnTCCn, which is found in the promoters of a number of genes involved in the response of plants to different environmental stress conditions. Different Hsfs demonstrate complex interactions, can bind their own gene promoters, and can form homo- or heterotrimers, leading to alteration of their nuclear localization and either enhanced or repressed levels of transcription. In humans and *Drosophila*, Hsf1 has been shown to directly and reversibly sense H_2O_2 and form a disulphide linked homotrimer that subsequently becomes nuclear localized and activates gene expression. In yeast, $O_2^{\bullet -}$ anions promote the physical association of two DNA bound Hsf homotrimers, which then activate expression of their own and the gene encoding CuZnSOD, the enzyme which in turn depletes the cell of $O_2^{\bullet -}$ (Miller and Mittler 2006). Thus, a circular regulatory process exists in which $O_2^{\bullet -}$ induces a system for its own removal and that is then downregulated after it has been removed. In plants, the promoter of the gene encoding the H_2O_2-scavenging enzyme ascorbate peroxidase 1 (Apx1) and those of the genes encoding a number of proteins and transcription factors involved in H_2O_2 signaling and defense, contain an Hsf-binding motif (Miller and Mittler 2006). Apx1 promoter analysis and its activity in *Arabidopsis* plants overexpressing AtHsfA1b suggest that this binding site is functional. Apx1 knockout mutant plants accumulate higher levels of H_2O_2 during light stress, and, correlated with this, a number of transcripts encoding Hsfs also show elevated levels in these plants and in wildtype plants treated with H_2O_2 (Miller and Mittler 2006). Interestingly, a point mutation in the gene encoding rice HsfA4a, which prevents this transcription factor binding DNA, results in a disease lesion mimic phenotype and suggests an antiapoptotic role for this factor. Overproduction of H_2O_2 and $O_2^{\bullet -}$ is known to be associated with such PCD and lesion formation. Thus, a similar mechanism of Hsf sensing of H_2O_2 to that found in other eukaryotes may also exist in plants (Miller and Mittler 2006).

Many proteins contain zinc finger domains and are grouped into classes dependent on the number and order of the Cys and His residues that bind the zinc ion. Recent studies have revealed that some of those in the C2H2 class may function as key transcriptional repressors involved in the defense and acclimation responses of plants to conditions of environmental stress (Ciftci-Yilmaz and Mittler 2008). For example, overexpression of ZAT10 results in transgenic plants that are more tolerant of drought, salt, heat, and osmotic stress and enhances the transcription of genes encoding Apx1 and 2 and iron superoxide dismutase (FSD1), which are involved in scavenging ROS. Similarly, ZAT12 is also required for the expression of Apx1 and the two important stress response proteins, ZAT7 and WRKY25, during oxidative stress. Again, this suggests that ZAT12 expression is an essential component of the processes underlying ROS metabolism. Interestingly, ZAT12 expression is reduced in plants expressing a dominant negative construct of HsfA4a. Thus, ZAT12 expression may be regulated by an Hsf that may directly sense H_2O_2. Whichever is the case, the various zinc finger proteins appear to confer differential effects during the application of different stresses (Ciftci-Yilmaz and Mittler 2008), suggesting that a complex, overlapping regulatory network exists. Again, it will require a concerted systems biology approach to unravel this network.

RS and Signal Crosstalk

RS and Signal Crosstalk in Response to Stress

Figure 7.2 outlines potential mechanisms and routes through which RS could act as nodes for signal crosstalk in response to various stresses. Exposure to abiotic and biotic stresses and signaling molecules involved in such challenge (e.g. microbial elicitors, hormones such as ABA) alter the RS landscape by changing the rates of RS generation and removal. Excessive RS may well impose oxidative/nitrosative stress within cells to the extent that cell metabolism is fatally impaired and either PCD or necrosis results. However, the altered RS landscape can result in a shift in intracellular signaling that is ultimately beneficial and that brings about ameliorative responses. Nevertheless, given that RS can have such global effects on cell activity through their widespread modifications of the transcriptome, proteome and metabolome and interactions with central signal transduction pathways, it is inevitable that there will be some disturbance of normal activities. Under stress situations, the imperative is cell and organism survival, and these negative effects may be a small price to pay for the initiation of protective mechanisms. The precise repertoire of cellular responses will depend, of course, on many factors such as the timing, location, strength, and duration of any "RS burst(s)." Stress stimuli undoubtedly induce a range of intracellular responses of which some are RS-dependent, and importantly, some RS-independent. As cells are exposed to several stresses concurrently, the final outcome for the cell will be dependent on this complex suite of interactions. Exposure to one stress that occurs before subsequent exposure to another is likely to induce cross-protection if it induces a set of responses that are protective against the second stress. This may occur, for example, by activating RS generation, the initiation of signaling pathways and expression of antioxidant enzymes.

Stress-related interactions of RS with jasmonate, ethylene, and ABA have been well-documented. This is, perhaps, not surprising because these hormones are themselves key factors in many stress responses including microbial challenge, ozone exposure, and drought stress. NO is induced by wounding and jasmonate in *Arabidopsis*, but although it can induce expression of jasmonate biosynthesis genes, it does not induce jasmonate biosynthesis per se (Huang *et al.* 2004). Exposure to ozone induces oxidative and nitric oxide bursts reminiscent of those activated during the hypersensitive response (Ederli et al. 2006; Mahalingham et al. 2006; Zaninotto et al. 2006). In *Arabidopsis,* transcriptomic changes affected by ozone overlap with those induced or repressed by ethylene and jasmonate (Mahalingham *et al.* 2006). In tobacco, increased production of ethylene but not jasmonate is induced by ozone, and a similar increase in ethylene can be induced by exogenous NO (Ederli et al. 2006). All three signals can induce the expression of the gene encoding the alternative oxidase, *AOX1a,* which is important for stress alleviation, but only NO is indispensable with regard to this response (Ederli et al., 2006). Other negative interactions between NO and ethylene have been reported. For example, NO reduces ethylene emanation from strawberries and delays ripening (Zhu and Zhou 2006). NO nitrosylates and inhibits the ethylene biosynthesis enzyme methionine adenosyl transferase (Lindermayr *et al.* 2006), suggesting that this enzyme is a point of NO-ethylene crosstalk.

Interactions between ABA and RS are likely to underpin several stress responses including those initiated by pathogen challenge as well as abiotic stresses such as drought and salt stress (Fujita et al. 2006). ABA induces the expression of many genes, and a number of these are also induced by ROS. Conversely, ROS induce the expression of genes that are involved in ABA and ethylene signaling. For example, the expression of genes encoding a number of ABA- and ethylene-responsive transcription factors is induced by ROS (Fujita et al. 2006). Induction of ROS (and NO) generation may be a common response to ABA, with the subsequent activation of MAP kinases and enhanced expression and activity of antioxidant enzymes (Neill et al. 2008).

ABA-induced RS contribute to the vital adaptive stress responses that lead to the induction of stomatal closure (see subsequent discussion) and may also regulate the gene expression and antioxidant status of guard cells.

RS Signaling Crosstalk in Stomatal Guard Cells

Stomatal movement is an excellent paradigm for plant cell signaling research in general and signal crosstalk in particular (Figure 7.6). Stomatal guard cells respond to many varied signals, and most, if not all, plant intracellular signaling pathways have been demonstrated to function within them. Indeed, guard cells have emerged as an exemplar system in which to construct increasingly complex systems biology models of plant cell signaling (Li et al. 2006).

Probably, the best studied of the signaling pathways that affect guard cell movements is that downstream of the perception of ABA (Hirayama and Shinozaki 2007). This vital hormone modulates a number of developmental responses to environmental stresses such as drought, high salinity, and reduced temperatures that all reduce the effective availability of water. ABA acts through reasonably well-understood signaling pathways that regulate stomatal apertures to mediate the rate of water loss through transpiration. A number of ABA receptors have been identified, two of which,

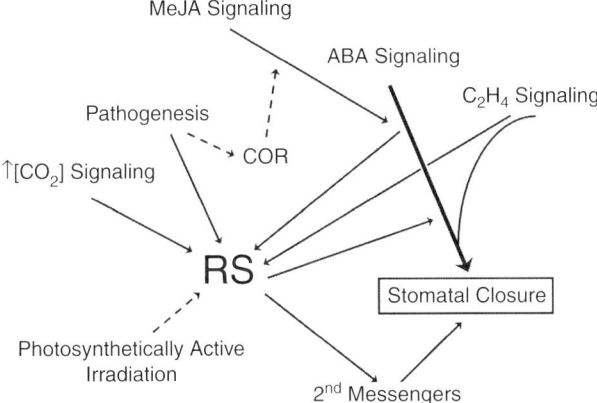

Fig. 7.6 RS (reactive oxygen species [ROS] and nitric oxide [NO] collectively) crosstalk in mediating stomatal closure. Stomata close in response to many stimuli that induce RS synthesis. Although the stomatal closure resulting from prolonged water stress may occur as a result of RS-independent abscisic acid (ABA) signaling, much of that which occurs in well-hydrated leaves may well require RS synthesis and signaling. RS-induced stomatal closure appears to require a functional C_2H_4 signaling pathway, and there is some evidence to suggest this may impinge on downstream components of the ABA signaling cascade. Stomatal RS signaling itself requires a functional ABA signaling pathway, and there is evidence to suggest that methyl-jasmonate (MeJA) induces RS synthesis and signaling through the ABA signaling pathway. High levels of CO_2 induce RS synthesis in guard cells, as does the transition from the light to dark. Thus, in well-hydrated leaves, RS signaling may mediate the degree of stomatal aperture in relation to light and CO_2 levels. Pathogen signals cause RS synthesis in the guard cells of inoculated leaves and subsequent stomatal closure. More virulent strains manage to overcome this and cause stomata to reopen by producing the opening signal coronatine (COR), which may also inhibit the MeJA signaling pathway. Thus, it is clear that in an environment of multiple stimuli, RS play a central role in mediating stomatal aperture.

ABAR/CHLH and GCR2, appear to be involved in guard cell movements. Components of the downstream signaling cascade include heterotrimeric G-proteins, protein kinases and phosphatases, and other proteins involved in lipid signaling, transcriptional regulation, RNA processing, and posttranslational protein modifications (Hirayama and Shinozaki 2007). Ca^{2+} and the RS, H_2O_2 and NO are also important components in normal stomatal responses to ABA. In guard cells, the synthesis of H_2O_2 is induced by ABA in a manner dependent on the normal upstream function of the protein phospatase 2C encoded by ABA insensitive 1 (ABI1) and plays a role in the required activation of plasma membrane Ca^{2+} channels during stomatal closure. Genetic evidence suggests that in *Arabidopsis*, this involves the NOX isoforms AtrbohD and AtrbohF and requires the normal downstream function of the protein phosphatase 2C encoded by ABA insensitive 2 (ABI2) (Hirayama and Shinozaki 2007) and the MAP kinase AtMPK3 (Gudesblat et al. 2007). Recent studies have also suggested that H_2O_2 synthesized by copper amine oxidase may similarly be involved in ABA-induced stomatal closure (An et al. 2008) and that the downstream signaling of H_2O_2 may require the function of other P38-like MAP kinases (Jiang et al. 2008). Interestingly, the closure of stomata induced by ethylene is also mediated through AtrbohD/F associated H_2O_2 synthesis (Desikan et al. 2006). Ethylene induces guard cell H_2O_2 synthesis and the ethylene receptor mutants, *etr1-1* and *etr1-3* are insensitive to ethylene in terms of both stomatal closure and H_2O_2 production. Interestingly, although ABA will induce H_2O_2 synthesis in *etr1-1*, in contrast to *etr1-3*, the *etr1-1* mutant is insensitive to this ROS (Desikan et al. 2006). In *etr1-1* the mutation affects the Cys residue known to associate with the Cu^{2+} ion required for ethylene binding. Thus, there is a suggestion that H_2O_2 may directly interact with this part of the ETR1 histidine kinase receptor. Histidine kinases have been implicated as ROS sensors in other systems (Desikan et al. 2006). Stomata of the *Arabidopsis* ethylene signaling mutants, *ein2-1* and *arr2*, do not close in response to either ethylene or H_2O_2 but do generate H_2O_2 following ethylene stimulation. The inference is, therefore, that a branch from the ethylene signaling pathway involving as yet unknown proteins and upstream of EIN2 and ARR2 leads to increased AtrbohD/F activity. Presumably, ethylene signaling downstream of EIN2 and ARR2 is still required for the resulting increased H_2O_2 synthesis to be effective. Whether this is mediated by the effects of ethylene insensitivity on downstream components of the ABA signaling pathway is unknown. However, it would appear that the NOX-associated synthesis of H_2O_2 and its downstream signaling through ABI2 may well constitute one point of convergence of the ABA, and ethylene signaling pathways that are active during stomatal closure.

ABA also induces the synthesis of NO in guard cells in a manner dependent on the prior synthesis of H_2O_2 (Bright et al. 2006), and this is required for proper stomatal closure in response to this hormone. ABA-induced stomatal closure is also impaired in the *Atnoa1* and *nia1*, *nia2* mutants, which fail to make NO in response to such hormonal challenge (Bright et al. 2006). NO is certainly made in guard cells in response to water stress, and this results as a consequence of increased ABA synthesis (Zhang et al. 2007). However, whether NO plays a signaling role in mediating drought-induced stomatal closure is still debatable. The stomata of rapidly wilted leaves of both the *nia1* and *nia1*, *nia2* mutants close with equal rapidity to those of wildtype plants undergoing similar treatment, whereas those of the *abi1* mutant remain open (Neill et al. 2008). Thus, stomatal closure appears to be both active and ABA-dependent but does not require NO. Perhaps NO has a more significant role in mediating stomatal movements in well-hydrated plants undergoing more gradual water loss. Moreover, dehydration is likely to induce the synthesis of several signaling molecules including ABA, H_2O_2, NO, and ethylene. Given that, in some circumstances, ethylene inhibits ABA- and NO-induced stomatal closure (Desikan et al. 2006), it is likely that the actual stomatal apertures achieved during drought stress result from considerable signaling crosstalk.

Both H_2O_2 and NO play a role during the stomatal closure that results from the light-to-dark transition. Inhibiting the guard cell synthesis of NO when leaves are transferred from the light to dark prevents stomatal closure (She et al. 2004). Also, pretreating leaves with ascorbate or catalase to remove their H_2O_2 also prevents stomatal closure on the transfer of leaves to the dark. Stomata of the *nia1* mutant also fail to close when leaves are transferred to the dark (Ribeiro et al. 2009) and the reopening of wildtype *Arabidopsis* stomata on transfer from the dark to the light can be inhibited by ABA or NO (Yan et al. 2007). This ABA-induced inhibition of stomatal reopening can be diminished by removing the NO. Recent studies also indicate that NO inhibits the blue light–induced reopening of stomata and that this involves the ABA signaling pathway (Zhang et al. 2007) and downstream Ca^{2+} signaling (Garcia-Mata and Lamattina 2007).

RS may modulate stomatal closure induced by elevated CO_2 levels. High levels of bicarbonate (HCO_3^-) ions have long been known to induce stomatal closure, and it has recently been shown that this correlates well with the synthesis of both H_2O_2 (Kolla et al. 2007) and NO (Kolla and Raghavendra 2007) in the guard cells of *Arabidopsis* and *Pisum sativum* respectively. In each case, removal of the RS effectively prevents this closure. How CO_2 initiates this increase in RS synthesis remains unknown, and a specific CO_2 receptor and signaling pathway has still not been identified. However, any protein that interacts with HCO_3^- may be a candidate for involvement in CO_2 sensing. In mammalian systems, such an enzyme is soluble adenylyl cyclase (Kamenetsky et al. 2006). Whichever is the case, it is intriguing to consider that RS signaling may be an important feature of the crosstalk that occurs to balance the degree of stomatal opening in relation to the available levels of photosynthetically useful irradiation and the atmospheric concentration of CO_2.

RS mediate the guard cell signaling crosstalk that occurs during pathogenesis. It has generally been assumed that pathogenic bacteria and fungi use stomata as passive ports of entry to the subepidermal tissues of leaves. However, in what has been termed a pathogen-associated molecular pattern (PAMP)-induced basal defense mechanism, it has recently become apparent that active stomatal closure constitutes part of the innate immunity of plants to some bacteria (Underwood et al. 2007). For example, *Arabidopsis* guard cells can perceive molecules such as the lipopolysaccharides (LPS) and flagellin on the cell surface of *Pseudomonas syringae*. Upon inoculation, these compounds trigger guard cell NO synthesis, which subsequently results in stomatal closure. Inhibiting the NO synthesis prevents this *P. syringae*–induced stomatal closure and, contrasting with those of wildtype, stomata of either the ABA-insensitive stomata mutant *ost1-2* or the ABA-deficient mutant *aba3-1*, fail to close in response to LPS and flagellin. Thus, it would appear that the ABA signaling pathway that leads to increased NO synthesis and signaling is required for this pathogen-induced response. Interestingly, some of the more virulent bacteria appear to be able to overcome this guard cell response and are able to cause stomata to reopen. Knockout mutants of *P. syringae* exist that are unable to elicit stomatal reopening, and these have been used to identify the reopening signal as the phytotoxin coronatine (COR; Underwood et al. 2007). COR effectively prevents ABA-induced stomatal closure in wildtype *Arabidopsis* leaves but has no effect on the guard cells of the coronatine insensitive mutant, *coi1*. COR does not appear to prevent ABA-induced NO biosynthesis in wildtype *Arabidopsis* guard cells, suggesting that its site of action may occur downstream of ABI2. In more recent studies, it has also been demonstrated that in methyl-jasmonate (MeJA)-induced stomatal closure (Suhita et al. 2003, 2004), the associated NO biosynthesis and Ca^{2+} channel activity are impaired in leaves of the *coi1* mutant (Munemasa et al. 2007). This suggests that COI1 plays a direct role in MeJA signaling in guard cells. In this study, MeJA also failed to cause stomatal closure in leaves of the *abi2* mutant, although the induction of NO synthesis was as in wildtype plants. Thus, it would seem that MeJA may signal through a branch point of the ABA signaling pathway upstream of ABI2 and induces ROS and NO production to effect stomatal closure. However, this observation does raise the issue as to whether the protein encoded by COI1 plays a direct role in the ABA signaling that leads to stomatal closure. Perhaps the effect of COR in guard cells is actually to promote stomatal opening rather than inhibiting ABA-induced closure and that an additional effect of its presence in guard cells is the inhibition of MeJA signaling. Certainly the predominant effect of light signaling appears to be to one of promoting stomatal opening (Shimazaki et al. 2007), and it will be interesting to see whether COI1 plays a role in this respect.

The extent to which stomata open clearly results from the ability of the guard cells to simultaneously perceive, transduce, and balance the effect of a number of input signals. It is also clear that many of the signaling pathways involved have common components and that RS such as H_2O_2 and NO may play a key central role in balancing the overall effect of the crosstalk between these multiple signaling inputs. However, fully unraveling the crosstalk that operates during guard cell movements and determining exactly how these signaling pathways impinge on each other will require a concerted systems biology approach.

Conclusions

It is clear that RS are essential components of many developmental and physiological processes in plants. Altered amounts and intracellular distributions of RS lead to altered signaling profiles within cells accompanied by widespread changes in gene expression and protein activities. Various abiotic and biotic stresses alter the RS landscape of cells by altering their turnover and distribution. Although the mechanisms by which this is achieved are likely to vary depending on the stress in question, many of the downstream effects of RS appear to be common to each of these different stresses. Inevitably then, RS act as nodes of signaling crosstalk between the different stresses and "normal" cell metabolism. Continued research into RS turnover and responses may identify key "RS sensors" and a better holistic picture of the positive and negative effectors of RS responses. Given the cross-stress effects of RS, it is to be hoped that this might facilitate the production of plants with simultaneous tolerance to several varieties of stresses.

References

An, Z., Jing, W., Liu, Y., and Zhang, W. 2008. Hydrogen peroxide generated by copper amine oxidase is involved in abscisic acid-induced stomatal closure in *Vicia faba*. *J Exp Botany* 59:815–825.

Apel, K., and Hirt, H. 2004. Reactive oxygen species: metabolism, oxidative stress and signal transduction. *Annu Rev Plant Biol* 55:373–399.

Arasimowicz, M., and Floryszak-Wieczorek, J. 2007. Nitric oxide as a bioactive signalling molecule in plant stress responses. *Plant Sci* 172:876–887.

Bailey-Serres, J., and Mittler, R. 2006. The roles of reactive oxygen species in plant cells. *Plant Physiol* 141:311.

Barroso, J., Corpas, F.J., Carreras, A., Rodréguez-Serrano, M., Esteban, F.J., Fernández-Ocaña, A., Chaki, M., Romero-Puertas, M.C., Valderrama, R., Sandalio, L.M., and del Río, L.A. 2006. Localization of S-nitrosoglutathione and expression of S-nitrosoglutathione reductase in pea plants under cadmium stress. *J Exp Botany* 57:1785–1793.

Belenghi, B., Romero-Puertas, M.C., Vercammen, D., Brackenier, A., Inzé, D., Delledonne, M., and Van Breusegem, F. 2007. Metacaspase activity of *Arabidopsis thaliana* is regulated by S-nitrosylation of a critical cysteine residue. *J Biol Chem* 282:1352–1358.

Besson-Bard, A., Courtois, C., Gauthier, A., Dahan, J., Dobrowolska, G., Jeandroz, S., Pugin, A., and Wendehenne, D. 2008. Nitric oxide in plants: production and cross-talk with Ca^{2+} signalling. *Mol Plant* 1:218–228.

Bindschedler, L.V., Dewdney, J., Blee, K.A., Stone, J.M., Asai, T., Plotnikov, J., Denoux, C., Hayes, T., Gerrish, C., Davies, D.R., Ausubel, F.M., and Bolwell, G.P. 2006. Peroxidase-dependent apoplastic oxidative burst in *Arabidopsis* required for pathogen resistance. *Plant J* 47:851–863.

Bright, J., Desikan, R., Hancock, J.T., Weir, I.S., and Neill, S.J. 2006. ABA-induced NO generation and stomatal closure in *Arabidopsis* are dependent on H_2O_2 synthesis. *Plant J* 45:113–122.

Brodersen, P., Petersen, M., Bjørn Nielsen, H., Zhu, S., Newman, M.A., Shokat, K.M., Rietz, S., Parker, J., and Mundy, J. 2006. *Arabidopsis* MAP kinase 4 regulates salicylic acid- and jasmonic acid/ethylene-dependent responses via EDS1 and PAD4. *Plant J* 47:532–546.

Ciftci-Yilmaz, S., and Mittler, R. 2008. The zinc finger network of plants. *Cell Mol Life Sci* 65:1150–1160.

de Pinto, M.C., Paradiso, A., Leonetti, P., and De Gara, L. 2006. Hydrogen peroxide, nitric oxide and cytosolic ascorbate peroxidase at the crossroad between defence and cell death. *Plant J* 48:784–795.

Delaunay, A., Pflieger, D., Barrault, M.B., Vinh, J., and Toledano, M.B. 2002. A thiol peroxidase is an H_2O_2 receptor and redox-transducer in gene activation. *Cell* 111:471–481.

Desikan, R., Last, K., Harrett-Williams, R., Tagliavia, C., Harter, K., Hooley, R., Hancock, J.T., and Neill, S.J. 2006. Ethylene-induced stomatal closure in *Arabidopsis* occurs *via* AtrbohF-mediated hydrogen peroxide synthesis. *Plant J* 47:907–916.

Dixon, D.P., Skipsey, M., Grundy, N.M., and Edwards, R. 2005. Stress-induced protein S-glutathionylation in *Arabidopsis*. *Plant Physiol* 138:2233–2244.

Durner, J., Wendehenne, D., and Klessig, D.F. 1998. Defense gene induction in tobacco by nitric oxide, cyclic GMP, and cyclic ADP-ribose. *Proc Natl Acad Sci U S A* 95:10328–10333.

Ederli, L., Morettini, R., Borgogni, A., Wasternack, C., Miersch, O., Reale, L., Ferranti, F., Tosti, N., and Pasqualini, S. 2006. Interaction between nitric oxide and ethylene in the induction of alternative oxidase in ozone-treated tobacco plants. *Plant Physiol* 142:595–608.

Fobert, P.R., and Despres, C. 2005. Redox control of systemic acquired resistance. *Curr Opin Plant Biol* 8:378–382.

Foyer, C.H., and Noctor, G. 2005. Redox homeostasis and antioxidant signalling: a metabolic interface between stress perception and physiological responses. *Plant Cell* 17:1866–1875.

Fujita, M., Fujita, Y., Noutoshi, Y., Takahashi, F., Narusaka, Y., Yamaguchi-Shinozaki, K., and Shinozaki, K. 2006. Crosstalk between abiotic and biotic stress responses: a current view from the points of convergence in the stress signalling networks. *Curr Opin Plant Biol* 9:436–442.

Gadjev, I., Vanderauwera, S., Gechev, T.S., Laloi, C., Minkov, I.N., Shulaev, V., Apel, K., Inze, D., Mittler, R., and Van Breusegem, F. 2006. Transcriptomic footprints disclose specificity of reactive oxygen species signaling in *Arabidopsis*. *Plant Physiol* 141:436–445.

Gapper, C., and Dolan, L. 2006. Control of plant development by reactive oxygen species. *Plant Physiol* 141:341–345.

Garcia-Mata, C., Gay, R., Sokolovski, S., Hills, A., Lamattina, L., and Blatt, M.R. 2003. Nitric oxide regulates K^+ and Cl^- channels in guard cells through a subset of abscisic acid-evoked signalling pathways. *Proc Natl Acad Sci U S A* 100:11116–11121.

Garcia-Mata, C., and Lamattina, L. 2007. Abscisic acid (ABA) inhibits light-induced stomatal opening through calcium- and nitric oxide-mediated signaling pathways. *Nitric Oxide* 17:143–151.

Gomi, K., Ogawa, D., Katou, S., Kamada, H., Nakajima, N., Saji, H., Soyano, T., Sasabe, M., Machida, Y., Mitsuhara, I., Ohashi, Y., and Seo, S. 2005. A mitogen-activated protein kinase NtMPK4 activated by SIPKK is required for jasmonic acid signaling and involved in ozone tolerance *via* stomatal movement in tobacco. *Plant Cell Physiol* 46:1902–1914.

Gudesblat, G.E., Iusem, N.D., and Morris, P.C. 2007. Guard cell-specific inhibition of *Arabidopsis* MPK3 expression causes abnormal stomatal responses to abscisic acid and hydrogen peroxide. *New Phytologist* 173:713–721.

Guo, F., Okamoto, M., and Crawford, N.M. 2003. Identification of a plant nitric oxide synthase gene involved in hormonal signaling. *Science* 302:100–103.

Halliwell, B. 2006. Reactive species and antioxidants. Redox biology is a fundamental theme of aerobic life. *Plant Physiol* 141:312–322.

Hancock, J., Desikan, R., Harrison, J., Bright, J., Hooley, R., and Neill, S. 2006. Doing the unexpected: proteins involved in hydrogen peroxide perception. *J Exp Botany* 57:1711–1718.

Hancock, J.T., Henson, D., Nyirenda, M., Desikan, R., Harrison, J., Lewis, L., Hughes, J., and Neill, S.J. 2005. Proteomic identification of glyceraldehyde 3-phosphate dehydrogenase as an inhibitory target of hydrogen peroxide in *Arabidopsis*. *Plant Physiol Biochem* 43:828–835.

He, J.M., Xu, H., She, X.P., Song, X.G., and Zhao, W.M. 2005. The role and the interrelationship of hydrogen peroxide and nitric oxide in the UV-B-induced stomatal closure in broad bean. *Funct Plant Biol* 32:237–247.

Hirayama, T., and Shinozaki, K. 2007. Perception and transduction of abscisic acid signals: keys to the function of the versatile plant hormone ABA. *Trends Plant Sci* 12:343–351.

Holtgrefe, S., Gohlke, J., Starmann, J., Druce, S., Klocke, S., Altmann, B., Wojtera, J., Lindermayr, C., and Scheibe, R. 2008. Regulation of plant cytosolic glyceraldehyde 3-phosphate dehydrogenase isoforms by thiol modifications. *Physiol Plantarum* 133:211–228.

Hong, J.K., Yun B-W., Kang, J-G., Raja, M.U., Kwon, E., Sorhagen, K., Chu, C., Wang, Y., and Loake, G. J. 2007. Nitric oxide function and signaling in plant disease resistance. *J Exp Botany* 59:147–154.

Hu, X., Bidney, D.L., Yalpani, N., Duvick, J.P., Crasta, O., Folkerts, O., and Lu, G. 2003. Overexpression of a gene encoding hydrogen peroxide-generating oxalate oxidase evokes defense responses in sunflower. *Plant Physiol* 133:170–181.

Hu, X., Jiang, M., Zhang, J., Zhang, A., Lin, F., and Tan, M. 2007. Calcium-calmodulin is required for abscisic acid-induced antioxidant defense and functions both upstream and downstream of H_2O_2 production in leaves of maize (*Zea mays*) plants. *New Phytol* 173:27–38.

Huang, X., Stettmaier, K., Michel, C., Hutzler, P., Mueller, M.J., and Durner, J. 2004. Nitric oxide is induced by wounding and influences jasmonic acid signalling in *Arabidopsis thaliana*. *Planta* 218:938–946.

Jiang, J., An, G., Wang, P., Wang, P., Han, J., Jia, Y., and Song, C. 2003. MAP kinase specifically mediates the ABA-induced H_2O_2 generation in guard cells of *Vicia faba* L. *Chin Sci Bull* 18:1919–1926.

Jiang, J., Wang, P., An, G., Wang, P., and Song, C.P. 2008. The involvement of a P38-like MAP kinase in ABA-induced and H_2O_2-mediated stomatal closure in *Vicia faba* L. *Plant Cell Reporter* 27:377–385.

Kalbina, I., and Strid, A. 2006. The role of NADPH oxidase and MAP kinase phosphatase in UV-B-dependent gene expression in *Arabidopsis*. *Plant Cell Environ* 29:1783–1793.

Kamenetsky, M., Middelhaufe, S., Bank, E.M., Levin, L.R., Buck, J., and Steegborn, C. 2006. Molecular details of cAMP generation in mammalian cells: a tale of two systems. *J Mol Biol* 362:623–639.

Klotz, L.O. 2002. Oxidant-induced signaling: effects of peroxynitrite and singlet oxygen. *J Biol Chem* 383:443–456.

Kobayashi, M., Ohura, I., Kawakita, N., Fujiwara, M., Shimamoto, K., Doke, N., and Yoshioka, H. 2007. Calcium-dependent protein kinases regulate the production of reactive oxygen species by potato NADPH oxidase. *Plant Cell* 19:1065–1080.

Kolla, V.A., and Raghavendra, A.S. 2007. Nitric oxide is a signaling intermediate during bicarbonate-induced stomatal closure in *Pisum sativum*. *Physiol Plantarum* 130:91–98.

Kolla, V.A., Vavasseur, A., and Raghavendra, A.S. 2007. Hydrogen peroxide production is an early event during bicarbonate induced stomatal closure in abaxial epidermis of *Arabidopsis*. *Planta* 225:1421–1429.

Kwak, J.M., Nguyen, V., and Schroeder, J.I. 2006. The role of reactive oxygen species in hormonal responses. *Plant Physiology* 141:323–329

Laloi, C., Stachowiak, M., Pers-Kamczyc, E., Warzych, E., Murgia, I., and Apel, K. 2007. Cross-talk between singlet oxygen- and hydrogen peroxide-dependent signalling of stress responses in *Arabidopsis thaliana*. *Proc Natl Acad Sci U S A*. 104:672–677.

Lamattina, L., Garcia-Mata, C., Graziano, M., and Pagnussat, G. 2003. Nitric oxide: the versatility of an extensive signal molecule. *Annu Rev Plant Biol* 54:109–136.

Li, S., Assmann, S.M., and Albert, R. 2006. Predicting essential components of signal transduction networks: a dynamic model of guard cell abscisic acid signaling. *Public Library Sci Biol* 4:1732–1748.

Lindermayr, C., Saalbach, G., Bahnweg, G., and Durner, J. 2006. Differential inhibition of *Arabidopsis* methionine adenosyltransferases by protein S-nitrosylation. *J Biol Chem* 281:1–7.

Lindermayr, C., Saalbach, G., and Durner, J. 2005. Proteomic identification of S-nitrosylated proteins in *Arabidopsis*. *Plant Physiol* 137:921–930.

Ludidi, N., and Gehring, C. 2003. Identification of a novel protein with guanylyl cyclase activity in *Arabidopsis thaliana*. *J Biol Chem* 278:6490–6494.

Mahalingham, R., Jambunathan, N., Gunjan, S.K., Faustin E., Weng, H., and Ayoubi, P. 2006. Analysis of oxidative signalling induced by ozone in *Arabidopsis thaliana*. *Plant Cell Environ* 29:1357–1371.

Miao, Y., Dong, L.V., Wang, P., Wang, X.C., Chen, J., Miao, C., and Song, C.P. 2006. An *Arabidopsis* glutathione peroxidase functions as both a redox transducer and a scavenger in abscisic acid and drought stress responses. *Plant Cell* 18:2749–2766.

Miles, G.P., Samuel, M.A., Zhang, Y., and Ellis, B.E. 2005. RNA interference-based RNAi suppression of AtMPK6, an *Arabidopsis* mitogen-activated protein kinase, results in hypersensitivity to ozone and misregulation of AtMPK3. *Environm Pollution* 138:230–237.

Miller, G., and Mittler, R. 2006. Could heat shock transcription factors function as hydrogen peroxide sensors in plants? *Ann Botany* 98:279–288.

Mou, Z., Fan, W., and Dong, X. 2003. Inducers of plant systemic acquired resistance regulate NPR1 function through redox changes. *Cell* 113:935–944.

Munemasa, S., Oda, K., Watanabe-Sugimoto, M., Nakamura, Y., Shimoishi, Y., and Murata, Y. 2007. The coronatine-insensitive 1 mutation reveals the hormonal signaling interaction between abscisic acid and methyl jasmonate in *Arabidopsis* guard cells. Specific impairment of ion channel activation and second messenger production. *Plant Physiol* 143:1398–1407.

Neill, S., Barros, R., Bright, J., Desikan, R., Hancock, J., Harrison, J., Morris, P., Ribeiro, D., and Wilson, I. 2008. Nitric oxide, stomatal closure and abiotic stress. *J Exp Botany* 59:165–176.

Perazzolli, M., Romero-Puertas, M.C., and Delledonne, M. 2006. Modulation of nitric oxide bioactivity by plant haemoglobins. *J Exp Botany* 57:479–488.

Pérez-Ruiz, J.M., Spínola, M.C., Kirchsteiger, K., Moreno, J., Sahrawy, M., and Cejudo, F.J. 2006. Rice NTRC is a high-efficiency redox system for chloroplast protection against oxidative damage. *Plant Cell* 18:2356–2368.

Pitzschke, A., and Hirt, H. 2006. Mitogen-activated protein kinases and reactive oxygen species signaling in plants. *Plant Physiol* 141:351–356.

Planchet, E., and Kaiser, W.M. 2006. Nitric oxide production in plants. *Plant Signal Behav* 1:46–51.

Queval, G., Issakidis-Bourguet, E., Hoeberichts, F.A., Vandorpe, M., Gakière, B., Vanacker, H., Miginiac-Maslow, M., Van Breusegem F., and Noctor, G. 2007. Conditional oxidative stress responses in the *Arabidopsis* photorespiratory mutant cat2 demonstrate that redox state is a key modulator of daylength-dependent gene expression, and define photoperiod as a crucial factor in the regulation of H_2O_2-induced cell death. *Plant J* 52:640–657.

Rentel, M.C., Lecourieux, D., Quaked, F., Usher, S.L., Petersen, L., Okamoto, H., Knight, H., Peck, S.C., Grierson, C.S., Hirt, H., and Knight, M.R. 2004. *OXI1* kinase is necessary for oxidative burst-mediated signalling in *Arabidopsis*. *Nature* 427:858–861.

Ribeiro, D.M., Desikan, R., Bright, J., Confraria, A., Harrison, J., Hancock, J.T., Barros, R.S., Neill, S.J. and Wilson, I.D. 2009. Differential requirement for NO during ABA-induced stomatal closure in turgid and wilted leaves. *Plant Cell Environ* 32:46–57.

Romero-Puertas, M.C., Campostrini, N., Mattè, A., Righetti, P.G., Perazzolli, M., Zolla, L., Roepstorff, P., and Delledonne, M. 2008. Proteomic analysis of S-nitrosylated proteins in *Arabidopsis thaliana* undergoing hypersensitive response. *Proteomics* 8:1459–1469.

Sadanandom, A., Poghosyan, Z., Fairbairn, D.J., and Murphy, D.J. 2000. Differential regulation of plastidial and cytosolic isoforms of peptide methionine sulfoxide reductase in *Arabidopsis*. *Plant Physiol* 123:255–264.

Sagi, M., and Fluhr, R. 2006. Production of reactive oxygen species by plant NADPH oxidases. *Plant Physiol* 141:336–340.

Schafer, F.Q., and Buettner, G.R. 2001. Redox environment of the cell as viewed through the redox state of the glutathione disulfide/glutathione couple. *Free Radical Biol Med* 30:1191–1212.

She, X.P., Song, X.G., and He, J.M. 2004. Role and relationship of nitric oxide and hydrogen peroxide in light/dark-regulated stomatal movement in *Vicia faba*. *Acta Botanica Sinica* 46:1292–1300.

Shimazaki, K., Doi, M., Assmann, S.M., and Kinoshita, T. 2007. Light regulation of stomatal movement. *Annu Rev Plant Biol* 58:219–247.

Stohr, C., and Stremlau, S. 2006. Formation and possible roles of nitric oxide in plant roots. *J Exp Botany* 57:463–470.

Suhita, D., Kolla, V.A., Vavasseur, A., and Raghavendra, A.S. 2003. Different signaling pathways involved during the suppression of stomatal opening by methyl jasmonate or abscisic acid. *Plant Sci* 164:481–488.

Suhita, D., Raghavendra, A.S., Kwak, J.M., and Vavasseur, A. 2004. Cytoplasmic alkalization precedes reactive oxygen species production during methyl jasmonate- and abscisic acid-induced stomatal closure. *Plant Physiol* 134:1536–1545.

Taylor, J.E., and McAinsh, M. 2004. Signalling crosstalk in plants: emerging issues. *J Exp Botany* 55:147–149.

Torres, M.A., Jones, J.D.G., and Dangl, J.L. 2006. Reactive oxygen species signaling in response to pathogens. *Plant Physiol* 141:373–378.

Tun, N.N., Santa-Catarina, C., Begum, T., Silveira, V., Handro, W., Floh, E.I.S., and Scherer, G.F.E. 2006. Polyamines induce rapid biosynthesis of nitric oxide NO in *Arabidopsis thaliana* seedlings. *Plant Cell Physiol* 47:346–354.

Underwood, W., Melotto, M., and He, S.Y. 2007. Role of plant stomata in bacterial invasion. *Cell Microbiol* 9:1621–1629.

Van Breusegem, F., and Dat, J.F. 2006. Reactive oxygen species in plant cell death. *Plant Physiol* 141:384–390.

Wang, Y., Yun, B.W., Kwon, E.J., Hong, J.K., Yoon, J.Y., and Loake G.J. 2006. S-nitrosylation: an emerging redox-based post-translational modification in plants. *J Exp Botany* 57:1777–1784.

Wendehenne, D., Pugin, A., Klessig, D.F., and Durner, J. 2001. Nitric oxide: comparative synthesis and signaling in animal and plant cells. *Trends Plant Sci* 6:177–183.

Wilson, I.D., Neill, S.J., and Hancock, J.T. 2008. Nitric oxide synthesis and signalling in plants. *Plant Cell Environ* 31:622–631.

Xing, Y., Jia, W., and Zhang, J. 2007. AtMEK1 mediates stress-induced gene expression of CAT1 catalase by triggering H_2O_2 production in *Arabidopsis*. *J Exp Botany* 58:2969–2981.

Yamasaki, H., and Cohen, M.F. 2006. NO signal at the crossroads: polyamine-induced nitric oxide synthesis in plants. *Trends Plant Sci* 11:522–524.

Yan, J., Tsuichihara, N., Etoh, T., and Iwai, S. 2007. Reactive oxygen species and nitric oxide are involved in ABA inhibition of stomatal opening. *Plant Cell Environ* 30:1320–1325.

Zago, E., Morsa, S., Dat, J.F., Alard, P., Ferrarini, A., Inzé, D., Delledonne, M., and Van Breusegem, F. 2006. Nitric oxide- and hydrogen peroxide-responsive gene regulation during cell death induction in tobacco. *Plant Physiol* 141:404–411.

Zaninotto, F., La Camera, S., Polverani, A., and Delledonne, M. 2006. Crosstalk between reactive nitrogen and oxygen species during the hypersensitive disease resistance response. *Plant Physiol* 141:379–383.

Zhang, A., Jiang, M., Zhang, J., Ding, H., Xu, S., Hu, X., and Tan, M. 2007. Nitric oxide induced by hydrogen peroxide mediates abscisic acid-induced activation of the mitogen-activated protein kinase cascade involved in antioxidant defense in maize leaves. *New Phytol* 175:36–50.

Zhang, X., Takemiya, A., Kinoshita, T., and Shimazaki, K. 2007. Nitric oxide inhibits blue light-specific stomatal opening *via* abscisic acid signaling pathways in *Vicia* guard cells. *Plant Cell Physiol* 48:715–723.

Zhu, S-H., and Zhou, J. 2006. Effect of nitric oxide on ethylene production in strawberry fruit during storage. *Food Chem* 100:1517–1522.

8 TORing with Cell Cycle, Nutrients, Stress, and Growth

Desh Pal S. Verma and Jayanta Chatterjee

Introduction

TOR (Target Of Rapamycin) signaling pathway has been subjected to almost three decades of intense investigation, beginning in the early 1970s with the discovery of rapamycin (Vezina et al. 1975; Heitman et al. 1991). This lipophilic macrolide was isolated from soil bacteria *Streptomyces hygroscopius* in the Pacific on Easter Island (locally called "rapa nui") as an anti-fungal agent (Vezina et al. 1975; Singh et al. 1979). This bacterial metabolite was purified and found to be a macrocyclic lactone, named "rapamycin." Rapamycin was found to inhibit proliferation of mammalian cells and to possess immunosuppressive properties. These intriguing observations prompted further investigation into the mode of action of rapamycin (Wullschleger et al. 2006). Research involving this antifungal agent gained momentum with the genetic identification of *TOR* genes in yeast (Heitman et al. 1991). Transcriptional profiling of rapamycin treatment of yeast, *Drosophila*, and mammalian cells showed that approximately 5% of the genes in the genome are affected, indicating that the TOR pathway has a broad impact on cellular functions (Hardwick et al. 1999; Peng et al. 2002; Guertin et al. 2006; Reiling and Sabatani 2006). TOR in yeast couples transcription, ribosome biogenesis, translation initiation, nutrient uptake, and autophagy to the abundance and quality of available nutrients. Thus, it functions as a temporal regulator of cell growth (Jacinto and Hall 2003; Jacinto et al. 2004; Corradetti and Guan 2006). TOR pathway also controls many aspects of cellular physiology including mitochondrial signaling and determination of the life span in animals, yeast, and *Caenorhabditis elegans* (Schieke and Finkel 2006).

Plants also contain TOR kinase pathway (Menand et al. 2002; Mahfouz et al. 2006), but its sensitivity to rapamycin varies in different species. In this chapter, we first discuss the caveat of plant resistance against rapamycin. A discussion of plant cell cycle, cellular differentiation, and growth in relation to TOR is followed by discussion of the relationship of plant TOR pathway with nutrients and stress (biotic and abiotic).

Conservation of TOR Pathway in Eukaryotes

TOR is an evolutionarily conserved protein present in all eukaryotes, and all eukaryotic genomes tested so far have at least one copy of TOR gene (Arsham and Neufeld 2006; Reiling and Sabatini 2006). TOR proteins are large (250–300 kD) serine/threonine protein kinases, which share 40% to 60% identity among different organisms. TOR belongs to the large family of PI3K-related kinases (PIKKs). Unlike yeast (*Schizosaccharomyces pombe* and *Saccharomyces cerevisiae*), the mammals, *Drosophila*, *C. elegans*, and *Arabidopsis* have only one TOR homolog (Mahfouz et al. 2006; Reiling and Sabatini 2006). Mammalian TOR (mTOR) has several characteristic

Fig. 8.1 General architecture of TOR (Target Of Rapamycin) protein

domains, viz., FRB-rapamycin domain (2019-2114 amino acid residue), PI3 Kinase domain (2153-2431 amino acid residue), FATC domain (2519-2549 amino acid residue), and several consecutive HEAT (**H**untington, **E**longation factor 3, **A** subunit of protein phosphatase 2A, and **T**OR1 proteins) repeats (Choi et al. 1996; Schmelzle and Hall 2000; Adami et al. 2007). Phosphorylation sites for human TOR have been mapped to tyr^{2446} and Ser$^{2448, 2481}$ (Stan et al. 1994; Choi et al. 1996). Figure 8.1 shows the general domain archetecture of TOR protein.

There are two distinct TOR complexes (TORC1 and TORC2) that constitute a primordial signaling network conserved in eukaryotic evolution to control the fundamental process of growth. TORC1 complex is rapamycin-sensitive, whereas TORC2 is not. Most of the key components of two TOR complexes are conserved in all eucaryotes. TORC1 complex contains TOR (TOR1 or TOR2 in yeast) and raptor/KOG1 and LST8/GβL, whereas TORC2 contains TOR (only TOR2 in yeast), rictor/AVO3, and LST8/GβL (Arsham and Neufeld 2006; Wullschleger et al. 2006). Rapamycin forms a complex with a peptidyl-prolyl cis/trans isomerase, FKBP12, an immunophilin. FKBP12–rapamycin complex mediates the drug's toxicity by binding to TOR complexes. Like yeast, a mammalian counterpart of TORC2 (mTORC2) also exists. mTORC2 contains mTOR, mLST8/GβL, and mAVO3/Rictor, but not raptor. mTORC2 is rapamycin-insensitive and seems to function upstream of Rho GTPases to regulate actin cytoskeleton, as does yeast TORC2 (Jacinto et al. 2004).

TORC2 mediates spatial control of cell growth by polarizing the actin cytoskeleton. In this way, it influences the secretory pathway towards the growth site. TORC2 signaling is rapamycin-insensitive owing to the inability of rapamycin to bind to TOR in TORC2 complex (Loewith et al. 2002; Jacinto et al. 2004). Besides TOR, LST8 (GβL), and KOG1 also have orthologs in all eukaryotic genomes examined to date. mTOR forms a rapamycin-sensitive TORC1 complex with mLST8 (GβL) and raptor (mKOG1; Loewith et al. 2002; Hara et al. 2002; Kim et al. 2002). Homologs of the AVOs are less obvious based on sequence homology (Jacinto et al. 2004). Figure 8.2 shows the current consensus model for TOR pathway regulation, which has been extensively reviewed (Wullschleger et al. 2006).

Presence of TOR Pathway in Plants and Caveat of Plant Resistance against Rapamycin

It is now well established that plants contain a functional TOR pathway, and this pathway responds to stress and growth signals (Menand et al. 2004; Mahfouz et al. 2006; Kim and Verma, unpublished data). Formation of the FKBP12–rapamycin–TOR^{FRB} complex is responsible for inhibition of TOR activity in a wide range of organisms, including *Drosophila*, mammals, *S. cerevisiae*, and *Cryptococcus neoformans* (Schmelzel

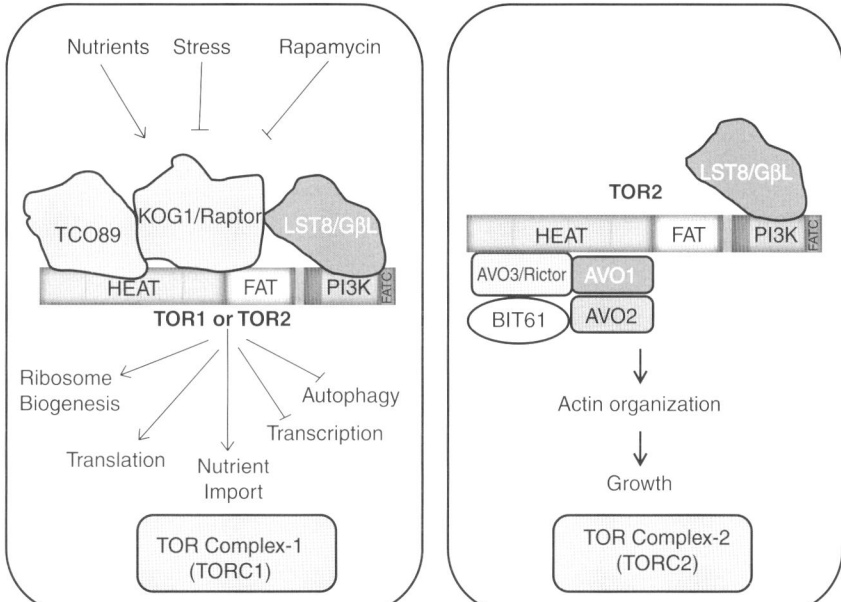

Fig. 8.2 TOR Complex 1 (TORC1) and TOR Complex 2 (TORC2) of *Saccharomyces cerevisiae*. Depicted are TOR-associated proteins (KOG1/Raptor, TCO89, LST8/GβL, AVO1-3, and BIT61) and the domains found in TOR protein (HEAT, FAT, FRB, kinase, and FATC). Both TORC1 and TORC2 are multimers, likely dimers. TORC1 mediates the rapamycin-sensitive signaling branch that couples growth cues to the accumulation of mass. Stimuli that positively regulate TORC1 and TORC1 outputs that promote the accumulation of mass are depicted with black arrows. Inputs that negatively regulate TORC1 and the stress- and starvation-induced processes that TORC1 regulates negatively are depicted with red bars. TORC2 signaling is rapamycin-insensitive and required for the organization of the actin cytoskeleton. Upstream regulators of TORC2 are not known. (Adapted from Wullschleger et al. 2006).

and Hall 2000; Zhang and Forde 2000). The unicellular algae *Chlamydomonas reinharditii* is susceptible to rapamycin treatment, whereas another algae, *Protheca segbwema*, is resistant (Baker et al. 1978). It is moderately active against filamentous fungi, including some *Dermatophytes* and weakly active against dimorphic fungi (Baker et al. 1978). In contrast, the growth of *Arabidopsis* was found to be insensitive to this drug even at concentrations up to 10 μM, which is 100 times the concentration inhibiting yeast growth (Menand et al. 2002; Mahfouz et al. 2006). In contrast, maize TOR kinase (ZmTOR) is responsive to rapamycin treatment (Reyes de la Cruz et al. 2004; Agredano-Moreno et al. 2007; Dinkova et al. 2007). Interestingly, rapamycin inhibits the growth of *C. reinhardtii*, but not of the bryophyte *Physcomitrella patens*. TOR kinase from monocot *Oryza sativa*, and the dicots *Nicotiana tabacum* and *Brassica napus* are also resistant to rapamycin treatment (Menand et al. 2002). These apparant contradictions suggest that the resistance to this drug may not be due to TOR but to other auxlliary proteins associated with the TOR pathway. This apparently involves plant FKBP12s rather than the TOR^{FRB} because it is the first level of interaction with rapamycin. The AtTOR^{FRB} is able to complex with rapamycin and yeast ScFKBP12, whereas the FKBP12 from *Vicia faba* (broad beans) does not restore the sensitivity of the *S. cerevisiae* FKBP12 mutant to rapamycin (Xu et al. 1998). Mutations in FKBP12 of *Chlamydomonas* alter the sensitivity with rapamycin (Crespo et al. 2005; Diaz-Troya et al. 2008).

The crucial residues involved in TOR^{FRB}–rapamycin–FKBP12 interaction are conserved in a wide range of organisms, including maize (Menand et al. 2002; Agredano-Moreno et al. 2005; Crespo et al. 2005). It has been found that TOR interacts with the triene arm of rapamycin, but only after the pipecolinyl moiety of rapamycin has bound to FKBP12 (Choi et al. 1996). Thus, the inhibition of TOR by rapamycin depends on the ability of FKBP12 to bind rapamycin (Agredano-Moreno et al. 2007). It has been suggested that the mutation in the conserved serine residue of TOR^{FRB} in yeast (*S. pombe*) leads to rapamycin resistance without affecting other functions of SpTOR. This mutant of SpTOR can also interact with SpFKBP12 protein (Weisman and Choder 2001). Agredano-Moreno and his colleagues argued that conservation of TOR^{FRB} is not enough to ensure sensitivity for ZmTOR kinase to rapamycin inhibition. Rather, FKBP12 protein contains structural features needed to acquire this characteristic, relevant to determination of TOR activity, which is regulated by TOR–Raptor/Rictor association (Agredano-Moreno et al. 2007).

One can argue that rapamycin does not enter plant cells or that a detoxification pathway exists in some plants (Menand et al. 2004). This seems unlikely, because rapamycin is able to enter *C. reinhardtii* cells (Menand et al. 2002). In *S. pombe*, SpTORs and SpFKBP12 proteins can interact with rapamycin without inhibiting some TOR functions (Weisman and Choder 2001).

Major Players of TOR Pathway in Plants

Plant TOR Kinase

Arabidopsis possesses a single TOR gene, AtTOR (At1G50030). It encodes a 279 kDa protein with several characteristic TOR domains. It is able to form a rapamycin-dependent complex with yeast 12-kDa FK506-binding protein (FKBP12) but does not interact with *Arabidopsis* homolog of FKBP12 (discussed later). AtTOR has 3 motifs (20–40 residues), named HEAT repeats, which are found in all TOR homologs. This domain may be involved in protein–protein interactions (Schmelzle and Hall 2000). In silico prediction of various domain structures of AtTOR is as follows: three HEAT repeats (204–240, 292–327, and 736–773 amino acid residues), FATC domain (1449–1815 amino acid residue), FRB domain (1920–2023 amino acid residue), PI3 kinase domain (2092–2417 amino acid residue). Protein sequence alignments also show that mTOR is the closest homolog of AtTOR, whereas AtTOR is closer to yeast TOR2 (Mennad et al. 2002; Mahfouz et al. 2006).

The regulatory-associated protein of TOR (RAPTOR) interacts with TOR and has HEAT and WD-40 protein-interacting domains. It recruits substrates for phosphorylation by TOR (Anderson and Hanson 2005; Mahfouz et al. 2006). Disruption of RAPTOR (encoded by two paralogous genes in *Arabidopsis*) results in arrest of seedling development while allowing normal embryonic development (Anderson and Hanson 2005). Increasing the level of RAPTOR expression in transgenic plants rendered S6K1 insensitive to osmotic stress (Mahfouz et al. 2006). Decreased TOR signaling activity may be associated with increased resistance to some stress conditions, suggesting that this pathway plays an important role in the adaptation of various stress conditions (Scott et al. 2002; Holzenberger et al. 2003; Reiling and Sabatini 2006).

S6 Kinase

Protein synthesis is the most energy-consuming anabolic process in a growing cell (Schmidt 1999). The major adenosine triphosphate (ATP)-consuming part in protein systhesis apparatus is translation machinery and ribosome biogenesis. Exponentially growing animal cells produce about 7,500 ribosomes per minute. If one considers that there are 80 distinct ribosomal proteins making up the 40S and 60S subunits of 80S ribosome and each is present as one protein in each 80S ribosome, this suggests that translational apparatus is producing approximately 600,000 ribosomal proteins per minute (Lewis and Tollervey 2000). In addition, there are a large number of other proteins including many initiation and elongation factors, kinases, rRNA helicase, pseudouridylating enzymes, sno rRNA, and different exo and endo nucleases involved in the ribosome biogenesis process (Lewis and Tollervey 2000). The whole process also needs to respond quickly to many environmental, nutritional, and growth signals. S6 kinase (S6K) is implicated as playing a crucial role in the translation process.

S6K is activated through a series of phosphorylations of key residues in its distinct regulatory domains in a hierarchical manner (Pullen and Thomas 1997). This kinase has a short amino terminus acidic region that is important for rapamycin sensitivity. This N-terminus acidic domain is followed by the catalytic domain, which contains 11 conserved residues found in all ser/thr kinases. A linker domain links the catalytic domain with the C-terminus auto-inhibitory domain (Pullen and Thomas 1997). Activation of S6K1 requires phosphorylation of four S/T residues in autoinhibitory domain (Pullen and Thomas 1997; Dennis and Thomas 2002). Phosphorylation of these four sites, in combination with phosphorylation of S371 in the linker domain, facilitates T389 phosphorylation. These phosphorylation events provide the docking site for 3-phosphoinositide-dependent protein kinase 1 (PDK1; Biondi et al. 2001). PDK1 docks on phosphorylated T389 and phosphorylate T229 in the activation loop of the S6K, leading to its activation. The critical event to activate S6K is phosphorylation of T389. It has been suggested that TOR kinase is responsible for phoshorylating T389 (Dennis et al. 2001). Plants also show PDK1-dependent phosphorylation of S6K (Mahfouz et al. 2006). Fig. 8.3 summarizes the modular structure and progressive phosphorylation events to activate the S6 kinase.

Drosophila cells express a single S6K protein (dS6K), whereas mammalian cells express two forms of this kinase, S6K1 and S6K2, also known as S6Kα and S6Kβ, respectively. Plants also have two S6K (Mahfouz et al. 2006). These two proteins are encoded by two genes that share a high level of sequence homology (Mahfouz et al. 2006; Ruvinsky and Meyuhas 2006). S6K1 has both cytosolic and nuclear isoforms, whereas the two isoforms of S6K2 are primarily nuclear (Martin et al. 2001; Mahfouz et al. 2006). *Drosophila* without dS6K activity shows a high degree of embryo lethality, with severely reduced body size due to reduced cell size rather than cell number. These surviving flies have a shorter life span (Ruvinsky and Meyuhas 2006). S6K1-null mice (S6K1$^{-/-}$) show similar phenotype of dSK-null *Drosophila*, whereas S6K2$^{-/-}$ mice

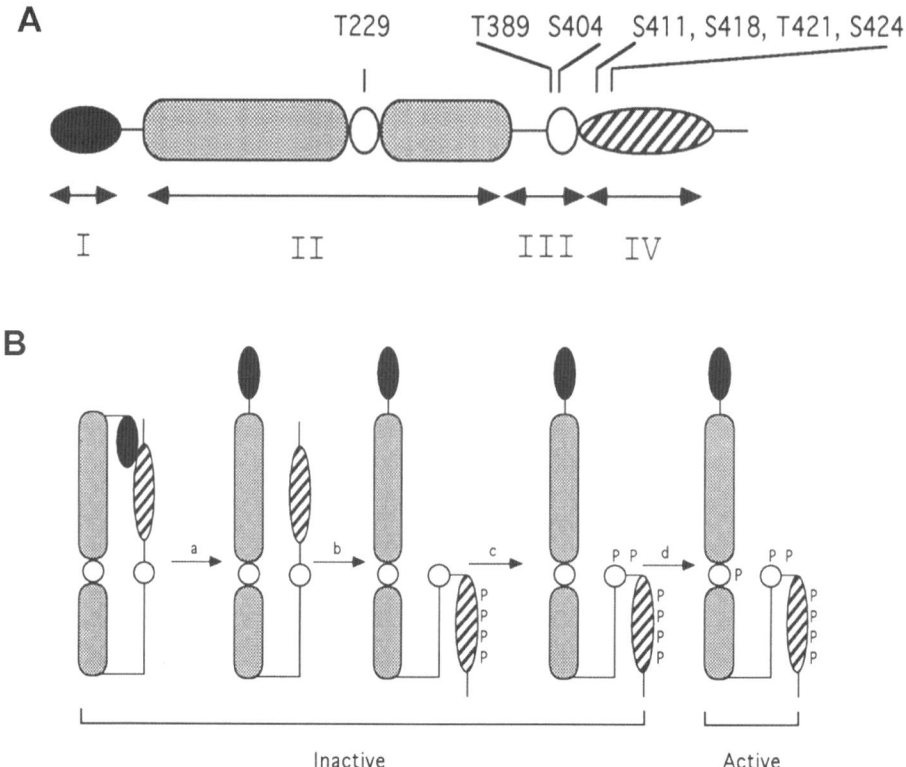

Fig. 8.3 (**A**) A modular representation of animal S6K (p70S6k). The primary structure of p70S6k has been functionally dissected into four modules, indicated as I–IV. Identified sites of phosphorylation are also indicated. (**B**) The modular activation of p70S6k. To generate p70S6k activity, the critical sites in Module II, T229, and III, T389, must become phosphorylated. The four-step model presented here predicts that a functional interaction between Modules I and IV is broken and stabilized (a) by phosphorylation of the S/T-P sites in Module IV (b). This event, possibly mediated by mTOR/FRAP, would allow phosphorylation of critical cognate site at T389 in Module III, thereby allowing the T-loop phosphorylation at T229 (c) to propagate full kinase activity (d). (From Pullen and Thomas 1997.)

have birth weight and size similar to that of wildtype. It has been shown that mTOR, in addition to its role as an activator of S6K, is also a substrate of S6K (Chiang and Abraham 2005; Holz and Blenis 2005; Ruvinsky and Meuhas 2006).

In addition to TOR, at least eight other substrates have been assigned to S6K in animals (Ruvinsky and Meyuhas 2006). One of those is small ribosomal protein 6, rpS6. This protein has attracted special attention because it was the first (and for many years the only) ribosomal protein and kinase substrate known to undergo inducible phosphorylation. It is worth mentioning here that the phosphorylated rpS6 can be detected in both nucleus and cytoplasm (Pende et al. 2004). Phosphorylation of rpS6 has been mapped to five residues: S235, 236, S240, S244, and S247 (Kreig et al. 1988). Phosphorylation of rpS6 in mice deficient in either S6K1 or S6K2 indicates that both of these S6K are required for complete phosphorylation of rpS6, yet it is predominantly carried out by S6K2. It has also been shown that S236 is the primary phosphorylation site (Flotow and Thomas 1992; Pende et al. 2004). Furthermore, substituting all the phosphorylatable serine residues by alanine (knock-in) has no apparent effect on yeast growth under various nutritional conditions. However, the yeast rpS6 is different compared with animals insofar as the number of phosphorylatable serine residues are concerned. Yeast also does not have any homologue for S6K1 and S6K2. Recent data shows that the proportion of ribosomes engaged in polysomes from liver cells of rpS6 knock-in mice (rpS6$^{p-/-}$) is similar to that of wildtype mice. The result indicate that phosphorylation of rpS6 is dispensable for recruitment of ribosomes to polysomes. However, the rate of global protein synthesis is significantly higher in mouse embryo fibroblast cells derived from rpS6$^{p-/-}$ mice (Ruvinsky et al. 2005). Conditional knockout of both alleles of rpS6 in adult mouse liver has demonstrated the requirement for this protein for ribosome assembly and cell proliferation but not for cell growth. Abrogation of rpS6 inhibits ribosome biogenesis and induces a checkpoint control that prevents cell cycle progression, but the cells grew (increased in size) in response to nutrients (Volarevic et al. 2000).

It has been proposed that S6K influences protein synthesis by phosphorylating eEF2K, which phosphorylates eukaryotic elongation factor 2 (eEF2) and/or eIF4B, while downregulating the same through rpS6 phosphorylation. These apparently reciprocal mechanisms seem to fine-tune the activation of protein synthesis after nutritional stimulation to ensure balanced protein synthesis with minimum energy waste (Ruvinsky and Meyuhas 2006).

The translational efficiency of mRNA with a 5′-terminal oligopyrimidine tract, commonly known as "TOP mRNAs," closely correlates with the levels of S6K and phosphorylation status of rpS6 (Hornstein et al. 2001). Such correlation led to the hypothesis that rpS6 phosphorylation increases the affinity of ribosomes to TOP mRNAs and thus facilitates translation of this particular class of mRNAs (Jeffries et al. 1994). It has been shown that translation of TOP mRNAs is regulated normally, at least in some cell types in S6K1$^{-/-}$/S6K2$^{-/-}$ double knockout mice (Pende et al. 2004) and in rpS6$^{p-/-}$ knock-in mice (Ruvinsky et al. 2005). Rapamycin-mediated translational efficiency of mTOR for TOP mRNAs varies significantly (Pende et al. 2004; Ruvinsky and Meyuhas 2006), and the exact role of TOR kinase to control TOP mRNA is yet to be fully ascertained (Ruvinsky and Meyuhas 2006).

S6K was first identified as a substrate of AGC kinase (cyclic adenosine monophosphate [cAMP]-dependent protein kinase A, cyclic guanosine monophosphate [cGMP]-dependent protein kinase G, phospholipid-dependent protein kinase C) protein in plants (Bogre et al. 2003). Two near-identical homologs of S6K gene were identified and cloned from Arabidopsis genome (Zhang et al. 1994; Mizoguchi et al. 1995; Mahfouz et al. 2006). These clones were termed as AtS6k1/Atpk1/AtPK6 and AtS6k2/Atpk2/AtPK19. Plant hormones such as auxin and cytokinin increase the activity of these two S6K proteins in Arabidopsis (Turck et al. 2004; Mahfouz et al. 2006). Transcript level of *AtS6K2* remains significantly higher in stationary phase or growth-arrested cells. AtS6K1 is translationally repressed in suspension cultured cells, compared with *AtS6K2*. Addition of fresh media to stationary phase cells increases S6K activity. The level of *S6K2* transcripts decreased, whereas that of *S6K1* remained constant under a growth-stimulating condition (Turck et al. 2004). Both *AtS6K1* and *AtS6K2* transcripts are induced under different stress conditions such as cold and higher salinity (Mizoguchi et al. 1995), whereas AtS6K1 activity is reduced to >70%

after 5% mannitol treatment (Mahfouz et al. 2006). *Arabidopsis* RAPTOR1, which interacts with HEAT repeats of AtTOR, also interacts with AtS6K1 in vivo, although it is not yet clear whether AtS6K2 also interacts with RAPTOR (Mahfouz et al. 2006). Overexpression of AtS6K1 renders *Arabidopsis* plant hypersensitive to osmotic stress (Mahfouz et al. 2006).

The C-terminal inhibitory domain, which is affected by mitogen-activated protein kinase (MAPK), as well as the N-terminal TOR signaling (TOS) motif are missing in AtS6K. However, TOR kinase and PDK1 kinase phosphorylation sites are conserved in AtS6K (Bogre et al. 2003). AtS6K2 is active in human cells. Its activity is increased by serum treatment, whereas it is negatively affected by PtdIns3K inhibitor, Wortmanin, but not affected by TOR inhibitor, rapamycin. Lack of rapamycin sensitivity is consistent with the fact that AtS6K2 contains no region homologous to the mammalian S6K amino terminus (autoinhibitory domain), which is required for rapamycin sensitivity. AtS6K2 can phosphorylate human rpS6 protein at $25°C$ but not at $37°C$. The plant S6K is rapidly inactivated in the human cell line at $37°C$, which is consistent with the observation that heat shock reduces rpS6 phosphorylatin in plants as well (Turck et al. 1998).

TOR-S6K signal transduction pathway is also functional in maize (Dinkova et al. 2007) and one S6K homolog from maize has been identified (Reyes de la Cruz et al. 2004). This 62 kDa maize protein can phosphorylate rpS6. The ZmS6K protein maintained its steady state level of expression during maize seed germination but its activity increased, consistent with Zm6SK phosphorylation (Reyes de la Cruz et al. 2004). Although insulin cannot enter the plant cell, adding insulin to germinating maize axes increases both ZmS6K activity and the extent of rpS6 phosphorylation. All these events were blocked by rapamycin (Reyes de la Cruz et al. 2004). *Zea mays* insulin-related peptide (ZmIGF) can also stimulate ZmS6K in germinating maize axis (Dinkova et al. 2007).

Unlike animals, plants have several isoforms of ribosomal proteins. *Arabidopsis* has three genes for ribosomal protein 6 (rpS6), one of the main downstream targets of TOR signaling. These genes (At5g10360, At3g17170, and At4g31700) encode cytosolic rpS6. EMBRYO DEFECTIVE (EMB3010) (At5g10360) and rpS6A (At3g17170) have very high sequence homology and are expressed in a similar manner. Genvestigator (https://www.genevestigator.ethz.ch) analysis indicates that plant hormones including cytokinin, gibberellin, and auxin induce higher levels of rpS6A. The root elongation zone and plants treated with flagellin also show increased levels of the same, whereas the transcript level of another isoform of rpS6, EMB3010 increases when the plant undergoes either abiotic (e.g., heat shock, cold treatment, nutrient/potassium deprivation, cell culture) or biotic stresses (e.g., *Agrobacterium* and nematode infection).

FKBP12

TOR-interacting partner, FKBP12, is a member of the Immunophilin family of proteins. Immunophilins are enzymes with a peptidyl-prolyl cis-trans isomerase activity (PPIase) involved in the folding of target proteins (Kay 1996; Schiene-Fischer and Yu 2001; Hur and Bruice 2002; Shaw 2002, 2007). This group of proteins has been studied extensively in humans. FK506-binding proteins (FKBPs) are immunophilins that bind two related drugs, FK506 (parvulins) and rapamycin. FKBP12-FK506 complex blocks Ca^{2+}-dependent signaling by binding to calcineurin (PP2B), a Ca^{2+}-calmodulin–regulated Ser/Thr protein phosphatase (Cardenas et al. 1999; Harrar et al. 2001), whereas rapamycin-FKBP12 complex binds not to calcineurin but to TOR kinase (Breiman and Camus 2002). FKBPs are found in most organisms, including prokaryotes, animals, and plants (Aghdasi et al. 2001; Harrar et al. 2001; Breiman and Camus 2002).

Although FKBPs differ in size, FKBP12 (12 kD) represents the smallest peptide, harboring the two main properties of FKBPs—namely, the PPIase activity and drug binding. FKBP12 is a ubiquitous and abundant protein localized in the cytosol (Maki et al. 1990). In mammals, FKBP12 is essential because knockout mice die during embryonic development (Shou et al. 1998), whereas in yeast (*S. cerevisiae*), the loss of FKBP12 function does not affect cell viability (Dolinski et al. 1997). In the absence of rapamycin, FKBP12 is associated with receptors such as the type II-TGFβ (transforming growth factor β) receptor or calcium channels such as the ryanodine receptor or the inositol-(1,4,5)-triphosphate receptor.

FKBP12 appears to be a regulator of cell cycle. Experiments on cells from FKBP12-deficient mice indicate that cells are arrested in G_1 because of an enhanced TGFβ signaling leading to the overactivation of p21, an inhibitor of the G_1/S transition (Aghdasi et al. 2001). In yeast two-hybrid screen for proteins interacting with AtFKBP12, Vespa and colleagues (2004) identified a 37kD FKBP12-interacting protein (AtFIP37). The drug FK506 disrupts AtFIP37 interaction with AtFKBP12 (Faure et al. 1998). AtFIP37 is a single copy gene in the *Arabidopsis* genome. These results demonstrate that AtFIP37 is critical for embryo and endosperm development and is involved in the endoreplicative cell cycle (Vespa et al. 2004). *Arabidopsis* FKBP12 does not complement yeast FKBP (Breiman and Camus 2002) but interact with human FIP37 (Faure et al. 1998; Vespa et al. 2004).

Inactivation of TOR by rapamycin is mediated by formation of a ternary complex in which rapamycin forms noncovalent link between FKBP12 and the FKBP-rapamycin-binding domain (FRB) of TOR protein, TOR^{FRB} (Choi et al. 1996; Menand et al. 2002). There are many FKBP-type proteins in *Arabidopsis* (Schubert et al. 2002; Romano et al. 2004; Harrar et al. 2001; Sormani et al. 2007; Geisler and Bailly 2007). Some of the FKBP-type immunophilins are also classified as cyclophilins (Harding et al. 1989; Romano et al. 2004). Cyclophilin (FKBP) expression has been shown to be influenced by many biotic and abiotic stresses, including viral infection, ethephon (an ethylene releaser), salisylic acid, salt stress, and heat and cold stress (Marivet et al. 1995; Scholze et al. 1999; Kullertz et al. 1999; Romano et al. 2004; Geisler and Bailly 2007), draught (Sharma and Singh 2003), wounding, fungal infection, absicic acid, and methyl jesmonate (Godoy et al. 2000; Kong et al. 2001) and light (Chou and Gasser 1997; Luan et al. 1996).

Besides FKBP12, there are few other FKBP paralogs that bind rapamycin, and many more immunophilins have PPIase activity. *Vicia faba* VfFKBP15 protein has rotamase activity inhibited by both FK506 and rapamycin with a K_i value of 30 nM and 0.9 nM, respectively, illustrating that VfFKBP15 binds rapamycin in preference over FK506 (Luan et al. 1996). Studies on VfFKBP15 homologs from *Arabidopsis* and rice have demonstrated that FKBP15 is encoded by a small gene family in higher plants, whereas only one gene is found in mammalian and yeast systems. The VfFKBP15 protein and mRNA are detected in all tissues, and its mRNA levels are regulated by heat shock, suggesting a possible role for this FKBP in the stress response of higher plants. Overexpression of wheat *FKBP73* in rice produced fertile plants, but overexpression of only the three PPIsae domains of wheat *FKBP73* produced male-sterile rice plants (Breiman and Camus 2002). Another FKBP member, FKBP25, has been characterized from mammalian cells to bind rapamycin preferably over FK506, with a more than 150-fold higher affinity for rapamycin than for FK506 (Luan et al. 1996). Human FKBP12 can also bind to *Arabidopsis* TOR kinase in a rapamycin-dependent manner (Mahfouz et al. 2006).

Eukaryotic Initiation Factor 4E (eIF4E) and Its Binding Proteins, 4E-BPs

Protein synthesis is energetically the most expensive process in the cell. Not surprisingly, the translation rates are tightly regulated, mainly at the level of initiation, predominantly by initiation factors (eIFs). Most eukaryotic mRNAs are translated through a m^7GTP cap-dependent mechanism involving the eIF4F complex, whereas other mRNAs may have a greater or lesser requirement for the eIF4F complex to be efficiently translated. For example, mRNAs possessing highly structured 5′-noncoding regions (e.g., those encoding c-*myc*, ornithine decarboxylase, and cyclin D1) exhibit a greater requirement for the eIF4F complex than do mRNAs with less structured 5′-noncoding regions (Pestova and Kolupaeva 2002). Under eIF4F limiting conditions, translation of mRNAs with highly structured 5′-noncoding regions is repressed to a greater extent than the general population of mRNAs. In part, this may reflect a requirement of such mRNAs for the RNA helicase activity of eIF4A as well as the helicase-stimulating activity of eIF4B to unwind secondary structure in the 5′-noncoding region and allow binding of the 40S ribosomal subunit to the mRNA, as well as its migration to the AUG start codon. Although little is known about the mechanism through which eIF4A and eIF4B are regulated, eIF4B is phosphorylated by S6K1 (Raught et al. 2004). This provides a possible link between hormone and nutrient signaling through TOR and eIF4A/eIF4B function (Kimball and Jefferson 2006).

Most mRNAs in eukaryotes are translated in a cap-dependent manner. The cap structure, m^7GpppN (where N is any nucleotide), is present at the 5′ end of all eukaryotic nuclear mRNAs (Gingras et al. 1999). The cap structure is bound by the eIF4F complex, which contains the 5′ cap-binding protein eIF4E and two other subunits, eIF4A (an ATP-dependent helicase) and eIF4G (a large scaffolding protein), which contains docking sites for the other proteins. eIF4F is directed to the 5′ end of the mRNA through eIF4E and acts through eIF4A along with eIF4B to unwind the mRNA 5′ secondary structure to facilitate ribosome binding (Gingras et al. 1999). The eIF4A, a DEAD-box RNA helicase, unwinds the mRNA 5′ UTR (Un-Translated Region) to facilitate the ribosome binding, and the eIF4G interacts with many other components of the translation machinery. The eIF4G also recruits the 40S small ribosomal subunit to the mRNA through the ribosome-associated eIF3 (Mamane et al. 2006). Overexpression of eIF4E leads to enhanced expression of many genes implicated in growth, proliferation, and survival (reviewed in Mamane et al. 2004). Expression and activity of eIF4E is regulated at many levels but primarily through transcription and phosphorylation (mainly by MAPK). Assembly of eIF4F complex is negatively affected by members of a family of repressors termed eIF4E-binding proteins (4E-BPs). 4EBPs and eIF4G compete for the same eIF4E-binding motif (Marcotrigiano et al. 1999). It has been shown that hypophosphorylated 4EBP binds strongly with eIF4E. Phosphorylation of 4EBP1 and probably other 4EBPs are rapamycin dependent and occur at multiple sites in a sequential manner (Gingras et al. 2001). After 4EBP1 is phosphorylated, it loses its affinity toward eIF4E and allows eIF4G to bind the complex, which initiates the cap-dependent translation of the mRNA (Mamane et al. 2006). Besides eIF4E, other translation initiation factors such as eIF4GI are implicated in TOR pathway. eIF4GI is regarded as a downstream target of mTOR because rapamycin treatment inhibits its phosphorylation (Raught et al. 2000). Figure 8.4 shows interaction of various factors for translation initiation. However, it is not known whether mTOR directly phosphorylates eIF4GI. *Arabidopsis* genome seems to contain at least four eIF4E homologs; viz. eIF4E1 (At4g18040), eIF4E2 (At1g29590) and eIF4E3 (At1g29550), and eIF(iso)4E (At5g35620) (Robaglia and Caranta 2006). So far there is no 4EBP reported from any

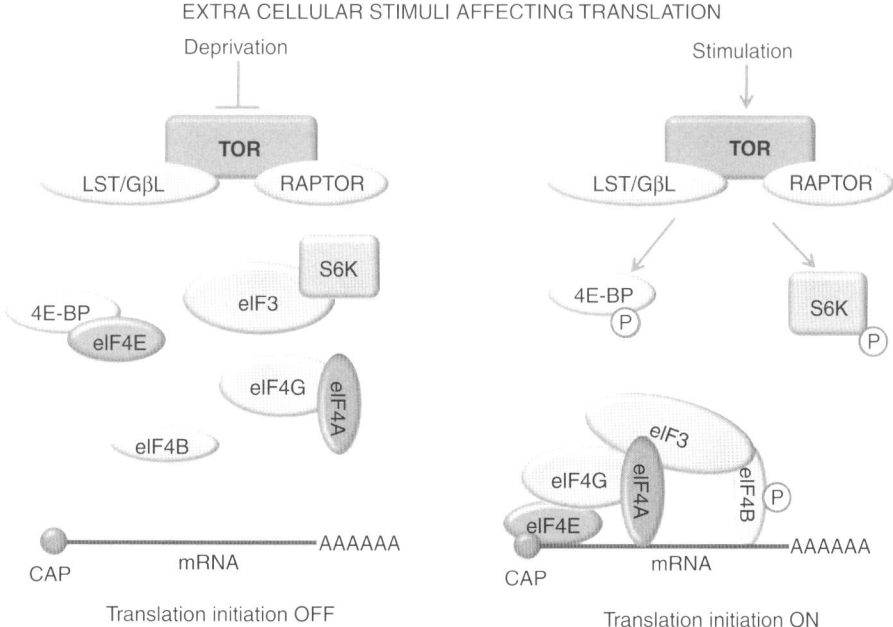

Fig. 8.4 Activation of translation initiation by mTOR. Nutrients, hormones, and growth factors activate the mTOR, which forms a complex with other proteins, including raptor and LST8/GβL. mTOR phosphorylates two major targets: 4E-BPs and S6Ks (most of the work has been performed with 4E-BP1 and S6K1). Phosphorylated 4E-BP is released from eIF4E. In the absence of extracellular stimuli, S6K1 is associated with eIF3. In response to extracellular stimuli such as growth factors or nutrients, the mTOR complex is recruited to eIF3 to phosphorylate S6K1 and 4E-BP1. Phosphorylation and activation of S6K1 leads to its dissociation from eIF3. Activated S6K1 then phosphorylates eIF4B and rpS6. Phosphorylation of eIF4B at S422 promotes its association with eIF3. mTOR also stimulates the association between eIF3 and eIF4G. These multiple interactions culminate in enhanced cap-dependent mRNA translation.

plant source. eIF4E and eIF(iso)4E also play a role in virus infection in plants (Nieto et al. 2006).

Other Eukaryotic Initiation Factors Implicated in TOR Signaling

eIF4G. In addition to eIF4E, other translation initiation factors have been implicated in tumorigenesis. eIF4G1 is overexpressed in squamous cell lung carcinomas (Brass et al. 1997; Keiper et al. 1999; Bauer et al. 2001, 2002), and eIF4GI transforms NIH 3T3 cells (Fukuchi-Shimogori et al. 1997). eIF4GI is a downstream target of mTOR as rapamycin treatment inhibits its phosphorylation (Raught et al. 2000). However, it is not known whether mTOR phosphorylates eIF4GI directly or indirectly (Mamane et al. 2006).

eIF3. It has been suggested that mTOR's role in translation initiation can be mediated through the eIF3 translation initiation complex (Lee et al. 2007). eIF3 is one of the largest initiation factors, with at least 12 subunits (Mayeur et al. 2003). eIF3 binds to the 40S ribosomal subunit, of which rpS6 is a component. Binding of eIF3 to the 40S subunit inhibits premature association with the 60S ribosomal subunit. In addition, eIF3 also enhances initiation by increasing the binding of the ternary complex (Gingras et al. 2001). Under serum starvation or rapamycin inhibition, S6K1 binds tightly to eIF3. S6K1 dissociates from eIF3 upon insulin stimulation. This association with eIF3 is disrupted by phosphorylation of the hydrophobic motif of S6K1 (T389). Thus, either phosphorylation by mTOR or phosphomimetic mutation seems to be sufficient to decrease the binding affinity between S6K1 and eIF3 (Holz and Blenis 2005). However, the association between eIF3 and mTOR changes with activation or inhibition of mTOR. Serum starvation and rapamycin treatment reduce the binding affinity between eIF3 and mTOR/raptor, whereas insulin stimulates binding between eIF3 and mTOR/raptor (Holz and Blenis 2005).

Protein Phosphatase A (TAP42)

Protein phosphatase A (PP2A) is a multimetric serine/threonine phosphatase that is highly conserved from budding yeast to humans and is regulated by TOR signaling (Cygnar et al. 2005; Inoki et al. 2005). It is involved in a variety of cellular functions such as cellular metabolism, DNA replication, transcription, translation, cell proliferation, and cell transformation (Mayer-Jaekel and Hemmings 1994). The PP2A holoenzyme is composed of a catalytic subunit C in complex with a scaffolding subunit A and a regulatory subunit B (Goldberg 1999). The regulatory subunit affect PP2A substrate specificity and localization. TAP42 is a yeast PP2A-interacting protein with a molecular mass of 42 kDa, which plays a critical role in regulating the PP2A activity (Wera and Hemmings 1995). Mutations in *tap42* block TOR-dependent phosphorylation of many substrates and mimic rapamycin treatment, including growth arrest, reduced translation, nutrient transporter degradation, and induction of stress response (Di Como and Arndt 1996; Duvel et al. 2003). *tap42* mutants show dominant rapamycin resistance (Di Como and Arndt 1996; Duvel et al. 2003; Cygnar et al. 2005). TAP42 is a positive effector of TOR signaling, although some TOR-regulated processes such as ribosomal protein genes transcription and autophagy are independent or negatively affected by TAP42 (Duvel et al. 2003; Cygnar et al. 2005). Current models predict that TAP42 dictate the substrate specificity of PP2A or phosphorylation of TAP42 may convert it from an activator to a repressor of PP2A activity (Duvel and Broach 2004; Cygnar et al. 2005).

Inhibitors of type 1 and type 2A phosphatases (e.g., calreticulin and okadaic acid) antagonize rapamycin-induced dephosphorylation of the TOR substrates S6K and 4E-BP1 and promote S6K activation and cell growth (Leicht et al. 1996; Lin and Lawrence 1997; Hara et al. 1998; Peterson et al. 1999; Krause et al. 2002). Furthermore, rapamycin treatment increases the activity of PP2A toward 4E-BP1 in vitro (Peterson et al. 1999). PP2A can directly bind and inhibit wildtype S6K, whereas a truncated version of S6K has reduced affinity for PP2A and does not require TOR for its activity (Peterson et al. 1999; Westphal et al. 1999; Gonzalez-Garcia et al. 2002). The potential function of TAP42 homologs in TOR signaling in higher eukaryotes is less clear than in yeast. The mammalian homolog of TAP42, $\alpha 4$, binds to type 2A-related phosphatases (Inoki et al. 2005; Cygnar et al. 2005). However, there are conflicting reports as to whether rapamycin disrupts this association (Cygnar et al. 2005). As in yeast, $\alpha 4$ may act to modify PP2A specificity because it has been shown to inhibit in vitro phosphatase activity toward 4E-BP1 and increase it toward other substrates (Murata et al. 1997; Inui et al. 1998; Nanahoshi et al. 1998). It has been observed that mTOR directly phosphorylates PP2A and represses its functions in vitro, whereas rapamycin treatment activates PP2A in vivo (Peterson et al. 1998). Plant PP2A ortholog is implicated in stress response, root development, and auxin distribution (Blakeslee et al. 2008).

Dephosphorylation of S6K following rapamycin treatment or amino acid starvation or other stress signals is blocked by inhibitors of PP2A (Bielinski and Mumby 2007). These inhibitors cannot distinguish between PP2A-like phosphatases and specific phosphatase(s) that dephosphorylate S6K. It is worth mentioning here that the specific phosphatase(s) that dephosphorylates S6K has not yet been identified (Bielinski and Mumby 2007). Using *Arabidopsis* PP2Ac gene as bait in yeast two-hybrid screen, Harris and his colleagues identified TAP46, the plant homolog of yeast TAP42 and mammalian $\alpha 4$ proteins. TAP46 gene appears to be a single-copy gene and is expressed in all *Arabidopsis* organs. TAP46 might be involved in the chilling response (Harris et al. 1999), whereas the role of TAP46 in plant TOR signaling remains to be established.

PI3 Kinase and PDK1

The importance of PI3K and TOR interplay increases as both pathways converge on common downstream targets via S6 Kinase and 4EBPs (Richardson et al. 2004). A wide variety of growth-promoting extracellular signals activates PI3K and generate lipid second messengers by phosphorylation of the 3′-OH position of phosphatidylinositols (Rameh and Cantley 1999). The modified lipids, particularly phosphatidylinositol 3,4,5-trisphosphate (PI-3,4,5-P_3) and phosphatidylinositol 3,4-bisphosphate (PI-3,4-P_2), recruit signaling molecules to the membrane, where these molecules subsequently become activated by a variety of mechanisms including phosphorylation, conformational changes, and entry into protein complexes (Richardson et al. 2004). The two kinases, which are activated by PI3K and specifically important for TOR signaling, are the proto-oncogene Akt (also known as PKB) and the PDK1. In contrast to PI3K, the lipid phosphatase and tumor suppressor protein PTEN dephosphorylates PI-3,4,5-P_3 and PI-3,4-P_2. This leads to subsequent inactivation of downstream signaling molecules and a downregulation of the pathway. Similar results can be achieved with the PI3K inhibitors wortmannin and LY294002 (Ramch and Cantley 1999). TOR contains sequence homology to PI3K but has been shown to contain phosphotransferase activity towards serine and threonine residues rather than activity on lipids. TOR and PI3K coordinately regulate phosphorylation of multiple sites in both S6K and 4EBP1. These phosphorylation events are induced by nutrients and growth factors and are inhibited by PI3K inhibitor wortmannin, LY294002, and TOR inhibitor rapamycin (Gingras et al. 2001; Martin and Blenis 2002). Several PI3K effectors have been identified as mediator for S6K phosphorylation. Apparently, the main mediator is PDK1 (Richerdson et al. 2004). PI3K also exerts its effect on TOR pathway via AKT kinase, phosphorylating TOR (Richardson et al. 2004).

PDK1 is a large (556-amino acids) enzyme possessing a kinase domain at its N-terminus (residues 70–359) and a pleckstrin homology (PH) domain at its C-terminus (residues 459–550), which interacts with PI-3,4,5-P3 and one of its immediate breakdown products PI-3,4-P2 (Mora et al. 2004). It has been found to have a novel role in regulating cortical actomyosin and cell motility by phosphorylating myosin light chain (Pinner and Sahai 2008). PDK1 homologs have been identified in *Arabidopsis* and rice, and PDK1 is shown to phosphorylate S6K in *Arabidopsis* (Mahfouz et al. 2006). *Arabidopsis* PDK1 can complement both yeast and mammalian cells lacking PDK1 homologs (Deak and Malamy 2005). AtPDK1 can phosphorylate at least one S6K protein, AtS6K2, in vitro and is also able to autophosphorylate Ser-177, Ser-276, and Ser-382 residues, like its mammalian counterpart (Otterhag et al. 2006). It is also shown that auto-phosphorylation of Ser-276 is essential for AtPDK1 downstream activity. Other sites identified that may be of importance for AtPDK1 activity are Asp-167, Thr-176, and Thr-211. Nine *Arabidopsis thaliana* 14-3-3 isoforms positively regulate AtPDK1 autophosphorylation and activation of AtS6k2, whereas one *A. thaliana* 14-3-3 isoform, o, negatively regulates the kinase activity (Otterhag et al. 2006).

AtPDK1 is also implicated in oxidative stress response via OXI1/AGC2-1 protein, a member of AGC protein kinase family. AtPDK1 binds phosphotidic acid (PA), resulting in activation of OXI1/AGC2-1 kinase (Anthony et al. 2006). PA is produced in response to many stress signals, including abscisic acid (ABA)-induced stomatal closure, pathogen attack and oxidative stress (Testernik and Munnik 2005). However, the activity of AtPDK1 itself is not inhibited by osmotic stress (Mahfouz et al. 2006).

Ribosomal Proteins, Histone-Modifying Enzymes, Upstream Binding Factor, and CyclinD-CDK4 Complexes

In plants each ribosomal protein is encoded by multiple genes located on different chromosomes and are expressed in all tissues. Apart from affecting components of the transcriptional machinery, other effectors have been implicated in TOR-dependent regulation of RP gene expression by chromatin-based mechanisms (Cardenas et al. 1999; Mayer and Grummt 2006). For example, two histone H4 modifying enzymes—ESA1, a histone acetyltransferase, and RPD3, a histone deacetylase subunit of the SIN3 complex—are implicated in the activation and repression of RP genes, respectively. Under good nutrient conditions, TOR signaling activates the expression of RP genes by promoting maintenance of the ESA1 complex to RP gene promoters. Upon starvation or rapamycin treatment, ESA1 is rapidly released from RP gene promoters (Rohde and Cardenas 2004). Conversely, RPD3 binds and inhibits RP genes in a TOR-dependent manner, indicating that nutrient availability affects the chromatin structure of RP genes (Mayer and Grummt 2006). In this process histone and other chromatin-binding proteins may play important roles. We have observed that a histone deacetylase (HDA) binds to hyperphosphorylated AtRPS6 (S. Kim and D.P.S. Verma, unpublished data).

Serine/Arginine–Rich (SR) Proteins

This class of phylogenetically conserved proteins was first identified for their roles in RNA splicing (reviewed by Graveley 2000). They have a modular structure consisting of one or two RNA-recognition motifs (RR motifs or RRMs) and a C-terminal arginine/serine-rich domain (RS domain; Caceres et al. 1997). Later SR proteins were implicated in targeting many mRNAs with premature termination codons to nonsense-mediated mRNA decay (NMD) pathway (Zhang and Krainer 2004). It has also been shown that SR proteins are involved in highly efficient splicing of all RNA polymerase II (RNAPII) transcripts (Das et al. 2007). A subset of SR proteins continuously shuttles between nucleus and cytoplasm. mRNAs containing binding sites for one such SR protein, SF2/ASF, enhances translation by increased binding with polyribosomes (Sanford et al. 2004). SF2/ASF-mediated translational activation requires cytoplasmic cap binding protein eIF4E (Michlewski et al. 2008).

Recently, SF2/ASF protein has been shown to interact with mTOR and PP2A and has been implicated in regulating translational initiation by enhancing phosphorylation of 4E-BP1 (Michlewski et al. 2008). Plants do have homologs of this important class of protein. Its *Arabidopsis* homolog, AtSR1, is encoded by a single gene (Lazar and Goodman 2000). However, AtSR1 differs from other SR proteins because it has a proline-serine-lysine (PSK)–rich C-terminus. *AtSR1* itself is regulated by alternative splicing. Of five transcripts from the same gene, only one encodes full-length protein, and all the alternatively spliced versions lack the PSK domain. Accumulation of SR1B isoform above 22°C indicates that this form could play a role in adaptation to high temperature and other stress conditions (Lazar and Goodman 2000).

Nucleolus is the site for ribosome biogenesis in all eukaryotes. Nuclear proteome includes ribosomal RNA processing factors and ribosomal proteins (Andersen et al. 2002). SR proteins are organized within the nucleus and are localized in irregularly shaped "speckles" (or splicing factor compartments; Sacco-Bubulya and Spector, 2002). Subcellular localization and dynamics of SR proteins in plants have been found to have a similar pattern as their animal counterparts and have a close correlation with cell cycle. SR proteins and RNA polymerase II closely localize within speckles in plants as well (Tillemans et al. 2005, 2006). It has been shown that regulation of transcription and alternative splicing of *Arabidopsis* SR protein mRNAs are responsive to different stress conditions (Tanabe et al. 2007). Overexpression of an *Arabidopsis*

SR protein confers salt tolerance in yeast and transgenic plants (Forment et al. 2002). Taking this information together, it seems that *Arabidopsis* SR protein(s) are novel candidates to connect TOR directly with transcription, splicing, ribosome biogenesis, and abiotic stresses, such as salt and temperature stress.

So far we have discussed some major proteins implicated in TOR kinase pathway. There are many more proteins involved in TOR signaling pathway (e.g., Rheb, PKA, PKB/AKT/Sch9 etc.). These are not discussed here primarily because there has been no homologs of these proteins detected in plant systems to date. Next, we discuss some major cellular and physiological processes influenced by TOR signaling pathway and their relevance to plant metabolism.

TOR Connections with Plant Cell Cycle and Growth

In the meristem, cells must sense and integrate signals when they are required to divide and when division must cease to allow differentiation into specialized organs (Francis 1998). In other words, after every round of replication, each daughter cell in a multicellular organism needs to make a fundamental decision, whether to undergo another round of replication or exit the cell cycle and undergo differentiation (Durfee et al. 2000). Conservation of several key proteins (e.g., cyclins, cyclin-dependent kinases [CDKs], retinoblastoma-related proteins [RBR], HDA, etc.) in plants has established the fact that the core cell cycle machinery is conserved in higher eukaryotes (Francis 2007).

No orthologs of the mammalian G_1/S-specific *CDK4* and *CDK6* genes are present in plants. As such, CDKA is seemingly the only CDK active at the G_1 and S phases in plant cells, whereas the entry into mitosis is probably controlled by multiple CDKs (Inze and Veylder 2006). Plants possess a unique class of CDKs—the so-called B-type CDKs—that have not been described in any other organism (Hirayama et al. 1991; Joubes et al. 2000; Boudolf et al. 2001). Plant cyclins can be classified into nine groups: Cyclin A1, A2, A3, B1, B2, D1, D2, D3, and D4 (Renaudin et al. 1996). Most of these cyclins have multiple members; for example, in *Arabidopsis*, cyclin A1 has two members, A2 has 4 and A3 has 4 (Yu et al. 2003), whereas cyclin D3 has three members (Meijer and Murray 2000). Most of the cyclins have characteristic expression as well as degradation patterns during the cell cycle. The time of expression and degradation plays a major role in determining at which phase of cell cycle they can form a complex with and thus activate the corresponding CDKs (Stals et al. 2000).

Plants and animals independently evolved mechanisms of cellular differentiation and cell–cell communications. However, there seems to be a gradient of difference between plants and animals, from very similar housekeeping processes in nucleus and cytoplasm to biochemically distinct regulatory networks for development and control of cellular differentiation (Meyerowitz 1997). Integrated responses to diverse internal and external signals lead to proper plant development, as summarized in Figure 8.5.

Two meristematic cell populations arise during embryogenesis in opposite polar directions and are maintained throughout the plant life. The shoot apical meristem (SAM) generates the aerial parts of the plant, whereas the root apical meristem (RAM) generates the underground parts (Laufs et al. 1998; Jurgens 2001). A constant flow of cells through the meristem, where the input of dividing pluripotent stem cells offsets the

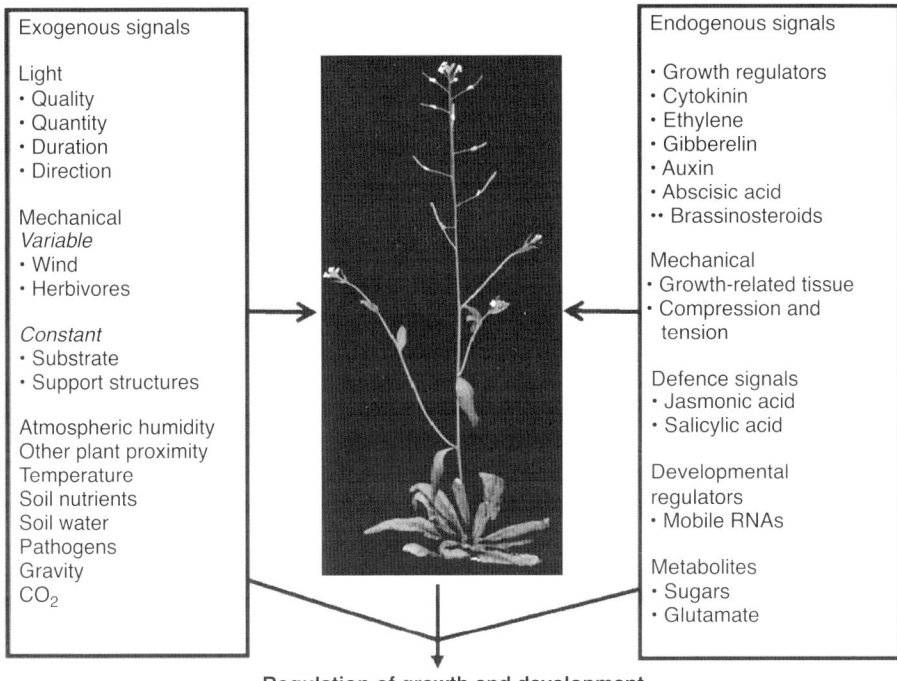

Fig. 8.5 Different internal and external signals that plants must respond to for growth and flowering.

output for the differentiating cells, must be miintained. This depends on extracellular signaling within the apical meristems, governed by a spatial regulatory feedback loop that maintains a reservoir of stem cells, and on factors that prevent meristem cells from differentiating prematurely (Carles and Fletcher 2003). Rapamycin targets have been implicated in signaling pathway leading to G_1-S progression in the cell cycle (Kunz et al. 1993; Luan et al. 1996). Yeast TOR1/TOR2 double disruption confers G_1 arrest, as does rapamycin (Kunz et al. 1993; Richardson et al. 2004). Delayed entry into S phase is observed by RNAi-mediated reduction of S6K or overexpression of 4EBP1, the two better known substrates of TOR kinase (Fingar and Blenis 2004). Another interesting connection between mTOR and G_1 cyclins has been observed. Ectopic expression of cyclin E, but not cyclin D, is able to rescue the G_1 block without affecting the RB-E2F pathway in animals (Leng et al. 1997; Lukas et al. 1997). Retinoblastoma protein (pRB) is another important protein involved in many cellular and physiological activities. It is one of the main checkpoints that controls G_1-S phase transition in eukaryotic cell cycle. One striking difference between animals and plants is the absence of E type cyclins and presence of more than one type of cyclin A in plants (animals have only one cyclin A), including *Arabidopsis* (Yu et al. 2003).

AtTOR is crucial for plant development. The embryos of *Arabidopsis to* knockout mutants arrest when cells begin to proliferate. It is consistent with its major role in growth control rather than cell division. In many respects, AtTOR null lines resembles *Arabidopsis* pRB (AtRBR1) null lines. Both are embryo-lethal (Ebel et al. 2004;

Mahfauz et al. 2006). Many of the TOR substrates, including RAPTOR, are conserved in plants (Anderson and Hanson 2005; Mahfouz et al. 2006; Ingram and Waites 2006). This indicates that molecular mechanisms of TOR-mediated growth control are conserved in plants. Phenotypic analysis of the two homologs of RAPTOR found in *Arabidopsis* (Anderson and Hanson 2005; Deprost et al. 2005) show that RAPTOR is necessary for normal plant development, although a double-knockout phenotype in which embryogenesis occurs normally but postembryonic growth is severely affected (Ingram and White 2006). This is consistent with TOR's ability to signal independently of RAPTOR, as is the case in other systems (Ingram and White 2006). Other reports indicate that early embryonic lethality is associated with knocking out one of the two RAPTOR genes (Deprost et al. 2005; Ingram and White 2006).

The transcription of rDNA is rapamycin-sensitive and apparently controlled by mTOR (Mahajan 1994; Mayer and Grummt 2006). Transcription initiation by Pol-I requires at least three basic factors, transcription initiation factor (TIF)-A, TIF-IB, and UBF (Grummt 2003). Results suggest that mTOR signaling leads to a cyclin D1–dependent increase in CDK4 activity, followed by phosphorylation-mediated dissociation of pRB from UBF with a concomitant increase in UBF availability (Hannan et al. 2003; Nader et al. 2005; Mayer and Grummt et al. 2006). Rapamycin treatment inhibits pRB phosphorylation and the increase in UBF availability without affecting UBF protein content. In addition, rapamycin prevented serum-stimulated increase of cyclin D1 protein expression and consequently the corresponding increase in CDK-4 kinase activity, with no detectable changes in CDK-4 or p21 protein (CDK inhibitor) levels or CDK-4/p21 functional interaction (Nader et al. 2005). These results suggest that TOR signaling may be downstream of cyclin E but upstream of cyclin D1 signaling. It also indicates that TOR activity is influenced by both of these cyclins and may have a link with pRB-E2F (UBF) signaling pathway. It is worth mentioning here that overexpression of pRB-associated transcription factor E2F in *Arabidopsis* (AtE2Fa) leads to overexpression of AtS6K2 (also known as AtPK19; He et al. 2004).

Drosophila cells having nonfunctional TOR homolog are about 4 times smaller than wildtype cells (Oldham et al. 2000; Zhang and Forde 2000; Inoki et al. 2005). Similarly, rapamycin treatment of mammalian cells reduces their size (Inoki et al. 2003, 2005). The effect on cell size by rapamycin is TOR-S6K mediated. Overexpression of rapamycin-resistant mTOR activates S6K, even in the presence of rapamycin, whereas knockdown mTOR expression through RNA interference represses S6K (Kim et al. 2002; Inoki et al. 2005). TOR phosphorylates S6K in multiple sites. Among them, Thr389 and Ser371 are the major sites as determined by in vitro experiments (Saitoh et al. 2002; Inoki et al. 2005). Phosphorylation on these two sites by mTOR is essential for S6K activation (Pearson et al. 1995; Inoki et al. 2005). Rapamycin was shown to inhibit phosphorylation of these sites (Han et al. 1995; Mahfouz et al. 2006). In addition, five amino acid residues (Phe-Asp-Ile-Asp-Ile) located close to the N terminus of S6K form the TOS motif (Sekito et al. 2002). Deletion or mutation of the TOS motif significantly blocked insulin-induced activation of S6K (Schalm and Blenis 2002). Deletion of the N terminus of S6K rendered it insensitive to rapamycin treatment, suggesting that the TOS motif is critical for mTOR-mediated activation of S6K. Importantly, S6K and 4EBP1 are two key elements of the TOR pathway that mediate regulation of cell size by TOR. Fruit flies deficient in S6K are smaller than wildtype flies and display decreased cell size (Montagne et al. 1999; Inoki and Guan 2006). S6K1 knockout mice had a body size approximately 80% of that of wildtype littermates (Shima et al. 1998). This phenotype is similar to those seen in *Drosophila* with *dTOR* mutations (Bodine et al. 2001; Inoki and Guan 2006). Furthermore, S6K overexpression rescues TOR knockout phenotypes in *Drosophila* (Zhang et al. 2000). These data suggest S6K as a major TOR target that mediates the effects of TOR on cell size regulation.

Plant Viral Infections and Possible TOR Connections

Resistance to cucumber mosaic virus (CMV) and turnip crinkle virus (TCV) have been shown to be associated with eIF4E and eIF4G, respectively (Yoshi et al. 2004), whereas eIF(iso)4G is implicated in rice yellow mottle virus (RYMV; Albar et al. 2006). *Arabidopsis* mutants lacking functional At-eIF4G genes showed that eIF4G factors are indispensable for potyvirus infection (Nicaise et al. 2007). In plants, there is a second eIF4F complex, eIF(iso)4F, containing the isoforms eIF(iso)4E and eIF(iso)4G (Browning 2004). Potyviruses recruit selectively eIF4E iso-forms. Disruption of At-eIF(iso)4E gene in *Arabidopsis* results in resistance to turnip mosaic virus (TuMV), tobacco etch potyvirus (TEV), lettuce mosaic virus (LMV), and plum pox virus (PPV), whereas the disruption of eIF4E gene results in resistance to clover yellow vein virus (ClYVV; Duprat et al. 2002; Lellis et al. 2002; Sato et al. 2005; Decroocq et al. 2006; Nicaise et al. 2007).

The molecular nature of recessive resistance for many viruses, particularly for potyvirus, is that the virus-encoded protein VPg (or its precursor, NIa) binds to the eukaryotic translation initiation factor 4E (eIF4E) or its isoform eIF(iso)4E in yeast two-hybrid and in vitro binding assays (Leonard et al. 2000). A mutation in VPg that abolishes the interaction with eIF(iso)4E in vitro prevents viral infection in planta (Leonard et al. 2000). The mutation of a single member of the eIF4E gene family is sufficient to ensure resistance against several potyviruses. This suggests that different members of eIF4E family are not redundant and cannot sustain viral replication in absence of one of the members (Robaglia and Caranta 2006). Recent results suggest that potyviruses differ in their ability to use eIF4E isoforms from a given host plant. In pepper, resistance to potato virus Y (PVY) and TEV is dependent on the mutation of a single eIF4E gene (the *pvr2* locus), whereas resistance to pepper veinal mottle virus (PVMV) requires both the pvr2 locus and another distinct locus, *pvr6*, corresponding to a natural knockout of an eIF(iso)4E gene (Ruffel et al. 2004; Robaglia and Caranta 2006). It appears that although some potyviruses require one specific eIF4E isoform to perform their replication cycle, others can use several of them.

Melon (*Cucumis melo*) necrotic spot virus (MNSV; belonging to the family Tombusviridae is an uncapped and nonpolyadenylated RNA virus. A single amino acid change at position 228 of Cm-eIF4E led to resistance to MNSV. Protein expression and cap-binding analysis showed that Cm-eIF4E encoded by a resistant plant was not affected in its cap-binding activity. These data, and data from other studies, suggest that translation initiation factors of the eIF4E family are universal determinants of plant susceptibility to RNA viruses (Nieto et al. 2006). Identification of 4EBP homologes in plants will shed more light on plausible TOR connections in plant viral infections.

Nutrient Sensing via TOR

The term "nutrient" is very general in nature. It includes all the molecules used as building blocks as well as those used in various cellular processes to generate ATP. Specific transporter proteins bring sugars, amino acids, lipids, and vitamins into cells. These "transporters" are of two types: 1) true nutrient transporter (e.g., sugar and amino acid transporters) forming a pore or channel across the membrane to carry specific nutrient and 2) nutrient receptors (e.g., transferring receptor [TfR], low-density lipoprotein [LDL], receptor, etc.). Expression of both types of transporters is positively regulated by various growth signals (Edinger 2007).

In 1972, Robert W. Holley (Nobel laureate for his work demonstrating that tRNA interprets the genetic code) suggested that mammalian cell growth is fundamentally regulated at the level of intracellular nutrient availability. He hypothesized that although the bloodstream supplies cells with constant levels of nutrients, hormones and/or growth factors regulate access to these nutrients by modulating the expression of nutrient transport systems in the plasma membrane (Holley 1972). It is well known in yeast and animals that a change in nutrients status influences TOR signaling and affects overall growth and development of the organism (Petersen and Nurse 2007). Although all cells depend on nutrients they acquire from the extracellular space, surprisingly little is known about how nutrient uptake is regulated in eukaryotic cells (Edinger 2007). In yeast, the expression and trafficking of a wide variety of nutrient transporters is controlled by TOR. Consistent with this, mTOR and the related protein PI3K play roles in coupling nutrient transporter expression to the availability of extrinsic signals. Also, a link between nutrient transporter expression and oncogenesis is revealed by several recent studies demonstrating that nutrient transporter expression drives, rather than parallels, cellular metabolism (Edinger 2007). Besides nutrient transporters, many kinases, phosphatases, and even spindle polar bodies (SPB) are involved in the process of sensing outside nutrient condition via TOR. TOR mediates nutrient-modulated control of mitotic entry via mitotic kinase cdc2 in yeast (Petersen and Nurse 2007).

Yeast also provided better understanding on how TOR protein localization depends on nutrient availability and the effect of rapamycin on the whole process. TOR1 is dynamically distributed in the cytoplasm and nucleus in yeast, as in many other organisms. Nuclear localization of TOR1 is nutrient-dependent and sensitive to rapamycin treatment. Nuclear localization of TOR1 is important for many of its activities. TOR1 is shown to be associated with 35S rRNA (rDNA) promoter via its HTH motif, although it may be that other protein(s) assist this interaction. TOR1 is dissociated from rDNA promoter and exits from nucleus on rapamycin treatment or starvation. This interaction is important for rDNA expression and cell growth (Li et al. 2006). No obvious nuclear localization signal is detected in TOR1. An intact TOR1 kinase domain is crucial for this nuclear localization as kinase-inactive TOR1-RR (D229E) is predominantly cytoplasmic (Li et al. 2006). In contrast to 35S rRNA genes, this cytoplasmic form of TOR1 is sufficient to regulate many other TOR-dependent Pol-II transcribed genes, including Gln3-dependent genes and ribosomal protein genes (Beck and Hall 1999; Duvel et al. 2003; Martin et al. 2004).

One of the roles of the two TOR kinases in *S. cerevisiae* is to control the transcription subject to nitrogen repression (Beck and Hall 1999; Cardenas et al. 1999). During nitrogen-rich conditions, TOR signaling prevents nuclear accumulation of Gln3p, maintaining it within a cytoplasmic complex with Ure2p, preventing activation of a suite of nitrogen utilization genes (Beck and Hall 1999; Fitzgibbon et al. 2005). Decreased TOR activity also promoted nuclear localization of the stress-related transcription factor Msn2. It has been proposed that upregulation of a highly conserved response to starvation-induced stress is important for life-span extension by decreased TOR signaling in yeast and higher eukaryotes (Powers et al. 2006). Filamentous fungus *Aspergillus nidulans* has one TOR homologue (Fitzgibbon et al. 2005), and mutation in this gene is consistant with the TOR signaling pathway having a role in nitrogen signaling in this fungus (Fitzgibbon et al. 2005).

Photosynthesis and Nitrogen Assimilation

Plants being protoprophic for carbon responds to light signals assimilating carbon to produce sugar for their own energy needs. Thus, plants must monitor their phosynthetic ability to produce sugar as a source of carbon for all their metabolic processes and to establish a link with nitrogen assimilation. A carbon–nitrogen sensing mechanism enables plants to activate genes involved in N-assimilation when carbon skeletons are available (Coruzzi and Zhou 2001). Plants stop N-assimilation when levels of photosynthate are low (Coruzzi and Zhou 2001). Apparently, 193 genes related to protein synthesis, including 40S ribosomal subunit proteins, are induced by carbon (Gutierrez et al. 2007).

Using photosynthetic gene promoter–reporter fusions, Sheen and her colleagues showed that seven maize photosynthetic genes were repressed by glucose or sucrose in a maize protoplast system (Sheen 1990; Fitzgibbon et al. 2005; Baena-Gonzalez et al. 2007). Genes encoding enzymes of the carbohydrate metabolism, glyoxylate cycle, defense responses, and even storage proteins have been shown to respond to sugars (reviewed by Halford and Paul 2003). One microarray experiment found that 444 *Arabidopsis* genes were upregulated by glucose, including those involved in biotic and abiotic responses, carbohydrate metabolism, N metabolism, lipid metabolism, inositol metabolism, secondary metabolism, nucleic acid–related activities, protein synthesis and degradation, transport, signal transduction (involving protein kinases, phosphatases, and transcription factors), hormone synthesis, and cell growth. A comparable number (534 genes) were downregulated by the same treatment (Price et al. 2004; Francis and Halford 2006). It is known that depriving meristems of sucrose results in cell cycle arrest in G_1 or G_2 (Van't Hof, 1968). Plant cells also arrest cell cycle if there is no phosphate in culture medium (Francis and Halford 2006). Genes known to regulate $G_0/G_1/S$ do respond to sugar availability. For example, sucrose and glucose can control the expression of Cyclin D2 and D3 genes in the G_1 phase of *Arabidopsis* (Riou-Khamlichi et al. 2000; Coruzzi and Zhou 2001; Dewitte and Murray 2003). Similar to the results in *Arabidopsis*, the expressions of three cyclin D genes in the apical meristems of snapdragon (*Antirrhinum majus*) are modulated in part by sucrose (Gaudin et al. 2000).

Sugars affect growth and development through crosstalk with hormone signaling pathways. A link between ABA and sugar signaling has been reviewed by Rook and Bevan (2003). Several *Arabidopsis* mutants that are impaired in their response to sugars (sugar response mutants) are also affected in their response to ABA, particularly with respect to germination and seedling growth. Alternatively, ABA and sugar signaling could be essentially separate but converge and crosstalk through specific factors (Halford and Paul 2003). Plant ICK1/2 (Inhibitor of CDK activity) is upregulated by ABA (Wang et al. 1998). Sugar–ABA–cell cycle signaling mechanism that would repress plant growth in unfavorable conditions and would crosstalk with the cell cycle in meristems (Francis and Halford 2006). *Arabidopsis* CDKA1, ortholog of yeast cdc2/cdc28 kinase (Iwakawa et al. 2006; Dissmeyer et al. 2007), appears to be the main kinase involved in both major check points: entry into S and M phases (Reichheld et al. 1999; Menges and Murray 2002). As described earlier, mTOR signaling leads to a cyclin D1–dependent increase in CDK4 activity, followed by phosphorylation-mediated dissociation of pRB from UBF with a concomitant increase in UBF availability (Hannan et al. 2003; Nader et al. 2005; Mayer and Grummt et al. 2006). Rapamycin prevents the upregulation of cyclin D1 as well as proteins acting downstream in the cell cycle in cultured rat hepatocytes cells. Rapamycin also inhibits the induction of global protein synthesis following growth factor treatment, whereas cyclin D1 stimulation overcomes this inhibition (Nelsen et al. 2003). These results suggest that cyclin D1 is a key mediator of increased protein synthesis and cell growth downstream of the TOR kinase pathway (Nelsen et al. 2003).

No results have been reported so far in plant systems to study TOR signaling regarding nutrient sensing. It seems plausible that TOR kinase is involved in sensing environmental signal(s) through cyclin D and related CDKs and may connect transcriptional repression (through pRB-E2F/UBF) and translational repression to nutrient status and cell cycle progression. In animals, cyclin D is expressed during G_1-S phase transition and integrates various internal and external signals into cell cycle (Sherr 1996). Plant *CYCD3.1* expression depends on the availability of sucrose and plant hormones (Riou-Khamlichi et al. 2000). The addition of sucrose to sucrose-deprived cell cultures results in the induction of *CYCD3.1* in late G_1 phase (Menges and Murray 2002), with the mRNA subsequently maintained at a relatively constant level during cell cycle (Ito 2000; Dewitte and Murray, 2003). Other mitogenic plant hormones such as auxin and gibberelin (Oakenfull et al. 2002), cytokinins (Riou-Khamlichi et al. 2000), and brassinosteriods (Hu et al. 2000; Boniotti and Griffith 2002; Sasse 2003) also have the ability to induce *CYCD3*. The observation that TORC1 signaling is required for the S-phase progression and viability of yeast cells in response to genotoxic stress (Shen et al. 2007) is consistent with its plausible connection with cyclin D, which plays a major role in G_1- to S-phase transition. It will be worth studying the effect of nutrients, mainly sugars, to connect growth with TOR and some major players in cell cycle in plants. It is likely that carbon (sugar) and nitrogen assimilation are directly linked since TOR independently senses carbon and amino acid levels (see below). Figure 8.6 summarizes the possible TOR involvement in carbon metabolism and nitrogen assimilation in plants.

Leucine and Other Amino Acid Connections

The translation step is regulated not only at initiation—that is, the recruitment of the initiator methionyl–transfer RNA (met-tRNA$_i$) and mRNA to the 40S ribosomal subunit followed by the joining of the 60S ribosomal subunit to complete assembly of

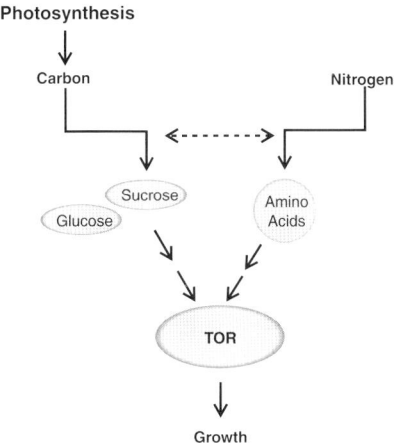

Fig. 8.6 TOR connecting carbon and nitrogen inputs to regulate plant growth

a translationally competent 80S ribosome—but also during termination. Many amino acids are known to influence gene expression through modulation of mRNA translation (Kimball and Jefferson 2006). The best characterized example of amino acid–induced regulation of signal transduction pathways involves TOR kinase, through which the branched-chain amino acids, particularly leucine, act to modulate both global mRNA translation and selection of specific mRNAs for translation (Kimball and Jefferson 2006). Leucine has been reported to have the most potent impact on the activation of S6K (Hara et al. 1998; Shigemitsu et al. 1999). Leucine in concert with other amino acids may regulate mTOR signaling through the tuberous sclerosis complex (TSC) (Tokunaga et al. 2004).

TORC1 is implicated in stress signal transduction arising from amino acid deficiency (Shen et al. 2007). The branched-chain amino acids leucine, isoleucine, and valine are the most abundant among the essential amino acids. In addition to being indispensable for growth, the branched-chain amino acids act as nutrient regulators of protein synthesis and degradation. Administration of leucine to food-deprived rats stimulates protein synthesis to 165% within 60 minutes, whereas administration of carbohydrate alone has no significant effect (Kimball and Jefferson 2006). Leucine plus carbohydrate has the same effect as leucine alone, and administration of either isoleucine or valine alone has no effect. Thus, leucine is unique among the branched-chain amino acids with regard to its effectiveness as a nutrient regulator of protein synthesis. Studies have shown that leucine has no effect on the met-tRNA$_i$ binding step in translation initiation, as assessed by the phosphorylation status of eIF2 on Ser51 of its α-subunit and by the guanine nucleotide exchange activity of eIF2B (Anthony et al. 2000). However, it has a stimulatory effect on assembly of the eIF4E complex, as assessed by the phosphorylation of the eIF4E-binding protein 4E-BP1 and by the association of eIF4E with 4E-BP1 and eIF4G. Leucine also has a stimulatory effect on the phosphorylation status of eIF4G, as well as S6K1 and its downstream substrate rpS6 (Bolster et al. 2004; Tokunaga et al. 2004).

Rapamycin blocks all effects of leucine on the downstream targets of mTOR signaling and prevents the stimulatory effect of this amino acid on protein synthesis.

Interestingly, some studies show that isoleucine, but not valine, mimics to some extent the effect of leucine on mTOR signaling. This is an unexpected result because isoleucine has no effect on protein synthesis. Leucine also has a stimulatory effect on the phosphorylation status of eIF4G, as well as S6K1 and its downstream substrate S6 (Bolster et al. 2004; Kimball and Jefferson 2006). Because phosphorylation of 4E-BP1, eIF4G, and S6K1 is mediated in part by mTOR, the results suggest that leucine stimulates a signaling pathway involving the serine/threonine protein kinase. This suggestion was confirmed in a separate study in which food-deprived rats were injected intravenously with rapamycin before leucine administration. Rapamycin blocks the effects of leucine on the downstream targets of mTOR signaling and attenuates the stimulatory effect of the amino acid on protein synthesis (Anthony et al. 2000).

Activity of mTORC1 is affected by amino acid level through Vps34 (vacuolar protein sorting 34), a class 3 PI3K (Edinger 2007; Backer 2008). Knocking down human Vps34 (hVps34) blocks insulin-stimulated phosphorylation of both S6K1 and 4E-BP1 (Byfield et al. 2005; Nobukuni et al. 2005). mTOR signaling inhibits the endocytic degradation of amino acid transporters but stimulates fluid phase uptake of bulk nutrients (Hennig et al. 2006). hVps34 knockdown also blocks amino acid stimulation of S6K1 (Byfield et al. 2005; Nobukuni et al. 2005; Backer 2008). Data suggest that hVps34 is upstream of mTOR mainly because Vps34 is not inhibited by rapamycin (Byfield et al. 2005). Yeast Vps34 seem to be an effector of $G\beta$ subunit assembly (Slessareva et al. 2006). In both yeast and mammalian cells, class 3 PI3K is implicated in TOR-mediated autophagy during nutrient starvation (Byfield et al. 2005).

$G\beta L$ is a positive regulator of mTOR because coexpression of $G\beta L$ and mTOR results in greatly increased kinase activity of mTOR toward 4E-BP1 and S6K1. Like raptor, $G\beta L$ coimmunoprecipitates with mTOR (Kim et al. 2002). Moreover, reducing $G\beta L$ expression using siRNA represses leucine- and serum-induced phosphorylation of S6K1, which suggests that $G\beta L$ is involved in hormone and amino acid signaling though mTOR (Kimball and Jefferson 2006). We have observed that the $G\beta L$ homolog of *Arabidopsis* responds to auxin and cytokinin treatments (S. Kim and D.P.S. Verma, unpublished data). Leucine induces phosphorylation of mTOR, on both Ser2448 and Ser2481, and its downstream effectors, S6K, rpS6, and 4E-BP1. Plants produce leucine primarily from valine inside mitochondria (Lucas et al. 2007). So far no study connects leucine or any other amino acid with TOR signaling in plants. Figure 8.7 summarizes the convergence of nutrient and energy sensing by TOR.

TOR as a Sensor of ATP

Oxidative phosphorylation generating ATP in plant cells also produces reactive oxygen species (ROS), the concentration of which increases under various stress conditions (Pastore et al. 2007). It has been demonstrated that TOR is a sensor of intracellular concentration of ATP, independent of the abundance of amino acids (Dennis et al. 1998). AMP-activated protein kinase (AMPK) is conserved from yeast to humans and plays a role in cellular energy homeostasis. The energy-sensing capability of AMPK is attributed to its ability to detect and react to fluctuations in the AMP:ATP ratio. It has been reported that ROS leads to the activation of AMP kinase (Hwang et al. 2004), and

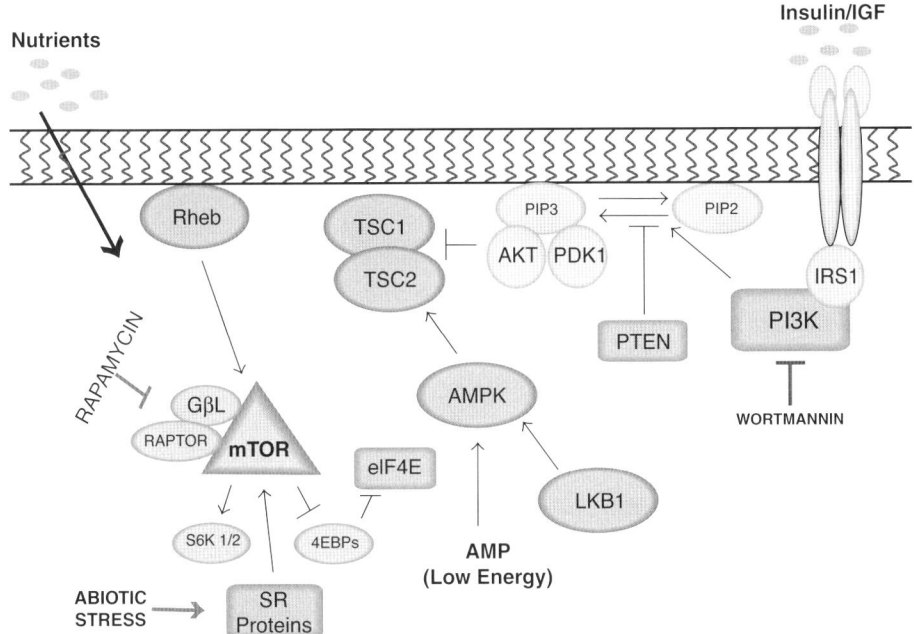

Fig. 8.7 Convergence of the nutrient- and energy-sensing mTOR pathway with the growth factor-sensing PI3K pathway. Activation of the mTOR pathway by the PI3K pathway occurs through the phosphorylation and inactivation of TSC2 by AKT. PIP3 complex. Inhibition of mTOR under low-energy conditions (when the AMP:ATP ratio rises) occurs through the phosphorylation and activation of TSC2 by AMPK. Abiotic stress-responsive SR protein (SF2/ASF) interacts with mTOR and mediate translation.

thus it is directly involved in cellular stress responses (Corton et al. 1994). AMPK is a homolog of SNF1 (Aguan et al. 1994). Plant SNF1 may be involved in the activation of gene expression in response to photosynthates, sucrose, and glucose (Coruzzi and Zhou 2001).

Even a small decrease in ATP can result in elevations of cellular AMP through the actions of adenylate kinase, which controls AMP concentration and is predominantly located in the mitochondrial intermembrane space rather than in the mitochondrial matrix (Schnaitman and Pedersen 1968), which leads to activation of AMPK and subsequent inhibition of mTOR-S6K signaling (Tokunaga et al. 2004). In yeast (*S. cerevisiae*), the SNF1 protein kinase is involved in various stress signaling pathways. It has been shown that activity of yeast SNF1 is negatively regulated by TOR via phosphorylation of Thr210 residue (Orlova et al. 2006).

Because plants are prototrophic with respect to carbon supply, they have significant reserves of carbohydrates (sugars), the supply of which is diuarnally regulated by the photosynthesis period. The rate of photosynthesis, however, is directly affected by stress conditions, as well as by light. If TOR senses the homeostasis of ATP, its role becomes extremely important in maintaining plant cellular metabolism in response to energy derived from photosynthates.

Osmotic Stress Signaling and TOR

Osmotic stress is the most critical abiotic factor for plants, because the soil moisture varies continuously with environmental conditions. Plants must adjust to their translation and transcription machineries in response to various biotic and abiotic stress conditions. Osmotic stress is challenging for almost all organisms, particularly unicellular organisms such as yeast and algae. It is therefore not surprising that osmotic stress influences cell growth in all organisms.

Arabidopsis S6K1 activity was found to be affected by osmotic stress (mannitol treatment). Mannitol treatment of tobacco or *Arabidopsis* leaves reduced S6K1 but not PDK1 activity by more than 70%, an effect that could be mitigated by overexpression of RAPTOR1 (Mahfouz et al. 2006). Down regulation of S6K activity appears to have a protective effect to sustain osmotic stress, as *Arabidopsis* overexpressing S6K is hypersensitive to osmotic stress (Mahfouz et al. 2006). This observation indicates that TOR may be involved in the regulation of the metabolic adjustment of plant cells to osmotic stress. Transgenic *Arabidopsis* seedlings overexpressing S6K1 were overly sensitive to the mannitol treatment. This result suggests that slowing down the S6K activity under unfavorable growth conditions may be one of the main mechanisms employed by plants to sustain harsh environmental conditions. The accumulation of hyperphosphorylated isoforms with phospho-Ser-238 RPS6 was reduced in response to oxygen deprivation and heat shock, whereas accumulation of these isoforms was elevated by cold stress. Salt and osmotic stress had no reproducible effect on RPS6 phosphorylation. The reduction in hyperphosphorylated isoforms under oxygen deprivation was blocked by okadaic acid, a Ser/Thr phosphatase inhibitor (Williams et al. 2003). In *Arabidopsis* cell culture, osmotic stress has almost no effect on the actively growing cells, whereas the S6K2 activity in the stationary-phase cells or starvation-induced cells is highly sensitive to osmotic stress (Mahfouz et al. 2006). Figure 8.8 shows the effect of osmotic stress (via mannitol treatment) on S6K1 kinase activity. These data suggest that various plants and tissues respond to osmotic stress differently and the involvement of TOR may also vary.

Although the overall amino acid sequences of plant S6Ks differ significantly from their mammalian counterparts, the catalytic domain is highly conserved with that of other AGC kinases. Work from our lab showed that the S6K1 is a substrate for PDK1 at least in vitro (Mahfouz et al. 2006). Phosphorylation of S6K by mammalian TOR is considered to be a prerequisite for PDK1 to make it active (Burnett et al. 1998; Dennis et al. 1998). Activation of animal PDK1 requires a second messenger phosphatidylinositol-3,4,5-phosphate, which is generated by the catalytic action of class-I PI-3K (Chan et al. 1999). No evidence for the existence of class-I PI-3K is available in plants. AtPDK1 activity was inhibited by wortmannin, an inhibitor of class-I PI-3K (Mahfouz et al. 2006). Plant PDK1 may be activated by a different type of phosphatidylinositol phosphate, as suggested by Deak et al. (1999), produced by a yet-to-be-identified lipid kinase. The observation that the PDK1 activity was not sensitive to osmotic stress suggests that the osmotic stress signal is likely to regulate S6K1 activity through the TOR pathway (Mahfouz et al. 2006). In yeast, *S. pombe*, a mutation in the TOR gene resulted in cells that are hypersensitive to osmotic stress (Weisman and Choder, 2001).

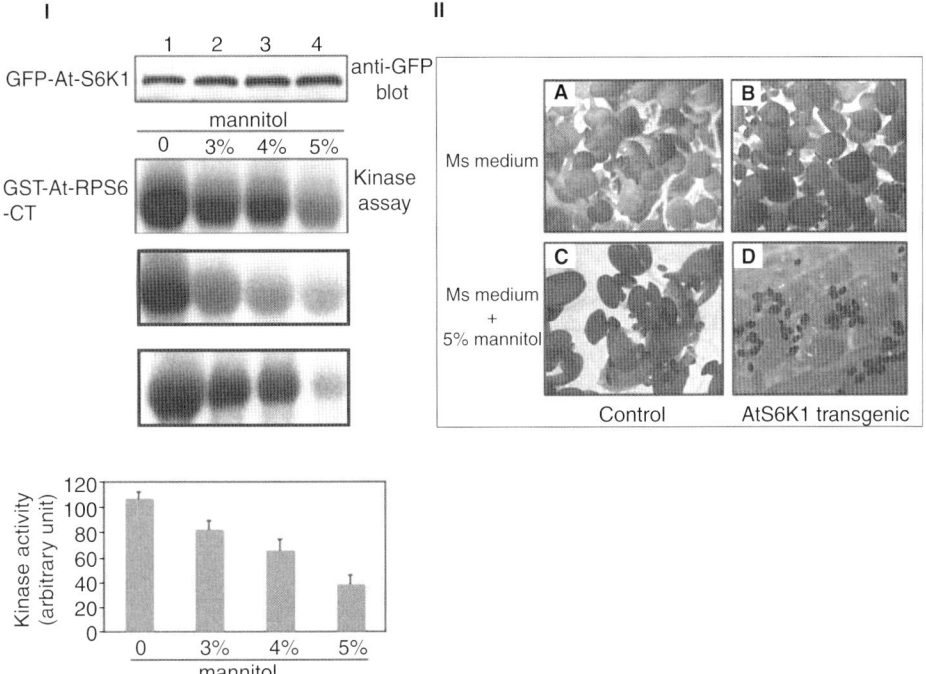

Fig. 8.8 Osmotic stress affects S6K1 kinase activity in plants. (**I**) Leaves of tobacco (*Nicotiana benthamiana*) transiently expressing the 35S::GFPS6K1 construct were treated with different concentration of mannitol. GFP-S6K1 protein was then immunoprecipitated with GFP antibody and subjected to in vitro kinase assay using GST-RPS6-CT as substrate. Top panel: protein gel blot showing equal loading of S6K1 in the kinase reaction. Bottom panel: autoradiograph showing the activity of S6K1 under different concentrations of mannitol treatment. (**II**) (**A**) *Arabidopsis* plants transformed with the empty vector construct and grown on Murashige and Skoog (MS) plates. (**B**) Transgenic *Arabidopsis* plants expressing S6K1 and grown on MS plates. (**C**) *Arabidopsis* plants transformed with the empty vector construct and grown on MS plates containing 5% mannitol. (**D**) Transgenic *Arabidopsis* plants expressing S6K1 and grown on MS plates containing 5% mannitol (note inhibition of seed germination). (From Mahfouz et al. 2006.)

Nitrogen starvation results in enhanced root growth, whereas an excess of nitrate inhibits root growth (Zhang et al. 1999; Zhang and Forde 2000). T-DNA insertion lines having elevated levels of AtTOR transcript did not modify root length under nitrogen-limiting conditions. Conversely, when plants were grown in the presence of excess nitrate, the AtTOR-overexpressing lines showed a longer primary root. This indicates that increased AtTOR expression relieves the inhibition of primary root growth by excess nitrogen (Deprost et al. 2007).

Plants with reduced AtTOR activity also resembled *Arabidopsis* mutants of KEEP ON GOING, an E3 ubiquitin ligase required for ABI5 degradation (Deprost et al. 2007; Stone et al. 2006). Furthermore, AtTOR RNAi lines were hypersensitive to high sugar concentrations, whereas plants overexpressing AtTOR were less affected by the same. High sugar concentrations can retard seedling growth through the action of ABA (Finkelstein and Gibson 2002). A positive correlation between AtTOR expression and sensitivity to ABA has also been proposed (Deprost et al. 2007). Together, these data indicate that ABA or osmotic stress invoke a decrease in AtTOR expression and/or activity. The observation that the AtTOR RNAi line is more sensitive to osmotic stress than to ABA can be explained by the fact that osmotic stress signaling is mediated only in part by ABA (Finkelstein and Gibson 2002, Deprost et al. 2007). These results indicate that AtTOR expression is required for postembryonic growth and might act as a relay for ABA signaling between environmental signals and the growth processes.

It has been shown that TOR1 in yeast is required for growth under high salt concentrations via regulation of GLN3 and GAT1 transcription factors (Crespo et al. 2001). Overexpression of SR proteins also confers salt tolerance in both yeast and plants (Forment et al. 2002) and some members of this protein family do interact with TOR (Michlewski et al. 2008). However, it is not known whether plant SR proteins accomplish that through TOR pathway.

Conclusion

Linking Stress to Growth

Plants form organs throughout their entire life spans, which can extend more than 1,000 years. For that purpose they maintain pluripotent somatic stem cells within the meristems, local pools of mitotically active cells (Weigel and Jurgens 2002; Wildwater et al. 2005). These meristems respond to nutrition and stress by controlling their growth. Although plants and animals evolved independently, it is surprising to see the conservation of TOR pathway in both kingdoms. This pathway links translation and transcription machineries to stress and nutrition, and hence, it is vital for the survival of the organism. Although most of the major players in this pathway have been identified in plants, it is likely that further studies would reveal other functions of TOR such as control of life span, autophagy, and cytoskeleton control and body size.

Plants need to successfully integrate external (abiotic and biotic) signals producing maximum vegetative mass and convert it into reproductive mass (seeds). This "harvest index" is the basis of plant breeding to improve crop productivity. Although plants are prototropic for carbon, they quickly respond to nitrogen inputs. Most legumes are prototropic even for nitrogen (via symbiotic nitrogen fixation), and hence they produce high protein-containing grains. In addition, plants are also totipotent and can grow from single cells, provided there are proper hormonal and nutritional inputs.

Abiotic stress immediately affects carbon metabolism by reducing the supply of photosynthates and hence ATP level. Because protein synthesis is a major energy-consuming cellular process, it is immediately affected by the supply of ATP. It is now well established that plants do use TOR kinase pathway to control both translation and transcription machineries as well as other cellular processes in response to nutrients status, hormones, and other environmental signals, such as osmotic stress to control their growth (Figure 8.9). Furthermore, the link between translation and transcription through histone deacetylase (HDA) suggests that chromatin modeling may be involved in this process, and sustained reduction in growth, either due to lack of nutrition or stress, may have long-term epigenetic consequences. Availability of TOR mutants is a major constraint in plant TOR research. It is hoped that the development of new

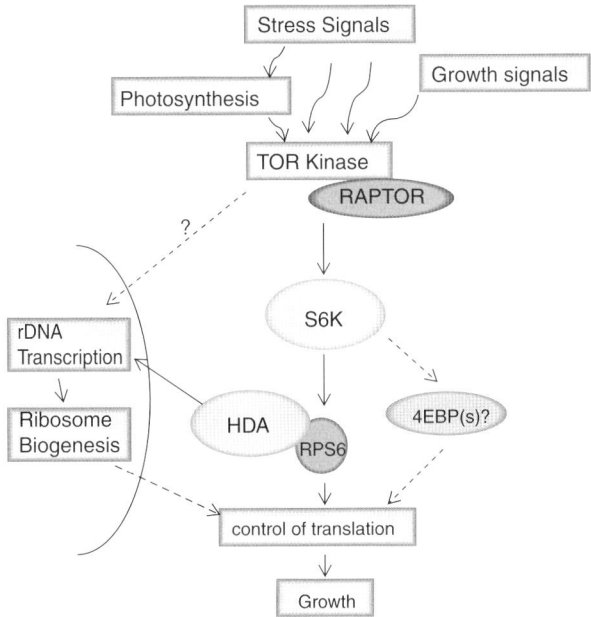

Fig. 8.9 Connection of TOR with stress, growth signals, and photosynthesis.

biochemical and genetic tools will help elucidate the complexity of this pathway, which is central in controlling plant growth. Expressing hFKBP12 or other FKBP12 ortholog that interacts with AtTOR in *Arabidopsis* to obtain rapamaycin-sensitive lines followed by chemical genomics experiments, as done in yeast to elucidate global functions of TOR protein (Chan et al. 2000), may allow us to identify positive and negative regulators of this pathway in plants. Identification and characterization of these components are of crucial importance to understanding the fundamental process of coordinated plant growth under a variety of environmental signals that a plant confronts to complete its life cycle.

Acknowledgment

The work on TOR kinase in our lab is supported by National Science Foundation, United States.

References

Adami, A., Garcia-Alvarez, B., Arias-Palomo, E., Barford, D., and Llorca, O. 2007. Structure of TOR and its complex with KOG1. *Mol Cell* 27:509–516.
Aghdasi, B., Ye, K., Resnick, A., Huang, A., Ha, H.C., Guo, X., Dawson, T.M., Dawson, V.L., and Snyder, S.H. 2001. FKBP12, the 12-kDa FK506-binding protein, is a physiologic regulator of the cell cycle. *Proc Natl Acad Sci USA* 98:2425–2430.

Agredano-Moreno, L.T., Reyes de la Cruz, H., Martinez-Castilla, L.P., and Sanchez de Jimenez, E. 2007. Distinctive expression and functional regulation of the maize (*Zea mays* L.) TOR kinase ortholog. *Mol Biosyst* 3:794–802.

Aguan, K., Scott, J., See, C.G., and Sarkar, N.H. 1994. Characterization and chromosomal localization of the human homologue of a rat AMP-activated protein kinase-encoding gene: a major regulator of lipid metabolism in mammals. *Gene* 149:345–350.

Albar, L., Bangratz-Reyser, M., Hebrard, E., Ndjiondjop, M.N., Jones, M., and Ghesquiere, A. 2006. Mutations in the eIF(iso)4G translation initiation factor confer high resistance of rice to rice yellow mottle virus. *Plant J* 47:417–26.

Andersen, J.S., Lyon, C.E., Fox, A.H., Leung, A.K., Lam, Y.W., Steen, H., Mann, M., and Lamond, A.I. 2002. Directed proteomic analysis of the human nucleolus. *Curr Biol* 12:1–11.

Anderson, G.H., and Hanson, M.R. 2005. The *Arabidopsis* Mei2 homologue AML1 binds AtRaptor1B, the plant homologue of a major regulator of eukaryotic cell growth. *BMC Plant Biol* 5:2.

Anthony, R.G., Khan, S., Costa, J., Pais, M.S., and Bogre, L. 2006. The *Arabidopsis* protein kinase PTI1-2 is activated by convergent phosphatidic acid and oxidative stress signaling pathways downstream of PDK1 and OXI1. *J Biol Chem* 281:37536–37546.

Anthony, T.G., Fabian, J.R., Kimball, S.R., and Jefferson, L.S. 2000. Identification of domains within the epsilon-subunit of the translation initiation factor eIF2B that are necessary for guanine nucleotide exchange activity and eIF2B holoprotein formation. *Biochim Biophys Acta* 1492:56–62.

Arsham, A.M., and Neufeld, T.P. 2006. Thinking globally and acting locally with TOR. *Curr Opin Cell Biol* 18:589–597.

Backer, J.M. 2008. The regulation and function of Class III PI3Ks: novel roles for Vps34. *Biochem J* 410:1–17.

Baena-Gonzalez, E., Rolland, F., Thevelein, J.M., and Sheen, J. 2007. A central integrator of transcription networks in plant stress and energy signalling. *Nature* 448:938–942.

Baker, H., Sidorowicz, A., Sehgal, S.N., and Vezina, C. (1978). Rapamycin (AY-22,989), a new antifungal antibiotic. III. In vitro and in vivo evaluation. *J Antibiot (Tokyo)* 31:539–545.

Bauer, C., Brass, N., Diesinger, I., Kayser, K., Grasser, F.A., and Meese, E. 2002. Overexpression of the eukaryotic translation initiation factor 4G (eIF4G-1) in squamous cell lung carcinoma. *Int J Cancer* 98:181–185.

Bauer, C., Diesinger, I., Brass, N., Steinhart, H., Iro, H., and Meese, E.U. 2001. Translation initiation factor eIF-4G is immunogenic, overexpressed, and amplified in patients with squamous cell lung carcinoma. *Cancer* 92:822–829.

Beck, T., and Hall, M.N. 1999. The TOR signalling pathway controls nuclear localization of nutrient-regulated transcription factors. *Nature* 402:689–692.

Bielinski, V.A., and Mumby, M.C. 2007. Functional analysis of the PP2A subfamily of protein phosphatases in regulating *Drosophila* S6 kinase. *Exp Cell Res* 313:3117–3126.

Biondi, R.M., Kieloch, A., Currie, R.A., Deak, M., and Alessi, D.R. 2001. The PIF-binding pocket in PDK1 is essential for activation of S6K and SGK, but not PKB. *Embo J* 20:4380–4390.

Blakeslee, J.J., Zhou, H.W., Heath, J.T., Skottke, K.R., Barrios, J.A., Liu, S.Y., and Delong, A. 2008. Specificity of RCN1-mediated protein phosphatase 2A regulation in meristem organization and stress response in roots. *Plant Physiol* 146:539–553.

Bodine, S.C., Stitt, T.N., Gonzalez, M., Kline, W.O., Stover, G.L., Bauerlein, R., Zlotchenko, E., Scrimgeour, A., Lawrence, J.C., Glass, D.J., et al. 2001. Akt/mTOR pathway is a crucial regulator of skeletal muscle hypertrophy and can prevent muscle atrophy in vivo. *Nat Cell Biol* 3:1014–1019.

Bogre, L., Okresz, L., Henriques, R., and Anthony, R.G. 2003. Growth signalling pathways in *Arabidopsis* and the AGC protein kinases. *Trends Plant Sci* 8:424–431.

Bolster, D.R., Vary, T.C., Kimball, S.R., and Jefferson, L.S. 2004. Leucine regulates translation initiation in rat skeletal muscle via enhanced eIF4G phosphorylation. *J Nutr* 134:1704–1710.

Boniotti, M.B., and Griffith, M.E. 2002. "Cross-talk" between cell division cycle and development in plants. *Plant Cell* 14:11–16.

Boudolf, V., Rombauts, S., Naudts, M., Inze, D., and De Veylder, L. 2001. Identification of novel cyclin-dependent kinases interacting with the CKS1 protein of Arabidopsis. *J Exp Bot* 52:1381–1382.

Brass, N., Heckel, D., Sahin, U., Pfreundschuh, M., Sybrecht, G.W., and Meese, E. 1997. Translation initiation factor eIF-4gamma is encoded by an amplified gene and induces an immune response in squamous cell lung carcinoma. *Hum Mol Genet* 6:33–39.

Breiman, A., and Camus, I. 2002. The involvement of mammalian and plant FK506-binding proteins (FKBPs) in development. *Transgenic Res* 11:321–335.

Browning, K.S. 2004. Plant translation initiation factors: it is not easy to be green. *Biochem Soc Trans* 32:589–591.

Burnett, P.E., Barrow, R.K., Cohen, N.A., Snyder, S.H., and Sabatini, D.M. 1998. RAFT1 phosphorylation of the translational regulators p70 S6 kinase and 4E-BP1. *Proc Natl Acad Sci USA* 95:1432–1437.

Byfield, M.P., Murray, J.T., and Backer, J.M. 2005. hVps34 is a nutrient-regulated lipid kinase required for activation of p70 S6 kinase. *J Biol Chem* 280:33076–33082.

Caceres, J.F., Misteli, T., Screaton, G.R., Spector, D.L., and Krainer, A.R. 1997. Role of the modular domains of SR proteins in subnuclear localization and alternative splicing specificity. *J Cell Biol* 138:225–238.

Cardenas, M.E., Cutler, N.S., Lorenz, M.C., Di Como, C.J., and Heitman, J. 1999. The TOR signaling cascade regulates gene expression in response to nutrients. *Genes Dev* 13:3271–3279.

Carles, C.C., and Fletcher, J.C. 2003. Shoot apical meristem maintenance: the art of a dynamic balance. *Trends Plant Sci* 8:394–401.

Carlson, M. 1999. Glucose repression in yeast. *Curr Opin Microbiol* 2:202–207.

Chan, T.F., Carvalho, J., Riles, L., and Zheng, X.F. 2000. A chemical genomics approach toward understanding the global functions of the target of rapamycin protein (TOR). *Proc Natl Acad Sci USA* 97, 13227–13232.

Chan, T.O., Rittenhouse, S.E., and Tsichlis, P.N. 1999. AKT/PKB and other D3 phosphoinositide-regulated kinases: kinase activation by phosphoinositide-dependent phosphorylation. *Annu Rev Biochem* 68:965–1014.

Chiang, G.G., and Abraham, R.T. 2005. Phosphorylation of mammalian target of rapamycin (mTOR) at Ser-2448 is mediated by p70S6 kinase. *J Biol Chem* 280:25485–25490.

Choi, J., Chen, J., Schreiber, S.L., and Clardy, J. 1996. Structure of the FKBP12-rapamycin complex interacting with the binding domain of human FRAP. *Science* 273:239–242.

Chou, I.T., and Gasser, C.S. 1997. Characterization of the cyclophilin gene family of *Arabidopsis thaliana* and phylogenetic analysis of known cyclophilin proteins. *Plant Mol Biol* 35:873–892.

Corradetti, M.N., and Guan, K.L. 2006. Upstream of the mammalian target of rapamycin: do all roads pass through mTOR? *Oncogene* 25:6347–6360.

Corton, J.M., Gillespie, J.G., and Hardie, D.G. 1994. Role of the AMP-activated protein kinase in the cellular stress response. *Curr Biol* 4:315–324.

Coruzzi, G.M., and Zhou, L. 2001. Carbon and nitrogen sensing and signaling in plants: emerging "matrix effects." *Curr Opin Plant Biol* 4:247–253.

Crespo, J.L., Diaz-Troya, S., and Florencio, F.J. 2005. Inhibition of target of rapamycin signaling by rapamycin in the unicellular green alga *Chlamydomonas reinhardtii*. *Plant Physiol* 139:1736–1749.

Cygnar, K.D., Gao, X., Pan, D., and Neufeld, T.P. 2005. The phosphatase subunit tap42 functions independently of target of rapamycin to regulate cell division and survival in *Drosophila*. *Genetics* 170:733–740.

Das, R., Yu, J., Zhang, Z., Gygi, M.P., Krainer, A.R., Gygi, S.P., and Reed, R. 2007. SR proteins function in coupling RNAP II transcription to pre-mRNA splicing. *Mol Cell* 26:867–881.

Deak, K.I., and Malamy, J. 2005. Osmotic regulation of root system architecture. *Plant J* 43:17–28.

Deak, M., Casamayor, A., Currie, R.A., Downes, C.P., and Alessi, D.R. 1999. Characterisation of a plant 3-phosphoinositide-dependent protein kinase-1 homologue which contains a pleckstrin homology domain. *FEBS Lett* 451:220–226.

Decroocq, V., Sicard, O., Alamillo, J.M., Lansac, M., Eyquard, J.P., Garcia, J.A., Candresse, T., Le Gall, O., and Revers, F. 2006. Multiple resistance traits control Plum pox virus infection in *Arabidopsis thaliana*. *Mol Plant Microbe Interact* 19:541–549.

Dennis, P.B., Jaeschke, A., Saitoh, M., Fowler, B., Kozma, S.C., and Thomas, G. 2001. Mammalian TOR: a homeostatic ATP sensor. *Science* 294:1102–1105.

Dennis, P.B., Pullen, N., Pearson, R.B., Kozma, S.C., and Thomas, G. 1998. Phosphorylation sites in the autoinhibitory domain participate in p70(s6k) activation loop phosphorylation. *J Biol Chem* 273, 14845–14852.

Dennis, P.B., and Thomas, G. 2002. Quick guide: target of rapamycin. *Curr Biol* 12:R269.

Deprost, D., Truong, H.N., Robaglia, C., and Meyer, C. 2005. An *Arabidopsis* homolog of RAPTOR/KOG1 is essential for early embryo development. *Biochem Biophys Res Commun* 326:844–850.

Deprost, D., Yao, L., Sormani, R., Moreau, M., Leterreux, G., Nicolai, M., Bedu, M., Robaglia, C., and Meyer, C. 2007. The *Arabidopsis* TOR kinase links plant growth, yield, stress resistance and mRNA translation. *EMBO Rep* 8:864–870.

Dewitte, W., and Murray, J.A. 2003. The plant cell cycle. *Annu Rev Plant Biol* 54:235–264.

Di Como, C.J., and Arndt, K.T. 1996. Nutrients, via the Tor proteins, stimulate the association of Tap42 with type 2A phosphatases. *Genes Dev* 10:1904–1916.

Dinkova, T.D., de la Cruz, A.F., Garcia-Flores, C., Aguilar, R., Jimenez-Garcia, L.P., and de Jimenez, E.S. 2007. Dissecting the TOR–S6K signal transduction pathway in maize seedlings: relevance on cell growth regulation. *Physiol Plantarum* 130:1–10.

Diaz-Troya, S., Florencio, F.J., and Crespo, J.L. 2008. Target of rapamycin and LST8 proteins associate with membranes from the endoplasmic reticulum in the unicellular green alga *Chlamydomonas reinhardtii*. *Eukaryot Cell* 7:212–222.

Dissmeyer, N., Nowack, M.K., Pusch, S., Stals, H., Inze, D., Grini, P.E., and Schnittger, A. 2007. T-loop phosphorylation of *Arabidopsis* CDKA;1 is required for its function and can be partially substituted by an aspartate residue. *Plant Cell* 19:972–985.

Dolinski, K., Muir, S., Cardenas, M., and Heitman, J. 1997. All cyclophilins and FK506 binding proteins are, individually and collectively, dispensable for viability in *Saccharomyces cerevisiae*. *Proc Natl Acad Sci USA* 94:13093–13098.

Duprat, A., Caranta, C., Revers, F., Menand, B., Browning, K.S., and Robaglia, C. 2002. The *Arabidopsis* eukaryotic initiation factor (iso)4E is dispensable for plant growth but required for susceptibility to potyviruses. *Plant J* 32:927–934.

Durfee, T., Feiler, H.S., and Gruissem, W. 2000. Retinoblastoma-related proteins in plants: homologues or orthologues of their metazoan counterparts? *Plant Mol Biol* 43:635–642.

Duvel, K., and Broach, J.R. 2004. The role of phosphatases in TOR signaling in yeast. *Curr Top Microbiol Immunol* 279:19–38.

Duvel, K., Santhanam, A., Garrett, S., Schneper, L., and Broach, J.R. 2003. Multiple roles of Tap42 in mediating rapamycin-induced transcriptional changes in yeast. *Mol Cell* 11:1467–1478.

Ebel, C., Mariconti, L., and Gruissem, W. 2004. Plant retinoblastoma homologues control nuclear proliferation in the female gametophyte. *Nature* 429:776–780.

Edinger, A.L. 2007. Controlling cell growth and survival through regulated nutrient transporter expression. *Biochem J* 406:1–12.

Faure, J.D., Gingerich, D., and Howell, S.H. 1998. An *Arabidopsis* immunophilin, AtFKBP12, binds to AtFIP37 (FKBP interacting protein) in an interaction that is disrupted by FK506. *Plant J* 15:783–789.

Fingar, D.C., and Blenis, J. 2004. Target of rapamycin (TOR): an integrator of nutrient and growth factor signals and coordinator of cell growth and cell cycle progression. *Oncogene* 23:3151–3171.

Finkelstein, R.R., and Gibson, S.I. 2002. ABA and sugar interactions regulating development: cross-talk or voices in a crowd? *Curr Opin Plant Biol* 5:26–32.

Fitzgibbon, G.J., Morozov, I.Y., Jones, M.G., and Caddick, M.X. 2005. Genetic analysis of the TOR pathway in *Aspergillus nidulans*. *Eukaryot Cell* 4:1595–1598.

Flotow, H., and Thomas, G. 1992. Substrate recognition determinants of the mitogen-activated 70K S6 kinase from rat liver. *J Biol Chem* 267:3074–3078.

Forment, J., Naranjo, M.A., Roldan, M., Serrano, R., and Vicente, O. 2002. Expression of *Arabidopsis* SR-like splicing proteins confers salt tolerance to yeast and transgenic plants. *Plant J* 30:511–519.

Francis, D. 1998. Cell Size and Organ Development in Higher Plants. London: Portland Press.

Francis, D. 2007. The plant cell cycle—15 years on. *New Phytol* 174:261–278.

Francis, D., and Halford, N.G. 2006. Nutrient sensing in plant meristems. *Plant Mol Biol* 60:981–993.

Fukuchi-Shimogori, T., Ishii, I., Kashiwagi, K., Mashiba, H., Ekimoto, H., and Igarashi, K. 1997. Malignant transformation by overproduction of translation initiation factor eIF4G. *Cancer Res* 57:5041–5044.

Gaudin, V., Lunness, P.A., Fobert, P.R., Towers, M., Riou-Khamlichi, C., Murray, J.A., Coen, E., and Doonan, J.H. 2000. The expression of D-cyclin genes defines distinct developmental zones in snapdragon apical meristems and is locally regulated by the Cycloidea gene. *Plant Physiol* 122:1137–1148.

Geisler, M., and Bailly, A. 2007. Tete-a-tete: the function of FKBPs in plant development. *Trends Plant Sci* 12:465–473.

Gingras, A.C., Raught, B., Gygi, S.P., Niedzwiecka, A., Miron, M., Burley, S.K., Polakiewicz, R.D., Wyslouch-Cieszynska, A., Aebersold, R., and Sonenberg, N. 2001. Hierarchical phosphorylation of the translation inhibitor 4E-BP1. *Genes Dev* 15:2852–2864.

Gingras, A.C., Raught, B., and Sonenberg, N. 1999. eIF4 initiation factors: effectors of mRNA recruitment to ribosomes and regulators of translation. *Annu Rev Biochem* 68:913–963.

Godoy, A., Lazzaro, A., Casalongue, C., and San Segundo, B. 2000. Expression of a *Solanum tuberosum* cyclophilin gene is regulated by fungal infection and abiotic stress conditions. *Plant Sci* 152:123–134.

Goldberg, Y. 1999. Protein phosphatase 2A: who shall regulate the regulator? *Biochem Pharmacol* 57:321–328.

Gonzalez-Garcia, A., Garrido, E., Hernandez, C., Alvarez, B., Jimenez, C., Cantrell, D.A., Pullen, N., and Carrera, A.C. 2002. A new role for the p85-phosphatidylinositol 3-kinase regulatory subunit linking FRAP to p70 S6 kinase activation. *J Biol Chem* 277:1500–1508.

Graveley, B.R. 2000. Sorting out the complexity of SR protein functions. *RNA* 6:1197–1211.

Grummt, I. (2003). Life on a planet of its own: regulation of RNA polymerase I transcription in the nucleolus. *Genes Dev* 17:1691–1702.

Guertin, D.A., Guntur, K.V., Bell, G.W., Thoreen, C.C., and Sabatini, D.M. 2006. Functional genomics identifies TOR-regulated genes that control growth and division. *Curr Biol* 16:958–970.

Guertin, D.A., Stevens, D.M., Thoreen, C.C., Burds, A.A., Kalaany, N.Y., Moffat, J., Brown, M., Fitzgerald, K.J., and Sabatini, D.M. 2006. Ablation in mice of the mTORC components raptor, rictor, or mLST8 reveals that mTORC2 is required for signaling to Akt-FOXO and PKCalpha, but not S6K1. *Dev Cell* 11: 859–871.

Gutierrez, R.A., Lejay, L.V., Dean, A., Chiaromonte, F., Shasha, D.E., and Coruzzi, G.M. 2007. Qualitative network models and genome-wide expression data define carbon/nitrogen-responsive molecular machines in Arabidopsis. *Genome Biol* 8:R7.

Halford, N.G., and Paul, M.J. 2003. Carbon metabolite sensing and signalling. *Plant Biotechnol J* 1:381–398.

Han, J.W., Pearson, R.B., Dennis, P.B., and Thomas, G. 1995. Rapamycin, wortmannin, and the methylxanthine SQ20006 inactivate p70s6k by inducing dephosphorylation of the same subset of sites. *J Biol Chem* 270:21396–21403.

Hannan, K.M., Brandenburger, Y., Jenkins, A., Sharkey, K., Cavanaugh, A., Rothblum, L., Moss, T., Poortinga, G., McArthur, G.A., Pearson, R.B., et al. 2003. mTOR-dependent regulation of ribosomal gene transcription requires S6K1 and is mediated by phosphorylation of the carboxy-terminal activation domain of the nucleolar transcription factor UBF. *Mol Cell Biol* 23:8862–8877.

Hara, K., Maruki, Y., Long, X., Yoshino, K., Oshiro, N., Hidayat, S., Tokunaga, C., Avruch, J., and Yonezawa, K. (2002). Raptor, a binding partner of target of rapamycin (TOR), mediates TOR action. *Cell* 110:177–189.

Hara, K., Yonezawa, K., Weng, Q.P., Kozlowski, M.T., Belham, C., and Avruch, J. 1998. Amino acid sufficiency and mTOR regulate p70 S6 kinase and eIF-4E BP1 through a common effector mechanism. *J Biol Chem* 273:14484–14494.

Harding, M.W., Galat, A., Uehling, D.E., and Schreiber, S.L. 1989. A receptor for the immunosuppressant FK506 is a cis-trans peptidyl-prolyl isomerase. *Nature* 341:758–760.

Hardwick, J.S., Kuruvilla, F.G., Tong, J.K., Shamji, A.F., and Schreiber, S.L. 1999. Rapamycin-modulated transcription defines the subset of nutrient-sensitive signaling pathways directly controlled by the Tor proteins. *Proc Natl Acad Sci USA* 96:14866–14870.

Harrar, Y., Bellini, C., and Faure, J.D. 2001. FKBPs: at the crossroads of folding and transduction. *Trends Plant Sci* 6:426–431.

Harris, D.M., Myrick, T.L., and Rundle, S.J. 1999. The *Arabidopsis* homolog of yeast TAP42 and mammalian alpha4 binds to the catalytic subunit of protein phosphatase 2A and is induced by chilling. *Plant Physiol* 121:609–617.

He, S.S., Liu, J., Xie, Z., O'Neill, D., and Dotson, S. 2004. *Arabidopsis* E2Fa plays a bimodal role in regulating cell division and cell growth. *Plant Mol Biol* 56:171–184.

Heitman, J., Movva, N.R., and Hall, M.N. 1991. Targets for cell cycle arrest by the immunosuppressant rapamycin in yeast. *Science* 253:905–909.

Hennig, K.M., Colombani, J., and Neufeld, T.P. 2006. TOR coordinates bulk and targeted endocytosis in the *Drosophila melanogaster* fat body to regulate cell growth. *J Cell Biol* 173:963–974.

Hirayama, T., Imajuku, Y., Anai, T., Matsui, M., and Oka, A. 1991. Identification of two cell-cycle-controlling cdc2 gene homologs in *Arabidopsis thaliana*. *Gene* 105:159–165.

Holley, R.W. (1972). A unifying hypothesis concerning the nature of malignant growth. *Proc Natl Acad Sci USA* 69:2840–2841.

Holz, M.K., and Blenis, J. 2005. Identification of S6 kinase 1 as a novel mammalian target of rapamycin (mTOR)-phosphorylating kinase. *J Biol Chem* 280:26089–26093.

Holzenberger, M., Dupont, J., Ducos, B., Leneuve, P., Geloen, A., Even, P.C., Cervera, P., and Le Bouc, Y. 2003. IGF-1 receptor regulates lifespan and resistance to oxidative stress in mice. *Nature* 421:182–187.

Hornstein, E., Tang, H., and Meyuhas, O. 2001. Mitogenic and nutritional signals are transduced into translational efficiency of TOP mRNAs. *Cold Spring Harb Symp Quant Biol* 66:477–484.

Hu, Y., Bao, F., and Li, J. 2000. Promotive effect of brassinosteroids on cell division involves a distinct CycD3-induction pathway in *Arabidopsis*. *Plant J* 24:693–701.

Hur, S., and Bruice, T.C. 2002. The mechanism of cis-trans isomerization of prolyl peptides by cyclophilin. *J Am Chem Soc* 124:7303–7313.

Hwang, J.T., Lee, M., Jung, S.N., Lee, H.J., Kang, I., Kim, S.S., and Ha, J. 2004. AMP-activated protein kinase activity is required for vanadate-induced hypoxia-inducible factor 1alpha expression in DU145 cells. *Carcinogenesis* 25:2497–2507.

Ingram, G.C., and Waites, R. 2006. Keeping it together: co-ordinating plant growth. *Curr Opin Plant Biol* 9:12–20.

Inoki, K., and Guan, K.L. 2006. Complexity of the TOR signaling network. *Trends Cell Biol* 16:206–212.

Inoki, K., Li, Y., Xu, T., and Guan, K.L. 2003. Rheb GTPase is a direct target of TSC2 GAP activity and regulates mTOR signaling. *Genes Dev* 17:1829–1834.

Inoki, K., Ouyang, H., Li, Y., and Guan, K.L. 2005. Signaling by target of rapamycin proteins in cell growth control. *Microbiol Mol Biol Rev* 69:79–100.

Inui, S., Sanjo, H., Maeda, K., Yamamoto, H., Miyamoto, E., and Sakaguchi, N. 1998. Ig receptor binding protein 1 (alpha4) is associated with a rapamycin-sensitive signal transduction in lymphocytes through direct binding to the catalytic subunit of protein phosphatase 2A. *Blood* 92:539–546.

Inze, D., and De Veylder, L. 2006. Cell cycle regulation in plant development. *Annu Rev Genet* 40:77–105.

Ito, M. 2000. Factors controlling cyclin B expression. *Plant Mol Biol* 43:677–690.

Iwakawa, H., Shinmyo, A., and Sekine, M. 2006. *Arabidopsis* CDKA;1, a cdc2 homologue, controls proliferation of generative cells in male gametogenesis. *Plant J* 45:819–831.

Jacinto, E., and Hall, M.N. 2003. Tor signalling in bugs, brain and brawn. *Nat Rev Mol Cell Biol* 4:117–126.

Jacinto, E., Loewith, R., Schmidt, A., Lin, S., Ruegg, M.A., Hall, A., and Hall, M.N. 2004. Mammalian TOR complex 2 controls the actin cytoskeleton and is rapamycin insensitive. *Nat Cell Biol* 6:1122–1128.

Jefferies, H.B., Reinhard, C., Kozma, S.C., and Thomas, G. 1994. Rapamycin selectively represses translation of the "polypyrimidine tract" mRNA family. *Proc Natl Acad Sci USA* 91, 4441–4445.

Joubes, J., Chevalier, C., Dudits, D., Heberle-Bors, E., Inze, D., Umeda, M., and Renaudin, J.P. 2000. CDK-related protein kinases in plants. *Plant Mol Biol* 43:607–620.

Joubes, J., Lemaire-Chamley, M., Delmas, F., Walter, J., Hernould, M., Mouras, A., Raymond, P., and Chevalier, C. 2001. A new C-type cyclin-dependent kinase from tomato expressed in dividing tissues does not interact with mitotic and G1 cyclins. *Plant Physiol* 126:1403–1415.

Jurgens, G. 2001. Apical-basal pattern formation in *Arabidopsis* embryogenesis. *Embo J* 20:3609–3616.

Kay, J.E. 1996. Structure-function relationships in the FK506-binding protein (FKBP) family of peptidylprolyl cis-trans isomerases. *Biochem J* 314:361–385.

Keiper, B.D., Gan, W., and Rhoads, R.E. 1999. Protein synthesis initiation factor 4G. *Int J Biochem Cell Biol* 31:37–41.

Kim, D.H., Sarbassov, D.D., Ali, S.M., King, J.E., Latek, R.R., Erdjument-Bromage, H., Tempst, P., and Sabatini, D.M. 2002. mTOR interacts with raptor to form a nutrient-sensitive complex that signals to the cell growth machinery. *Cell* 110:163–175.

Kimball, S.R., and Jefferson, L.S. 2006. New functions for amino acids: effects on gene transcription and translation. *Am J Clin Nutr* 83:500S–507S.

Kong, H., Lee, S., and Hwang, B. 2001. Expression of pepper cyclophilin gene is differentially regulated during the pathogen infection and abiotic stress conditions. *Physiol Mol Plant Pathol* 59:189–199.

Krause, U., Bertrand, L., and Hue, L. 2002. Control of p70 ribosomal protein S6 kinase and acetyl-CoA carboxylase by AMP-activated protein kinase and protein phosphatases in isolated hepatocytes. *Eur J Biochem* 269:3751–3759.

Krieg, J., Hofsteenge, J., and Thomas, G. 1988. Identification of the 40 S ribosomal protein S6 phosphorylation sites induced by cycloheximide. *J Biol Chem* 263:11473–11477.

Kullertz, G., Liebau, A., Rucknagel, P., Schierhorn, A., Diettrich, B., Fischer, G., and Luckner, M. 1999. Stress-induced expression of cyclophilins in proembryonic masses of *Digitalis lanata* does not protect against freezing/thawing stress. *Planta* 208:599–605.

Kunz, J., Henriquez, R., Schneider, U., Deuter-Reinhard, M., Movva, N.R., and Hall, M.N. 1993. Target of rapamycin in yeast, TOR2, is an essential phosphatidylinositol kinase homolog required for G1 progression. *Cell* 73:585–596.

Laufs, P., Grandjean, O., Jonak, C., Kieu, K., and Traas, J. 1998. Cellular parameters of the shoot apical meristem in *Arabidopsis*. *Plant Cell* 10:1375–1390.

Lazar, G., and Goodman, H.M. 2000. The *Arabidopsis* splicing factor SR1 is regulated by alternative splicing. *Plant Mol Biol* 42:571–581.

Lee, C.H., Inoki, K., and Guan, K.L. 2007. mTOR pathway as a target in tissue hypertrophy. *Annu Rev Pharmacol Toxicol* 47, 443–467.

Leicht, M., Simm, A., Bertsch, G., and Hoppe, J. 1996. Okadaic acid induces cellular hypertrophy in AKR-2B fibroblasts: involvement of the p70S6 kinase in the onset of protein and rRNA synthesis. *Cell Growth Differ* 7:1199–1209.

Lellis, A.D., Kasschau, K.D., Whitham, S.A., and Carrington, J.C. 2002. Loss-of-susceptibility mutants of *Arabidopsis thaliana* reveal an essential role for eIF(iso)4E during potyvirus infection. *Curr Biol* 12:1046–1051.

Leng, X., Connell-Crowley, L., Goodrich, D., and Harper, J.W. 1997. S-Phase entry upon ectopic expression of G1 cyclin-dependent kinases in the absence of retinoblastoma protein phosphorylation. *Curr Biol* 7: 709–712.

Leonard, S., Plante, D., Wittmann, S., Daigneault, N., Fortin, M.G., and Laliberte, J.F. 2000. Complex formation between potyvirus VPg and translation eukaryotic initiation factor 4E correlates with virus infectivity. *J Virol* 74:7730–7737.

Lewis, J.D., and Tollervey, D. 2000. Like attracts like: getting RNA processing together in the nucleus. *Science* 288:1385–1389.

Li, H., Tsang, C.K., Watkins, M., Bertram, P.G., and Zheng, X.F. 2006. Nutrient regulates Tor1 nuclear localization and association with rDNA promoter. *Nature* 442:1058–1061.

Lin, T.A., and Lawrence, J.C., Jr. 1997. Control of PHAS-I phosphorylation in 3T3-L1 adipocytes: effects of inhibiting protein phosphatases and the p70S6K signalling pathway. *Diabetologia* 40(Suppl 2):S18–S24.

Loewith, R., Jacinto, E., Wullschleger, S., Lorberg, A., Crespo, J.L., Bonenfant, D., Oppliger, W., Jenoe, P., and Hall, M.N. 2002. Two TOR complexes, only one of which is rapamycin sensitive, have distinct roles in cell growth control. *Mol Cell* 10:457–468.

Luan, S., Kudla, J., Gruissem, W., and Schreiber, S.L. 1996. Molecular characterization of a FKBP-type immunophilin from higher plants. *Proc Natl Acad Sci USA* 93, 6964–6969.

Lucas, K.A., Filley, J.R., Erb, J.M., Graybill, E.R., and Hawes, J.W. 2007. Peroxisomal metabolism of propionic acid and isobutyric acid in plants. *J Biol Chem* 282:24980–24989.

Lukas, J., Herzinger, T., Hansen, K., Moroni, M.C., Resnitzky, D., Helin, K., Reed, S.I., and Bartek, J. 1997. Cyclin E-induced S phase without activation of the pRb/E2F pathway. *Genes Dev* 11:1479–1492.

Mahajan, P.B. 1994. Modulation of transcription of rRNA genes by rapamycin. *Int J Immunopharmacol* 16:711–721.

Mahfouz, M.M., Kim, S., Delauney, A.J., and Verma, D.P. 2006. *Arabidopsis* TARGET OF RAPAMYCIN interacts with RAPTOR, which regulates the activity of S6 kinase in response to osmotic stress signals. *Plant Cell* 18:477–490.

Maki, N., Sekiguchi, F., Nishimaki, J., Miwa, K., Hayano, T., Takahashi, N., and Suzuki, M. 1990. Complementary DNA encoding the human T-cell FK506-binding protein, a peptidylprolyl cis-trans isomerase distinct from cyclophilin. *Proc Natl Acad Sci USA* 87:5440–5443.

Mamane, Y., Petroulakis, E., LeBacquer, O., and Sonenberg, N. 2006. mTOR, translation initiation and cancer. *Oncogene* 25:6416–6422.

Mamane, Y., Petroulakis, E., Rong, L., Yoshida, K., Ler, L.W., and Sonenberg, N. 2004. eIF4E—from translation to transformation. *Oncogene* 23:3172–3179.

Marcotrigiano, J., Gingras, A.C., Sonenberg, N., and Burley, S.K. 1999. Cap-dependent translation initiation in eukaryotes is regulated by a molecular mimic of eIF4G. *Mol Cell* 3:707–716.

Marivet, J., Frendo, P., and Burkard, G. 1995. DNA sequence analysis of a cyclophilin gene from maize: developmental expression and regulation by salicylic acid. *Mol Gen Genet* 247:222–228.

Martin, K.A., and Blenis, J. 2002. Coordinate regulation of translation by the PI 3-kinase and mTOR pathways. *Adv Cancer Res* 86:1–39.

Martin, K.A., Rzucidlo, E.M., Merenick, B.L., Fingar, D.C., Brown, D.J., Wagner, R.J., and Powell, R.J. 2004. The mTOR/p70 S6K1 pathway regulates vascular smooth muscle cell differentiation. *Am J Physiol Cell Physiol* 286:C507–C517.

Martin, K.A., Schalm, S.S., Romanelli, A., Keon, K.L., and Blenis, J. 2001. Ribosomal S6 kinase 2 inhibition by a potent C-terminal repressor domain is relieved by mitogen-activated protein-extracellular signal-regulated kinase kinase-regulated phosphorylation. *J Biol Chem* 276:7892–7898.

Mayer, C., and Grummt, I. 2006. Ribosome biogenesis and cell growth: mTOR coordinates transcription by all three classes of nuclear RNA polymerases. *Oncogene* 25:6384–6391.

Mayer-Jaekel, R.E., and Hemmings, B.A. 1994. Protein phosphatase 2A—a "menage a trois." *Trends Cell Biol* 4:287–291.

Mayeur, G.L., Fraser, C.S., Peiretti, F., Block, K.L., and Hershey, J.W. 2003. Characterization of eIF3k: a newly discovered subunit of mammalian translation initiation factor eIF3. *Eur J Biochem* 270:4133–4139.

Meijer, M., and Murray, J.A.H. 2000. The role and regulation of D-type cyclins in the plant cell cycle. *Plant Mol Biol* 43:621–633.

Menand, B., Desnos, T., Nussaume, L., Berger, F., Bouchez, D., Meyer, C., and Robaglia, C. 2002. Expression and disruption of the *Arabidopsis* TOR (target of rapamycin) gene. *Proc Natl Acad Sci USA* 99:6422–6427.

Menand, B., Meyer, C., and Robaglia, C. 2004. Plant growth and the TOR pathway. *Curr Top Microbiol Immunol* 279:97–113.

Menges, M., and Murray, J.A. 2002. Synchronous *Arabidopsis* suspension cultures for analysis of cell-cycle gene activity. *Plant J* 30:203–212.

Meyerowitz, E.M. 1997. Plants and the logic of development. *Genetics* 145:5–9.

Michlewski, G., Sanford, J.R., and Caceres, J.F. 2008. The splicing factor SF2/ASF regulates translation initiation by enhancing phosphorylation of 4E-BP1. *Mol Cell* 30:179–189.

Mizoguchi, T., Hayashida, N., Yamaguchi-Shinozaki, K., Kamada, H., and Shinozaki, K. 1995. Two genes that encode ribosomal-protein S6 kinase homologs are induced by cold or salinity stress in *Arabidopsis thaliana*. *FEBS Lett* 358:199–204.

Montagne, J., Stewart, M.J., Stocker, H., Hafen, E., Kozma, S.C., and Thomas, G. 1999. Drosophila S6 kinase: a regulator of cell size. *Science* 285:2126–2129.

Mora, A., Komander, D., van Aalten, D.M., and Alessi, D.R. 2004. PDK1, the master regulator of AGC kinase signal transduction. *Semin Cell Dev Biol* 15:161–170.

Murata, K., Wu, J., and Brautigan, D.L. 1997. B cell receptor-associated protein alpha4 displays rapamycin-sensitive binding directly to the catalytic subunit of protein phosphatase 2A. *Proc Natl Acad Sci USA* 94:10624–10629.

Nader, G.A., McLoughlin, T.J., and Esser, K.A. 2005. mTOR function in skeletal muscle hypertrophy: increased ribosomal RNA via cell cycle regulators. *Am J Physiol Cell Physiol* 289:C1457–C1465.

Nanahoshi, M., Nishiuma, T., Tsujishita, Y., Hara, K., Inui, S., Sakaguchi, N., and Yonezawa, K. 1998. Regulation of protein phosphatase 2A catalytic activity by alpha4 protein and its yeast homolog Tap42. *Biochem Biophys Res Commun* 251:520–526.

Nelsen, C.J., Rickheim, D.G., Tucker, M.M., Hansen, L.K., and Albrecht, J.H. 2003. Evidence that cyclin D1 mediates both growth and proliferation downstream of TOR in hepatocytes. *J Biol Chem* 278:3656–3663.

Nicaise, V., Gallois, J.L., Chafiai, F., Allen, L.M., Schurdi-Levraud, V., Browning, K.S., Candresse, T., Caranta, C., Le Gall, O., and German-Retana, S. 2007. Coordinated and selective recruitment of eIF4E and eIF4G factors for potyvirus infection in *Arabidopsis thaliana*. *FEBS Lett* 581:1041–1046.

Nieto, C., Morales, M., Orjeda, G., Clepet, C., Monfort, A., Sturbois, B., Puigdomenech, P., Pitrat, M., Caboche, M., Dogimont, C., *et al.* 2006. An eIF4E allele confers resistance to an uncapped and non-polyadenylated RNA virus in melon. *Plant J* 48:452–462.

Nobukuni, T., Joaquin, M., Roccio, M., Dann, S.G., Kim, S.Y., Gulati, P., Byfield, M.P., Backer, J.M., Natt, F., Bos, J.L. *et al.* 2005. Amino acids mediate mTOR/raptor signaling through activation of class 3 phosphatidylinositol 3OH-kinase. *Proc Natl Acad Sci USA* 102:14238–14243.

Oakenfull, E.A., Riou-Khamlichi, C., and Murray, J.A. 2002. Plant D-type cyclins and the control of G1 progression. *Philos Trans R Soc Lond B Biol Sci* 357:749–760.

Oldham, S., Montagne, J., Radimerski, T., Thomas, G., and Hafen, E. 2000. Genetic and biochemical characterization of dTOR, the *Drosophila* homolog of the target of rapamycin. *Genes Dev* 14:2689–2694.

Orlova, M., Kanter, E., Krakovich, D., and Kuchin, S. 2006. Nitrogen availability and TOR regulate the Snf1 protein kinase in *Saccharomyces cerevisiae*. *Eukaryot Cell* 5:1831–1837.

Otterhag, L., Gustavsson, N., Alsterfjord, M., Pical, C., Lehrach, H., Gobom, J., and Sommarin, M. 2006. *Arabidopsis* PDK1: identification of sites important for activity and downstream phosphorylation of S6 kinase. *Biochimie* 88:11–21.

Pastore, D., Trono, D., Laus, M.N., Di Fonzo, N., and Flagella, Z. 2007. Possible plant mitochondria involvement in cell adaptation to drought stress. A case study: durum wheat mitochondria. *J Exp Bot* 58:195–210.

Pearson, R.B., Dennis, P.B., Han, J.W., Williamson, N.A., Kozma, S.C., Wettenhall, R.E., and Thomas, G. 1995. The principal target of rapamycin-induced p70s6k inactivation is a novel phosphorylation site within a conserved hydrophobic domain. *Embo J* 14:5279–5287.

Pende, M., Um, S.H., Mieulet, V., Sticker, M., Goss, V.L., Mestan, J., Mueller, M., Fumagalli, S., Kozma, S.C., and Thomas, G. 2004. S6K1(-/-)/S6K2(-/-) mice exhibit perinatal lethality and rapamycin-sensitive 5′-terminal oligopyrimidine mRNA translation and reveal a mitogen-activated protein kinase-dependent S6 kinase pathway. *Mol Cell Biol* 24:3112–3124.

Peng, T., Golub, T.R., and Sabatini, D.M. 2002. The immunosuppressant rapamycin mimics a starvation-like signal distinct from amino acid and glucose deprivation. *Mol Cell Biol* 22:5575–5584.

Pestova, T.V., and Kolupaeva, V.G. 2002. The roles of individual eukaryotic translation initiation factors in ribosomal scanning and initiation codon selection. *Genes Dev* 16:2906–2922.

Petersen, J., and Nurse, P. 2007. TOR signalling regulates mitotic commitment through the stress MAP kinase pathway and the Polo and Cdc2 kinases. *Nat Cell Biol* 9:1263–1272.

Peterson, R.T., Desai, B.N., Hardwick, J.S., and Schreiber, S.L. 1999. Protein phosphatase 2A interacts with the 70-kDa S6 kinase and is activated by inhibition of FKBP12-rapamycinassociated protein. *Proc Natl Acad Sci USA* 96:4438–4442.

Pinner, S., and Sahai, E. 2008. PDK1 regulates cancer cell motility by antagonising inhibition of ROCK1 by RhoE. *Nat Cell Biol* 10:127–137.

Powers, R.W., 3rd, Kaeberlein, M., Caldwell, S.D., Kennedy, B.K., and Fields, S. 2006. Extension of chronological life span in yeast by decreased TOR pathway signaling. *Genes Dev* 20:174–184.

Price, J., Laxmi, A., St. Martin, S.K., and Jang, J.C. 2004. Global transcription profiling reveals multiple sugar signal transduction mechanisms in *Arabidopsis*. *Plant Cell* 16:2128–2150.

Pullen, N., and Thomas, G. 1997. The modular phosphorylation and activation of p70s6k. *FEBS Lett* 410:78–82.

Rameh, L.E., and Cantley, L.C. 1999. The role of phosphoinositide 3-kinase lipid products in cell function. *J Biol Chem* 274:8347–8350.

Raught, B., Gingras, A.C., Gygi, S.P., Imataka, H., Morino, S., Gradi, A., Aebersold, R., and Sonenberg, N. 2000. Serum-stimulated, rapamycin-sensitive phosphorylation sites in the eukaryotic translation initiation factor 4GI. *Embo J* 19:434–444.

Raught, B., Peiretti, F., Gingras, A.C., Livingstone, M., Shahbazian, D., Mayeur, G.L., Polakiewicz, R.D., Sonenberg, N., and Hershey, J.W. 2004. Phosphorylation of eukaryotic translation initiation factor 4B Ser422 is modulated by S6 kinases. *Embo J* 23:1761–1769.

Reichheld, J., Vernoux, T., Lardon, F., Van Montagu, M., and Inze, D. 1999. Specific checkpoints regulate plant cell cycle progression in response to oxidative stress. *Plant J* 17:647–656.

Reiling, J.H., and Sabatini, D.M. 2006. Stress and mTORture signaling. *Oncogene* 25:6373–6383.

Renaudin, J.P., Doonan, J.H., Freeman, D., Hashimoto, J., Hirt, H., Inze, D., Jacobs, T., Kouchi, H., Rouze, P., Sauter, M., *et al.* 1996. Plant cyclins: a unified nomenclature for plant A-, B- and D-type cyclins based on sequence organization. *Plant Mol Biol* 32:1003–1018.

Reyes de la Cruz, H., Aguilar, R., and Sanchez de Jimenez, E. 2004. Functional characterization of a maize ribosomal S6 protein kinase (ZmS6K), a plant ortholog of metazoan p70(S6K). *Biochemistry* 43:533–539.

Richardson, C.J., Schalm, S.S., and Blenis, J. 2004. PI3-kinase and TOR: PIKTORing cell growth. *Semin Cell Dev Biol* 15:147–159.

Riou-Khamlichi, C., Menges, M., Healy, J.M., and Murray, J.A. 2000. Sugar control of the plant cell cycle: differential regulation of *Arabidopsis* D-type cyclin gene expression. *Mol Cell Biol* 20:4513–4521.

Robaglia, C., and Caranta, C. 2006. Translation initiation factors: a weak link in plant RNA virus infection. *Trends Plant Sci* 11:40–45.

Rohde, J.R., and Cardenas, M.E. 2004. Nutrient signaling through TOR kinases controls gene expression and cellular differentiation in fungi. *Curr Top Microbiol Immunol* 279:53–72.

Romano, P.G., Horton, P., and Gray, J.E. 2004. The *Arabidopsis* cyclophilin gene family. *Plant Physiol* 134:1268–1282.

Rook, F., and Bevan, M.W. 2003. Genetic approaches to understanding sugar-response pathways. *J Exp Bot* 54:495–501.

Ruffel, S., Caranta, C., Palloix, A., Lefebvre, V., Caboche, M., and Bendahmane, A. 2004. Structural analysis of the eukaryotic initiation factor 4E gene controlling potyvirus resistance in pepper: exploitation of a BAC library. *Gene* 338:209–216.

Ruvinsky, I., and Meyuhas, O. 2006. Ribosomal protein S6 phosphorylation: from protein synthesis to cell size. *Trends Biochem Sci* 31:342–348.

Ruvinsky, I., Sharon, N., Lerer, T., Cohen, H., Stolovich-Rain, M., Nir, T., Dor, Y., Zisman, P., and Meyuhas, O. 2005. Ribosomal protein S6 phosphorylation is a determinant of cell size and glucose homeostasis. *Genes Dev* 19:2199–2211.

Sacco-Bubulya, P., and Spector, D.L. 2002. Disassembly of interchromatin granule clusters alters the coordination of transcription and pre-mRNA splicing. *J Cell Biol* 156:425–436.

Saitoh, M., Pullen, N., Brennan, P., Cantrell, D., Dennis, P.B., and Thomas, G. 2002. Regulation of an activated S6 kinase 1 variant reveals a novel mammalian target of rapamycin phosphorylation site. *J Biol Chem* 277:20104–20112.

Sanford, J.R., Gray, N.K., Beckmann, K., and Caceres, J.F. 2004. A novel role for shuttling SR proteins in mRNA translation. *Genes Dev* 18:755–768.

Sasse, J.M. 2003. Physiological actions of brassinosteroids: an update. *J Plant Growth Regul* 22:276–288.

Sato, M., Nakahara, K., Yoshii, M., Ishikawa, M., and Uyeda, I. 2005. Selective involvement of members of the eukaryotic initiation factor 4E family in the infection of *Arabidopsis* thaliana by potyviruses. *FEBS Lett* 579:1167–1171.

Schalm, S.S., and Blenis, J. 2002. Identification of a conserved motif required for mTOR signaling. *Curr Biol* 12:632–639.

Schieke, S.M., and Finkel, T. 2006. Mitochondrial signaling, TOR, and life span. *Biol Chem* 387:1357–1361.

Schiene-Fischer, C., and Yu, C. 2001. Receptor accessory folding helper enzymes: the functional role of peptidyl prolyl cis/trans isomerases. *FEBS Lett* 495:1–6.

Schmelzle, T., and Hall, M.N. 2000. TOR, a central controller of cell growth. *Cell* 103:253–262.

Schmidt, E.V. 1999. The role of c-myc in cellular growth control. *Oncogene* 18:2988–2996.

Schnaitman, C.A., and Pedersen, P.L. (1968). Localization of oligomycin-sensitive ADP-ATP exchange activity in rat liver mitochondria. *Biochem Biophys Res Commun* 30:428–433.

Scholze, C., Peterson, A., Diettrich, B., and Luckner, M. 1999. Cyclophilin isoforms from Digitalis lanata: sequences and expression during embryogenesis and stress. *J Plant Physiol* 155:212–219.

Schubert, M., Petersson, U.A., Haas, B.J., Funk, C., Schroder, W.P., and Kieselbach, T. 2002. Proteome map of the chloroplast lumen of *Arabidopsis thaliana*. *J Biol Chem* 277:8354–8365.

Scott, B.A., Avidan, M.S., and Crowder, C.M. 2002. Regulation of hypoxic death in *C. elegans* by the insulin/IGF receptor homolog DAF-2. *Science* 296:2388–2391.

Sekito, T., Liu, Z., Thornton, J., and Butow, R.A. 2002. RTG-dependent mitochondria-to-nucleus signaling is regulated by MKS1 and is linked to formation of yeast prion [URE3]. *Mol Biol Cell* 13:795–804.

Sharma, A., and Singh, P. 2003. Effect of water stress on expression of a 20 kD cyclophilin-like protein in drought susceptible and tolerant cultivars of Sorghum. *J Plant Biochem Biot* 12:77–80.

Shaw, P.E. 2002. Peptidyl-prolyl isomerases: a new twist to transcription. *EMBO Rep* 3:521–526.

Shaw, P.E. 2007. Peptidyl-prolyl cis/trans isomerases and transcription: is there a twist in the tail? *EMBO Rep* 8:40–45.

Sheen, J. 1990. Metabolic repression of transcription in higher plants. *Plant Cell* 2:1027–1038.

Shen, C., Lancaster, C.S., Shi, B., Guo, H., Thimmaiah, P., and Bjornsti, M.A. 2007. TOR Signaling is a determinant of cell survival in response to DNA damage. *Mol Cell Biol* 27:7007–7017.

Sherr, C.J. 1996. Cancer cell cycles. *Science* 274:1672–1677.

Shigemitsu, K., Tsujishita, Y., Miyake, H., Hidayat, S., Tanaka, N., Hara, K., and Yonezawa, K. 1999. Structural requirement of leucine for activation of p70 S6 kinase. *FEBS Lett* 447:303–306.

Shima, H., Pende, M., Chen, Y., Fumagalli, S., Thomas, G., and Kozma, S.C. 1998. Disruption of the p70(s6k)/p85(s6k) gene reveals a small mouse phenotype and a new functional S6 kinase. *Embo J* 17:6649–6659.

Shou, W., Aghdasi, B., Armstrong, D.L., Guo, Q., Bao, S., Charng, M.J., Mathews, L.M., Schneider, M.D., Hamilton, S.L., and Matzuk, M.M. 1998. Cardiac defects and altered ryanodine receptor function in mice lacking FKBP12. *Nature* 391:489–492.

Singh, K., Sun, S., and Vezina, C. (1979). Rapamycin (AY-22,989), a new antifungal antibiotic. IV. Mechanism of action. *J Antibiot (Tokyo)* 32:630–645.

Slessareva, J.E., Routt, S.M., Temple, B., Bankaitis, V.A., and Dohlman, H.G. 2006. Activation of the phosphatidylinositol 3-kinase Vps34 by a G protein alpha subunit at the endosome. *Cell* 126:191–203.

Sormani, R., Yao, L., Menand, B., Ennar, N., Lecampion, C., Meyer, C., and Robaglia, C. 2007. *Saccharomyces cerevisiae* FKBP12 binds *Arabidopsis thaliana* TOR and its expression in plants leads to rapamycin susceptibility. *BMC Plant Biol* 7:26.

Stals, H., Casteels, P., Van Montagu, M., and Inze, D. 2000. Regulation of cyclin-dependent kinases in *Arabidopsis thaliana*. *Plant Mol Biol* 43:583–593.

Stan, R., McLaughlin, M.M., Cafferkey, R., Johnson, R.K., Rosenberg, M., and Livi, G.P. 1994. Interaction between FKBP12-rapamycin and TOR involves a conserved serine residue. *J Biol Chem* 269:32027–32030.

Stone, S.L., Williams, L.A., Farmer, L.M., Vierstra, R.D., and Callis, J. 2006. KEEP ON GOING, a RING E3 ligase essential for *Arabidopsis* growth and development, is involved in abscisic acid signaling. *Plant Cell* 18:3415–3428.

Tanabe, N., Yoshimura, K., Kimura, A., Yabuta, Y., and Shigeoka, S. 2007. Differential expression of alternatively spliced mRNAs of *Arabidopsis* SR protein homologs, atSR30 and atSR45a, in response to environmental stress. *Plant Cell Physiol* 48:1036–1049.

Testerink, C., and Munnik, T. 2005. Phosphatidic acid: a multifunctional stress signaling lipid in plants. *Trends Plant Sci* 10:368–375.

Tillemans, V., Dispa, L., Remacle, C., Collinge, M., and Motte, P. 2005. Functional distribution and dynamics of *Arabidopsis* SR splicing factors in living plant cells. *Plant J* 41:567–582.

Tillemans, V., Leponce, I., Rausin, G., Dispa, L., and Motte, P. 2006. Insights into nuclear organization in plants as revealed by the dynamic distribution of *Arabidopsis* SR splicing factors. *Plant Cell* 18:3218–3234.

Tokunaga, C., Yoshino, K., and Yonezawa, K. 2004. mTOR integrates amino acid- and energy-sensing pathways. *Biochem Biophys Res Commun* 313:443–446.

Turck, F., Kozma, S.C., Thomas, G., and Nagy, F. 1998. A heat-sensitive *Arabidopsis thaliana* kinase substitutes for human p70s6k function in vivo. *Mol Cell Biol* 18:2038–2044.

Turck, F., Zilbermann, F., Kozma, S.C., Thomas, G., and Nagy, F. 2004. Phytohormones participate in an S6 kinase signal transduction pathway in *Arabidopsis*. *Plant Physiol* 134:1527–1535.

Vandepoele, K., Raes, J., De Veylder, L., Rouze, P., Rombauts, S., and Inze, D. 2002. Genome-wide analysis of core cell cycle genes in *Arabidopsis*. *Plant Cell* 14:903–916.

Van't Hof, J. (1968). Control of cell progression through the mitotic cycle by carbohydrate provision. I. Regulation of cell division in excised plant tissue. *J Cell Biol* 37:773–780.

Vespa, L., Vachon, G., Berger, F., Perazza, D., Faure, J.D., and Herzog, M. 2004. The immunophilin-interacting protein AtFIP37 from *Arabidopsis* is essential for plant development and is involved in trichome endoreduplication. *Plant Physiol* 134:1283–1292.

Vezina, C., Kudelski, A., and Sehgal, S.N. (1975). Rapamycin (AY-22,989), a new antifungal antibiotic. I. Taxonomy of the producing streptomycete and isolation of the active principle. *J Antibiot (Tokyo)* 28:721–726.

Volarevic, S., Stewart, M.J., Ledermann, B., Zilberman, F., Terracciano, L., Montini, E., Grompe, M., Kozma, S.C., and Thomas, G. 2000. Proliferation, but not growth, blocked by conditional deletion of 40S ribosomal protein S6. *Science* 288:2045–2047.

Wang, H., Qi, Q., Schorr, P., Cutler, A.J., Crosby, W.L., and Fowke, L.C. 1998. ICK1, a cyclin-dependent protein kinase inhibitor from *Arabidopsis thaliana* interacts with both Cdc2a and CycD3, and its expression is induced by abscisic acid. *Plant J* 15:501–510.

Weigel, D., and Jurgens, G. 2002. Stem cells that make stems. *Nature* 415:751–754.

Weisman, R., and Choder, M. 2001. The fission yeast TOR homolog, tor1+, is required for the response to starvation and other stresses via a conserved serine. *J Biol Chem* 276:7027–7032.

Wera, S., and Hemmings, B.A. 1995. Serine/threonine protein phosphatases. *Biochem J* 311:17–29.

Westphal, R.S., Coffee, R.L., Jr., Marotta, A., Pelech, S.L., and Wadzinski, B.E. 1999. Identification of kinase-phosphatase signaling modules composed of p70 S6 kinase-protein phosphatase 2A (PP2A) and p21-activated kinase-PP2A. *J Biol Chem* 274:687–692.

Wildwater, M., Campilho, A., Perez-Perez, J.M., Heidstra, R., Blilou, I., Korthout, H., Chatterjee, J., Mariconti, L., Gruissem, W., and Scheres, B. 2005. The RETINOBLASTOMA-RELATED gene regulates stem cell maintenance in *Arabidopsis* roots. *Cell* 123:1337–1349.

Williams, A.J., Werner-Fraczek, J., Chang, I.F., and Bailey-Serres, J. 2003. Regulated phosphorylation of 40S ribosomal protein S6 in root tips of maize. *Plant Physiol* 132:2086–2097.

Wullschleger, S., Loewith, R., and Hall, M.N. 2006. TOR signaling in growth and metabolism. *Cell* 124:471–484.

Xu, Q., Liang, S., Kudla, J., and Luan, S. 1998. Molecular characterization of a plant FKBP12 that does not mediate action of FK506 and rapamycin. *Plant J* 15:511–519.

Yoshii, M., Nishikiori, M., Tomita, K., Yoshioka, N., Kozuka, R., Naito, S., and Ishikawa, M. 2004. The Arabidopsis cucumovirus multiplication 1 and 2 loci encode translation initiation factors 4E and 4G. *J Virol* 78:6102–6111.

Yu, Y., Steinmetz, A., Meyer, D., Brown, S., and Shen, W.H. 2003. The tobacco A-type cyclin, Nicta;CYCA3;2, at the nexus of cell division and differentiation. *Plant Cell* 15:2763–2777.

Zhang, H., and Forde, B.G. 2000. Regulation of *Arabidopsis* root development by nitrate availability. *J Exp Bot* 51:51–59.

Zhang, H., Jennings, A., Barlow, P.W., and Forde, B.G. 1999. Dual pathways for regulation of root branching by nitrate. *Proc Natl Acad Sci USA* 96:6529–6534.

Zhang, H., Stallock, J.P., Ng, J.C., Reinhard, C., and Neufeld, T.P. 2000. Regulation of cellular growth by the *Drosophila* target of rapamycin dTOR. *Genes Dev* 14:2712–2724.

Zhang, S.H., Broome, M.A., Lawton, M.A., Hunter, T., and Lamb, C.J. 1994. atpk1, a novel ribosomal protein kinase gene from *Arabidopsis*. II. Functional and biochemical analysis of the encoded protein. *J Biol Chem* 269:17593–17599.

Zhang, Z., and Krainer, A.R. 2004. Involvement of SR proteins in mRNA surveillance. *Mol Cell* 16:597–607.

Index

-SOH. *See* Sulphenic acid (–SOH)
[Ca^{2+}]cyt. *See* Cytoplasmic calcium ([Ca^{2+}]cyt)
[Ca^{2+}]cyt
   elicitor-induced changes in, 65–66
   in Nod-factor signaling pathway, 64
   oscillations in, 62–63, 74–77
   role in pollen tube growth, 62–63
   spatial and temporal dynamics of, 61
   stimulus-induced changes in guard cell, 63–64
   transients in, 74–77
9-cis-epoxycarotenoid dioxygenase (*NCED*), 5

ABA. *See* Abscisic acid (ABA); abscisic acid (ABA); Hormone abscisic acid (ABA)
ABA insensitive 1 (ABI1), 154
ABA receptors, 153
ABA signaling
   alterations in secondary plant cell wall and, 7
   MYB transcription factors, 45–47
   MYC transcription factors, 45–47
   RLM1$_{Col}$ pathway and, 8
ABA-induced ROS production, 142
aba1-3, 8
aba2-1, 11
aba3, ABA-deficient mutant, 3
ABI1. *See* ABA insensitive 1 (ABI1)
Abiotic stress regulation, interaction with biotic stress regulation, 15–17
Abiotic stress-induced Ca^{2+} signals
   rapid transient in, 60–61
   signaling information, 61
Abiotic stress-signaling network, 43–54
Abiotic stresses, 136
Abscisic acid (ABA), 97–98, 137
   and auxin, 12
   and GA, 12
   and interaction between *Arabidopsis* and *Leptosphaeria maculans*, 8
   ascorbate deficiency and accumulation of, 6
   CALRR1 gene in pepper plants, 6
   concentration in tomato plant and *sitiens* mutation, 4–5
   crosstalk
      with JA/ET signaling pathway, 11–12
   crosstalk with NO signaling, 12–13
   defense and PAMP, 4
   downregulation of antifungal β-1,3-glucanase, 2
   in *Arabidopsis*-necrotrophic fungus interactions, 4
   in BABA-induced resistance against biotrophs and necrotrophs, 15
   in disease promotion during pathogenic infection of plants, 2–3
   in priming, 13–15
   inoculation of *Pst*, 3
   interaction between *Arabidopsis* and *Pythium irregulare*, 8
   leaf humidity and, 2

Abscisic acid (ABA) (Cont.)
  role in disease resistance, 1–10
    decreases, 1–5
    increases, 5–10
  role in plant-insect interaction, 2
  role in regulation of papillae
      formation, 6
  treatment of *Arabidopsis*, 2–3
  with BR, 13
  with SA, 10–11
Alternaria brassicicola, 4
  ABA against, 8
  BABA-IR against, 13, 14
Alternaria longipes, 16
ANP1, 30
Antifungal β-1,3-glucanase,
    downregulation by ABA, 2
Arabidopsis
  interaction with *Leptosphaeria maculans*, 8
  interaction with *Pythium irregulare*, 8
  treatment with ABA, 2–3
Arabidopsis coi1 mutants, 9
Arabidopsis genome
  cell suspension cultures in, 146
  NOS sequence in, 143
Arabidopsis lesion mimic, environmental sensitive, 117–118
Arabidopsis MAPK pathways, 24
Arabidopsis MKK3, 34–35
Arabidopsis MKK3-MPK6 pathway, 34
Arabidopsis MPK3/MPK6 pathways, 28
Arabidopsis mutants, pathogen resistant, 113–130
Arabidopsis thaliana
  Ca^{2+}-ATPases in, 72
  Ca^{2+}/H^+ antiport activity, 73
  CAMTAs in, 83
  CBLs and CIPKs in, 79
  cold-shock-induced changes in, 60–61

  elicitor-induced changes in $[Ca^{2+}]_{cyt}$, 65
  genes for ribosomal protein 6, 168
  guard cells, I_{Ca} channels in, 68
  pollen tubes, 62
  rapid Ca^{2+} transient in, 61
  S6K
    activity, 167, 168
    TOR kinase and PDK1 kinase phosphorylation sites in, 168
Arabidopsis-necrotrophic fungus interactions, in ABA, 4
ASc-GSH. *See* Ascorbate-glutathione (ASc-GSH)
Ascorbate deficiency, and accumulation of ABA and SA, 6
Ascorbate-glutathione (ASc-GSH), 144
Ascorbic acid (AsA) deficiency, impact on *NCED* levels, 5–6
AtCesA8/IRX1 gene, 7
AtFIP37, FKBP12 interacting protein, 169
AtMRP5 ATP-binding cassette (ABC) transmembrane regulator protein mutation, 68
AtMYB2, 17
ATMYC2, 46
AtMYC2, 11, 17
AtNOA1. *See* Nitric oxide associated 1 (AtNOA1)
AtSAC1/IBS2 gene, 5
Autoinhibitory domain, CDPKs, 81
Auxins, 99–100
  and ABA, 12
avrRpt2/RPS2, 9

β-1,3glucanase, 6
β-amino butyric acid (BABA)
  in callose priming, 13–15
  induced resistance against biotrophs, 15
BABA. *See* β-amino butyric acid (BABA)

BABA-induced resistance (BABA-IR), 13, 14
BABA-IR. See BABA-induced resistance (BABA-IR)
Basic leucine zipper (bZIP) proteins, 150
bin2. See Brassinosteroid-insensitive2 (bin2)
Biosynthesis, and transcription factors, 51–53
Biotic signals, and stomatal movements, 102
Biotic stress regulation, interaction with abiotic stress regulation, 15–17
Biotic stress-signaling network, 43–54
Biotic stresses, 136
Blumeria graminis, 4, 129
bon1 mutant, 120
BOS1. See Botrytis SUSCEPTIBLE1 *(BOS1)*
Botrytis cinerea, 4, 7
Botrytis SUSCEPTIBLE1 *(BOS1)*, 17
BR. *See* Brassinosteroids (BR)
Brassinosteroid-insensitive2 (bin2), 13
Brassinosteroids (BR), 13, 100–101
bZIP. *See* Basic leucine zipper (bZIP) proteins

Ca^{2+} channels, 66–67
    and CaM-activated kinases (CCaMK), 82
    hyperpolarization-activated, 68
Ca^{2+} efflux pathways
    Ca^{2+}-ATPases role in, 72–73
    Ca^{2+}/H^+ antiporters role in, 74
Ca^{2+} sensitivity priming hypothesis, 78
Ca^{2+} signaling, 127–128
    decoding
        CAMTAs role in, 82–83
        CBL–CIPK interactions, 79–81
        CDPKs role in, 82
    generation mechanism in plant cells, 59
    generation of
        efflux transporters, 72–73
        endomembrane, 69–72
        plasma membrane Ca^{2+} channels, 66–69
    in abiotic stress responses
        cold-shock-induced changes, 60–61
        signaling information, 61
    in guard cells, 63–64
    oscillations and transients in $[Ca^{2+}]_{cyt}$, specificity in, 74–77
    with specific spatio-temporal dynamics, 60
Ca^{2+} transport pathways in plant cells, 67
Ca^{2+}-ATPases, 72–73
Ca^{2+}-based signaling systems, crosstalk in, 59
Ca^{2+}-dependent protein kinases, 60
Ca^{2+}-permeable channels, 67
Ca^{2+}/CaM-domain protein kinases (CDPK)
    antisense repression of, 81
    biochemical studies of, 81
    functions of, 81–82
Ca^{2+}/H^+ antiporters, 73
Calcineurin B-like Ca^{2+}-binding proteins (CBL), 79
Callose priming, 13–15
Callose, role in disease resistance, 6
CALRR1 gene, in pepper plants, 6
CaM-binding transcription activators (CAMTA), 83
Cation exchanger (CAX) genes, 73
CAX genes. *See* Cation exchanger (CAX) genes
CBL-interacting protein kinases (CIPK), 79
CBL–CIPK interactions, 79–80
CDPK genes
    genetic loss-of-function studies in, 81
    mutants in, 81–82
Cell plate formation, in tobacco NPK1-NtMEK1-NTF6 cascade, 35–36

cGMP. *See* Cyclic guanosine monophosphate (cGMP)
Chlamydomonas reinhardtii, susceptible to rapamycin treatment, 163
Cladosporium cucumerinum, 2
Cladosporium fulvum, 106, 114–115
CNGC (cyclic necleotide-gated ion channel), 116, 119
Colletotrichum coccodes, 6
Colletotrichum gloesporoides, 6
Commelina communis
   stimulus-induced changes in guard cells, 63
   stomata opening by OGA, 103
   treatment with plant phospholipase C (U-73122), 5
Concentration-dependent effects, of JA, 100
copine gene, 124
cpr22, 123
Cucumber mosaic virus (CMV), resistance to, 179
Cuticular component regulation, transcription factors, 51–53
Cyclic guanosine monophosphate (cGMP), 146
Cyclophilin (FKBP) expression, 169
Cytokinins, 99–100
Cytoplasmic calcium ([Ca^{2+}]cyt), ABA-induced, 5

Defense-related mutants, 116–121
Dehydration-responsive element binding proteins/c-repeat binding factors (DREB/CBF), 48–50
Dietary components, 144

Effector triggered susceptibility (ETS), 9
eIF3, translation initiation factor, 172
eIF4G1, translation initiation factor, 172
Elicitation competency, 13
Elicitor-induced changes in [Ca^{2+}]$_{cyt}$ transients, 65

Endomembrane Ca^{2+}-permeable channels
   characterization of, 69
   ligand-gated, 70
   role in ABA responses, 69–70
   sources of Ca^{2+} release from, 71–72
   voltage-dependent, 70–71
Environmental sensitivity
   and lesion mimic mutants, 116–121
   in pathogen resistant arabidopsis mutants, 113–130
ERF subfamily transcription factors, 50–51
ERF/AP2 family transcription factors, 48–51
Erwinia chrysanthemi, 4–5
Erysiphe cichoracearum, 7
ET. *See* Ethylene (ET)
Ethylene (ET), 1, 98
Ethylene biosynthesis, in plant MAPK cascades, 30–31
Ethylene-responsive element binding factors (ERF), 48–50
ETS. *See* Effector triggered susceptibility (ETS)

FK506-binding proteins (FKBP), 169
FKBP12, immunophilin, 168
   and FRB of TOR proteins, noncovalent link between, 169
   paralogs, 169–170
   properties of, 169
FSD1. *See* Iron superoxide dismutase (FSD1)
Fusarium oxysporum, 4, 11

GA signaling pathways, and ABA, 12
GAPDH. *See* Glyceraldehyde-3-phosphate dehydrogenase (GAPDH)
GbERF2. *See Gossypium barbadense* transcription factor gene (*GbERF2*)

Genetic loss-of-function studies in
    CDPK genes, 81
Glyceraldehyde-3-phosphate
    dehydrogenase (GAPDH), 150
Gossypium barbadense, 16
Gossypium barbadense transcription
    factor gene *(GbERF2)*, 16
GSNO. See S-nitrosoglutathione (GSNO)
Guard cell signaling, 96–107

H_2O_2 accumulation, in tomato plants,
    4–5
Halobacterium halobium, 127
Heat shock transcription factors (Hsfs),
    150
HLM1, 119
Homeostasis, ion, 127–128
Hormonal regulation
    of guard cell signaling, 96–107
    of stomatal movements, 97–101
Hormone abscisic acid (ABA). See
    Abscisic acid (ABA)
HR (Hypersensitive Response), 113, 115
Hsfs. See Heat shock transcription
    factors (Hsfs)
humidity sensitivity 114–115
Hyaloperonospora arabidopsis, 3
    BABA-IR against, 16
Hyaloperonospora parasitica, 116
Hyperpolarization-activated Ca^{2+}
    channels, 68

*ibs2 (insensitive to BABA-induced
    sterility2)*, 5
ibs3. See *Insensitive to BABA-induced
    sterility3 (ibs3)*
I_{Ca} channels, 68–69
Initiation factors for TOR signaling
    pathway
    and 4EBPs, 171
    eIF4A, eIF4B, and eIF4E, 170
    eIF4E and eIF(iso)4E, 172
    eIF4G1, 172

    interaction of, 171
*Insensitive to BABA-induced sterility3
    (ibs3)*, 14
Ion channel regulation, 106–107
Ion homeostasis, 127–128
Ipomoea batatas, 16
Iron superoxide dismutase (FSD1), 151
Irregular xylem5 (irx5) mutant, 7
irx1, 7
irx5. See *Irregular xylem5 (irx5)* mutant
Isochorismate synthase 1 (ICS1), 124

JA. See Jasmonic acid (JA); jasmonic
    acid (JA)
JA signaling
    and *Arabidopsis* MKK3-MPK6
        pathway, 34
    MYB transcription factors, 45–47
    MYC transcription factors, 45–47
JA/ET signaling pathway, crosstalk with
    ABA, 11–12
Jasmonate and ethylene-responsive
    factor 3 *(JERF3)*, 16
Jasmonic acid (JA), 1, 98–99
JERF3. See Jasmonate and
    ethylene-responsive factor 3
    (JERF3)
jin1/myc2, 11

Kinase open stomata 1 (OST1), 142

L-arginine, 143
L-citrulline, 143
Leaf humidity, impact on insects, 2
Leaf vitamin C contents, role in disease
    resisitance, 5–6
Leaf wilting (lew) mutant, 7
Leptosphaeria maculans
    interaction with *Arabidopsis*, 8
    ABA and BABA priming, 14–15
Lesion mimic mutants, 116–121
Leucine-rich repeat (LRR) region, 114
lew. See *Leaf wilting (lew)* mutant

Lilium longiflorum, $[Ca^{2+}]_{cyt}$ in, 62
Lipopolysaccharides (LPS), 9
LPS. *See* Lipopolysaccharides (LPS)

Magnaporthe grisea, 2
MAP kinase, 142
MAP kinases, 125–126
MAPK. *See* mitogen-activated protein kinase (MAPK); Mitogen-activated protein kinase (MAPK) signaling
MAPK cascades. *See* mitogen-activated protein kinase (MAPK) cascades
MAPK kinase (MAPKK), 25
MAPK kinase kinase (MAPKKK), 25, 146
MAPKK. *See* MAPK kinase (MAPKK)
MAPKKK. *See* MAPK kinase kinase (MAPKKK); MAPK kinase kinases (MAPKKK)
mekk1 mutant, 33
MEKK1-MKK1/MKK2-MPK4 pathway, 31–33
Melon necrotic spot virus (MNSV), 180
Methionine sulfoxide (MetSO), 148
MetSO. *See* Methionine sulfoxide (MetSO)
Mitogen-activated protein kinase (MAPK) cascades, 23–38
Mitogen-activated protein kinase (MAPK) signaling, 145
MPK3, 25–29
mpk4 mutant, 33
MPK6, 25–29
mTOR, 161–162
  rapamycin-resistant, overexpression of, 178–179
  rRNA synthesis and, 175
MYB transcription factors, 45–47
MYC transcription factors, 45–47
Myzus nicotianae, 2

N gene mediated HR, 115–116
NAC transcription factors, 47–48
NADPH. *See* Nicotinamide adenine dinucleotide phosphate (NADPH) oxidase
NBS. *See* Nucleotide-binding site (NBS)
NCED. *See* 9-cis-epoxycarotenoid dioxygenase *(NCED)*
Neovossia indica, 5
Ni-NOR. *See* Nitrite-NO-oxidoreductase (Ni-NOR)
Nicotiana attenuata, 2
Nicotinamide adenine dinucleotide phosphate (NADPH) oxidase, 141
Nitrate reductases (NR), 143
Nitric oxide (NO), 136
  crosstalk with ABA, 12–13
  generation in plants, 143
Nitric oxide associated 1 (AtNOA1), 143
Nitric oxide synthases (NOS), 143
Nitrite-NO-oxidoreductase (Ni-NOR), 143
Nitrosonium cation (NO^+), 136
Nitroxyl anion (NO^-), 136
NO. *See* Nitric oxide (NO)
NO^+. *See* Nitrosonium cation (NO^+)
NO^-. *See* Nitroxyl anion (NO^-)
Nod-factor-induced Ca^{2+}-spikes, 64
Non Expressor of Pathogenesis Related Genes 1 (NPR1), 150
Nonsymbiotic hemoglobins (nsHbs), 144
NOS. *See* Nitric oxide synthases (NOS)
NOX enzymes, 141
NOX genes, 141
NPR1. *See* Non Expressor of Pathogenesis Related Genes 1 (NPR1)
NR. *See* Nitrate reductases (NR)
nsHb. *See* Nonsymbiotic hemoglobins (nsHb)
Nuclear proteome, 176
Nucleotide-binding site (NBS), 114

OST1. *See* Kinase open stomata 1 (OST1)
ost1 kinase, 9
OXI1. *See* Oxidative signal-inducible 1 (OXI1)
Oxidative signal-inducible 1 (OXI1), 146

PAL. *See* Phenylalanine ammonia lyase (PAL)
PAMP. *See* Pathogen-associated molecular patterns (PAMP)
Papaver rhoeas, 62
Papillae formation, ABA role in, 6
Pathogen
   and ERF subfamily transcription factors, 50–51
   and signaling crosstalk, 96–107
Pathogen infection, and *Arabidopsis* MKK3, 34–35
Pathogen modulation, of stomatal movements, 106–107
Pathogen recognition, R-gene-mediated, 122–123
Pathogen regulation, of stomatal regulation, 102–107
Pathogen resistance, R gene-mediated, 114–116
Pathogen resistant *arabidopsis* mutants, environmental sensitivity, 113–130
Pathogen virulence, mechanism, 104–106
Pathogen-associated molecular patterns (PAMP), 4
PCD. *See* Programmed cell death (PCD)
PDK1 (enzyme), 174
Pectobacterium chrysanthemi, 16
Phenylalanine ammonia lyase (PAL), 2, 10
Phytophthora capsici, 6
Phytophthora infestans, 2
Phytophthora sojae, 2
PI3 kinase, 173–174

Plant cells, 136
Plant defense response, 102–103
   to pathogen attack and changes in $[Ca^{2+}]_{cyt}$, 64–66
Plant MAPK cascades, 23–38
   ANP1, 30
   Arabidopsis MKK3, 34–35
   components, 24–25
   MPK3, 25–29
   MPK6, 25–29
   overview, 23–24
   plant MKK, 27
   plant MPK, 26
   protein phosphatases, 36–37
   tobacco NPK1-NtMEK1-NTF6 cascade, 35–36
Plant stress signaling, 136–137
Plant TOR kinase, 164–165
Plant-insect interaction, ABA role in, 2
Plasma membrane Ca^{2+} channels, classes of, 66–68
Plasmopara viticola, 9
   infection, 102
Plectosporium cucumerina
   ABA against, 4
   BABA-IR against, 13
   callose formation, 6
pmr4-1 mutant, 14–15
Pollen tube growth, $[Ca^{2+}]_{cyt}$ role in, 62–63
Priming, 13
Priming, ABA in, 13–15
Programmed cell death (PCD), 140
Protein kinases, MAPK, 23–38
Protein phosphatase A (PP2A), subunits, 172
Protein phosphatases, and plant MAPK cascades, 36–37
Pseudomonas syringae, 116
Pseudomonas syringae pv *tomato (Pst)*, 3
Pst. See Pseudomonas syringae pv *tomato (Pst)*

Pst cor, 9
Pst DC3000, in ABA content, 3
Pti5 gene in tomato, 5
Pythium irregulare, interaction with
    Arabidopsis, 8

R-gene-mediated pathogen recognition,
    122–123
R-gene-mediated pathogen resistance,
    114–116
Ralstonia solanacearum, 7
Rapamycin treatment
    Chlamydomonas reinharditii
        susceptibility to, 163
    of mammalian cells, 178
    transcriptional profiling of, 161
RAPTOR genes, 178
Reactive nitrogen species (RNS), 136
Reactive oxygen species (ROS),
    126–127, 136
    generation in plants, 141–142
Receptor-like protein kinases, 125
Recombinant aequorin technique, 60
Reversible protein phosphorylation,
    145–146
Ribosomal protein (RP) genes, 174
    expression, regulation of, 175
Ribosomal protein 6 (rpS6), genes for,
    168
Ribosomal proteins, 174
RLM1$_{Col}$ pathway, 8
RNS. *See* Reactive nitrogen species
    (RNS)
Root hairs and nodulation signals, 64
ROS. *See* Reactive oxygen species (ROS)
ROS signaling, 43–45
RS (reactive oxygen species and nitric
    oxide)
    as nodes for signaling crosstalk,
        139–141
    as signaling intermediates, 137–139
    intracellular movement of, 140
    removal of, 144

RS signaling, 145–151
    changes in gene expression, 149–151
    reversible protein phosphorylation,
        145–146
    second messengers, 146–147
    thiol (-SH) group modification in,
        147–149
Ryanodine receptor calcium channels
    (RYR), 146
RYR. *See* Ryanodine receptor calcium
    channels (RYR)

S-nitrosoglutathione (GSNO), 149
S6 kinase (S6K)
    activation of, 165
    activity of plant, 167
    as substrate of AGC kinase, 166–167
    dephosphorylation, 173
    drosophila and mammalian cells,
        165–166
    in mice, 166
    p70s6k, 167
    protein synthesis, 166–167
    ribosomal protein and kinase substrate,
        166
SA. *See* Salicylic acid (SA)
SA positive feedback loop, 123–128
Saccharomyces cerevisiae, 23
    FKBP12 mutant sensitivity, 164
    TORC1 and TORC2 of, 163
Salicylic Acid, 123–128
Salicylic acid (SA), 98
    ascorbate deficiency and accumulation
        of, 6
    with ABA, 10–11
Salinity signaling, 50–51
Schizaphis graminum–resistant sorghum
    plants, 2
Sclerotinia sclerotiorum infection, 102
Second messengers, 146–147
Serine/arginine–rich (SR) proteins,
    175–176
sGC. *See* Soluble guanylyl cyclase (sGC)

Signal transduction pathways
    Ca^{2+} role in, 59
Signal transduction, in plant MAPK
        cascades, 30–31
Signaling crosstalk
    in pathogen, 96–107
    in plant MAPK cascades, 23–38
    response to stress, 151–153
    RS nodes for, 139–141
    stomatal guard cells, 153–156
    transcription factors, 43–54
Sitiens mutation, in tomato plant, 4–5
SLH1, 120
Slow-activating vacuolar (SV) channel,
        70
Solanum chilense, 16
Soluble guanylyl cyclase (sGC), 146
Soybean-*Phytophthora megasperma f.
        sp. glycinea (Pmg)*, 10
Spodoptera exigua, 2, 2
Stomatal closure promotion, 102–103
Stomatal guard cells
    RS signalling crosstalk, 153–156
    stimulus-induced changes in, 63–64
Stomatal movements
    and hormonal regulation, 97–101
    and pathogen regulation, 102–107
    biotic signals, 102
    of pathogen modulation, 106–107
Stomatal opening promotion, 104–106
Stress
    abiotic, 136
    biotic, 136
    signaling crosstalk response, 151–153
Stress-signaling network, and
        transcription factors, 43–54
Stretch-activated Ca^{2+} channels, 68
Sulphenic acid (–SOH), 144

T3SS. *See* Type III secretory system
        (T3SS)
TaERF1. *See Triticum eastivum*
        ET-responsive factor 1 *(TaERF1)*

TAP42
    and PP2A activity, 173
    mutations in, 172
Temperature sensitivity, and N gene
        mediated HR, 115–116
Thiol (-SH) group modification, in RS
        signaling, 147–149
TINY, 50
Tobacco NPK1-NtMEK1-NTF6 cascade,
        35–36
TOP mRNA, regulation of translation,
        166
TOR (target of rapamycin) signaling
        pathway
    CDK4/cyclinD activity and, 175
    conservation in eukaryotes, 161–162
    FKBP12 role in, 168–170
    histone-modifying enzymes and, 175
    initiation factors
        and 4EBPs, 171
        eIF4A, eIF4B, and eIF4E, 170
        eIF4E and eIF(iso)4E, 172
        eIF4G1, 172
        interaction of, 171
    PI3 kinase and PDK1, 173–174
    plant cell cycle influenced by
        cyclins, 176–177, 178
        rapamycin targets, 177
    plant growth
        AtTOR role in, 178
        internal and external signals for,
            177
    plant TOR kinase role in, 164–165
    plant viral infections and, 179–180
    PP2A activity, 172–173
    ribosomal proteins, 174
    RP gene promoter activation by, 175
    S6 kinase role in
        as substrate of AGC kinase,
            166–167
        drosophila and mammalian cells,
            165–166
        in mice, 166

TOR (target of rapamycin) signaling
      pathway (*Cont.*)
   p70s6k, 167
   protein synthesis, 166–167
 SR proteins and, 175–176
 translation initiation factors implicated
      in, 172
TOR complexes, 162
TOR gene, 161
TOR proteins
 as sensor of ATP, 185–186
 general architecture of, 161, 162
 nutrient sensing via
   amino acids, 183–185
   in yeast, 180
   leucine, 183
   photosynthesis and nitrogen
      assimilation, 181–183
   TOR kinases role in, 181
   TOR protein localization,
      180–181
 osmotic stress signaling and
   AtTOR activity, 188
   growth signals and photosynthesis,
      188–190
   PDK1 activity, 187–188
   S6K1 activity, 186–187
 regulatory-associated protein of, 164
 signaling activity, 164–165
 TORC1 and TORC2
   functionality of, 162
   of *Saccharomyces cerevisiae*, 163
TOR-S6K signal transduction pathway,
      168
TOR^{FRB}–rapamycin–FKBP12
      interaction, 164–165
Transcription factors, and
      stress-signaling network, 43–54
 ERF/AP2 family transcription factors,
      48–51

in biosynthesis, 51–53
in cuticular component regulation,
      51–53
MYB transcription factors, 45–47
MYC transcription factors, 45–47
NAC transcription factors, 47–48
overview, 43
zinc finger transcription factors, 43–45
Triticum eastivum, 16
Triticum eastivum ET-responsive factor 1
      (*TaERF1*), 16
TSI1-interacting protein1 (TSIP1), 51
TSIP1. *See* TSI1-interacting protein1
      (TSIP1)
Turnip crinkle virus (TCV), resistance to,
      179
Type III secretory system (T3SS), 3

Vacuolar voltage-gated Ca^{2+} (VVCa)
      channel, 70
Vegetative storage protein2 (VSP2), 16
Verticillium daliae, 16
VfFKBP15 protein, 170
VSP2. *See* Vegetative storage protein2
      (VSP2)

Wax inducer1/shine1 (WIN1/SHN1), 52
Wax production1 (WXP1), 52
Wax production2 (WXP2), 52
WIN1/SHN1. *See* Wax inducer1/shine1
      (WIN1/SHN1)
WRK, 125
WXP1. *See* Wax production1 (WXP1)
WXP2. *See* Wax production2 (WXP2)

Xanthomonas campestris, 6
Xanthomonas oryzae, 114

Zinc finger transcription factors, 43–45
ZmS6K protein, 168